Social Stratification in Contemporary China:
Definitive Survey and Analysis

Social Stratification in Contemporary China: Definitive Survey and Analysis

By Li Qiang

Bridge21Publications

Distributed by Transaction Publishers

Social Stratification in Contemporary China:
Definitive Survey and Analysis
by Li Qiang

Distributed by Transaction Publishers
10 Corporate Place South, Suite 102
Piscataway, NJ 08854

Translation managed by E-Town Cultural Development Co. Limited and funded by Chinese Fund for the Humanities and Social Sciences

Chief Translator: He Gan
Language Editors: Adam Lorimer, Xiaodan Zheng, and Tait Bergstrom
Copyeditors: Amanda Brown and Melissa Peitsch
Layout Design: Sue Balcer
Cover Design: Lana Tara

For information contact Bridge21 Publications, LLC
16705 Huerta Road, Encino, California 91436

Published in the United States

ISBN: 978-1-62643-037-2 Hardcover
978-1-62643-044-0 eBook

Table of Contents

Prologue

The Beijing Normal University Press plans to issue a series of books called *The Contemporary Chinese Sociologists Library*, and I have been invited to publish some of my writings based on the findings of my sociological research. It was in the 1980s that I first entered the field of sociological study. At that time, I was conducting my undergraduate and postgraduate study at Renmin University of China, where the Department of Sociology had not yet been set up. There were hardly any areas of the curriculum relevant to sociology; therefore, the knowledge of sociology I have acquired was mainly obtained through self-study. The main method of my self-study was reading books during that period of time—I read many sociological works. My bachelor's thesis was on "collective bargaining" in the American labor movement, while my master's thesis dealt with the "white-collar class" in Western countries. Those theses were relevant to sociology, both theoretically and methodologically, although I had actually majored in International Politics. In 1985, when I graduated, I volunteered to work at the Institute of Sociology, which had just been established at Renmin University. Therefore, 1985 can be regarded as the starting point of my career as a sociologist.

At the time of writing of this prologue, twenty-five years have passed since 1985. Over that period, I have explored a wide range of interests in sociological studies and have conducted many studies on different social issues, among which my study of social stratification is of primary importance. Social stratification is a concept referring to the classification of people into different classes and strata, such as the poor, the rich, different interest groups, and people with different status. Is it vital to understand social stratification? My answer is an emphatic yes. I read a survey conducted by a journal in 2009 in which those polled were asked what contemporary social problem was the most urgent in China. Most of them considered the poverty gap to be the most serious problem, which is evidence that social stratification has received the Chinese public's interest.

The concept of social stratification informs us bluntly that people in a society are ranked into different strata in a hierarchy, just like a huge ladder. Some people are at the top of the ladder, while others are at lower strata of it. Many people doubt that the phenomenon of social stratification is fair. Although the concepts most easily accepted by the public to date are social equity and social justice, brutal social realities reveal that inequity and injustice can be found everywhere. How do we interpret this kind of social problem? This is the question the present book will deal with.

To answer the questions of why society is stratified, how society is stratified, and what kinds of stratification are fair and unfair, we should first determine our position, or "value orientation." That is to say, we have to clearly define social equity and social justice. It is necessary to clarify these two issues from two different perspectives. The first perspective concerns equality and inequality, which means to observe whether or not everyone gets—both in quality and quantity—the same share of property, income, or any other kind of social resource distributed in a society. However, we have not yet found a method by which property, income, or any kind of social resource can be equally distributed to everyone in any society. Differences always exist. We used to believe that the distribution in primitive tribes was completely equal, but research has shown that distribution between the tribal chiefs and tribe members was not completely equal; it is just that social stratification had not yet been institutionalized. Before the reforms and the opening up of China, and during the experimental period of land reform and socialist transformation, differences were apparent between city and countryside, cadres and workers, mental labor and manual labor. Since complete equality cannot be achieved, we must then decide what kind of inequality is reasonable and acceptable. This is the second perspective—what kind of inequality is reasonable and fair? In other words, what is reasonable and fair social stratification? What kind of inequality is acceptable to common people?

The explorations in most of the chapters of this book are drawn from these two perspectives. In some chapters, of course, basic theoretical and macro perspectives will be adopted—such as in chapters 1, 2, and 7. Other chapters, such as chapters 4, 5, and 6, focus on a historical analysis of the changes in social stratification over the past sixty years since the founding of the People's Republic of China in 1949, including the changes in equality and inequality and an analysis on what was seen to be reasonable and fair and what was seen to be unreasonable and unfair. Chapters 8, 9, 10, and 11 take pains to analyze the upper, middle, and lower classes. However, discussion in these chapters still focuses on exploring what is reasonable and fair and what is unequal and unfair. Chapters 13 and 14 focus on an analysis of equality and inequality, as well as justice and injustice in the allocation of housing resources. Essentially, we can sort out the basic ideas about social stratification from the two perspectives outlined above. In order to explain the basics of studying social stratification, chapter 3 provides a special introduction to the methodology of measuring social stratification. The application and demonstration of these methods can also be found in other chapters. Therein lies the origin of the subtitle of this book: "A Measurement and Analysis."

Some chapters are papers I have previously published. Because the findings of those papers are still applicable in explaining present social reality, I have revised them, made additions and deletions, and included them in this book. Some chapters are from my most recent papers. Also, I should explain that some chapters are papers written in cooperation with other researchers, so I would like to express my sincere thanks to the following collaborators: Liu Haiyang from the Department of Sociology at Tsinghua University, with whom I surveyed occupational reputation in 2009 and wrote a paper, part of which is adopted in the third section of chapter 3; Dr. Wang Meiqin from the Department of Sociology at Shandong University, with whom I wrote a paper which is used as chapter 14; Dr. Deng Jianwei who works in Changsha, with whom I wrote a paper which is used as chapter 15; Dr. Zhang Wanli of the Chinese Academy of Social Science's Institute of Sociology, with whom I wrote the papers which make up sections one, two, and three of chapter 16, and of which he was the chief author; and Liu Yang from the Department of Sociology at Tsinghua University, with whom I wrote the papers which are used as chapters 4 and 5. I am grateful to all these collaborators and their contributions to this book are highly appreciated.

I am particularly grateful to the editor of Beijing Normal University Press, Professor Yang Geng. It is under his invitation and encouragement that I publish this book. I am also grateful to the editor of this book, Li Xuejie, who was both supportive of my writing and dedicated to the compiling and collation of this work. I am also grateful to the translator Professor He Gan and the English language editor

Mr. Adam Lorimer, who each spared no effort and played a significant role in transforming the Chinese original into an English version. Other co-translators, a group of teachers from the College of Foreign Languages and Literature of Wuhan Donghu University, have worked hard on the first draft in their spare time and made it possible to complete the English version within a tight time limit of ten months.

I know well the complexity of social stratification—especially contemporary Chinese social stratification—and that different people have different views and opinions on this topic, so I sincerely welcome criticisms and suggestions from my readers.

Li Qiang
February 6, 2010
Lanqiying, Tsinghua University

Chapter 1

Understanding Social Structure and Social Stratification

1. What Are Social Structure and Social Stratification?

Although the term social structure (社会结构) sounds quite academic, it is not as complicated as it seems. How then should we understand it well?

First of all, we should address the question of how society is constituted. Some people believe that society is composed of people. This is, of course, true; people are the basic element from which a society is constituted. However, a society is formed of a great many individuals. For instance, there are 1.3 billion people in China. Therefore, if we study an entire society by focusing on individuals instead of structure, the scale of the task becomes impossible to handle. Social structure refers to stable relationships between individuals, so close scrutiny of those relationships may help us understand a society. If a definition must be given to *social structure*, we can say that *social structure* refers to the relatively stable pattern of relationships and interactions between the elements and parts of a society. For instance, social economic structures, social class structures, social occupational structures, and social organizational structures reflect respectively the stable relationship patterns of the social elements of economics, groups, occupations, and organizations.

Some people might argue that we can only see individuals, not the *social structures* mentioned above. As a matter of fact, social structures can be perceived, but only from a macroscopic perspective. It is rather like watching people paint a mural on a mountainside; one cannot see the picture when in the same position as the painters, only abstract colors. But, if one looks at it from a distance, then a grand picture is revealed. This is the same as the macroscopic perspective, which must be employed in the study of social structures.

In this sense, social structures are kinds of relationships that reach beyond individuals. They condition the behavior of individuals in a macroscopic mode. Social structures can be observed from various viewpoints. Generally speaking, the social position structures of individuals and groups receive more attention. The structures of social position have two aspects: *vertical structure* and *horizontal structure*.

The *vertical structure* of social positions is also known as the structure of social inequality, referring to the different social positions of individuals and groups in a society, such as with class structures, wealth structures formed by differences in income and corresponding assets, and power structures (among others). The *horizontal structure* refers to the lateral or transverse relationships between individuals and groups, such as spatial, geographical, industrial, and gender distributions—as well as the marital status of individuals. Of course, in fact the *vertical structure* and the *horizontal structure* are usually interwoven and hard to separate. For instance, geographical or residential distribution is superficially horizontal, but people inhabiting cities tend to have a higher status than those in the countryside, and those in big cities such as Beijing and Shanghai possess more educational resources than others. It is therefore hard to tell the *vertical structure* from the *horizontal structure*.

In short, with social structure, the primary issue is that a society is organized by ranking or classifying people according to their different incomes, statuses, and assets. *Social structure* is therefore also referred to as *social stratification structure*, or *social stratification*.

The term *social stratification* implies inequality of status and position, and one cannot deny that *social stratification* is a universal phenomenon. No country has yet been found completely free of *social stratification*.

How, then, should we academically define *social stratification* (社会分层)? A previously attempted definition states: *social stratification* refers to the phenomenon of stratification or difference generated by the possession of varying amounts of s*ocial resources* by different groups and members of society, especially an institutionalized system of social difference established on the basis of laws and regulations. One might have noticed the use of the term social resources. What are s*ocial resources*? *Social resources* is a generic term for any valuable resource, including political resources, economic resources, and cultural resources. Among these, the most critical are economic resources—including assets and income. Since social resources are varied, the stratification phenomena they produce are also varied. For example, possession of different economic resources produces a rich stratum and a poor stratum; an unequal distribution of educational resources produces well-educated groups and groups with poor educational backgrounds.

Moreover, in talking about social structure and social stratification, one cannot fail to mention *social mobility* (社会流动). Social structure and social stratification deal with the established positions of individuals and groups in a society. However, are those positions stable? Is it possible for people to climb from a lower social stratum to a higher one? Is it possible for those who possess great fortunes to go bankrupt? Here the problem of *social mobility* should be discussed. *Social mobility* refers to change in the status or position of individuals and groups—or, to be more specific, change in the position of individuals and groups in a social stratification structure, as well as change of location in a spatial and geographical structure. Of the two, sociologists place more emphasis on the former—the change, upward or downward, of one's social position. Rural farmers migrating to the cities, workers changing jobs, changes in the positions of one generation relative to another, cadre promotion, and the coalition of the rich are all good examples of social mobility. Social stratification and *social mobility* are two sides of one coin and are closely linked. One deals with the problem of how people are divided into different strata, the other with how people enter different strata. An analysis of the relationship between these concepts will be made below.

2. Who Gets What and Why

Regarding the two aspects of social structure mentioned above, we can describe them in the following standardized academic language: the first aspect refers to a social inequality structure or "status differentiation structure"—in other words, society is divided into structures of status inequality or networks of social status difference. This can be observed at first sight when making an investigation into any society or community, such as when we find income stratification, rank stratification, power stratification, etc.

The second aspect deals with the problem of how people enter into the status differentiation structure. For instance, how do graduates of schools at different levels come to stratified vocational positions? Why do some poor people later become rich, and vice versa? Sociologists call this entering procedure a "recruitment" system"[1](地位准入), or status attainment (地位获得), referring to how people obtain access to their positions.

Social stratification theorist, G. E. Lenski, gave a popular explanation of the two aspects of social stratification, saying that all theorists in this field have sought to answer one basic question: who gets what and why?[2] "Who gets what" concerns the result—who becomes rich and who does not—while "why" turns its attention to the process of entering into social strata. Why do some own huge quantities of assets? Are they being competitive in the market, or taking advantage by using monopoly power or through corrupt means?

Accordingly, we can find two very different notions of fairness. One believes that social stratification structures are in themselves unfair and that social equity can be realized in a society only when its stratification structure is eliminated. The other holds that the differences between each stratum are hard to eliminate. Social equity lies in ensuring that the process by which individuals or groups enter into social strata is fair. These two perspectives could be provisionally summarized as "equity based on status differentiation elimination" (or, "equity based on fair results") and "equity based on recruitment" (or, "equity based on equal opportunity" or, finally, "equity based on fair process").

As far as "recruitment" or "status attainment" is concerned, the following mechanisms have appeared in history.

First, "status attainment" can be based on consanguinity, or rank. There is no lack of examples of this, both in the succession to the imperial throne in ancient China and in the feudal system in precapitalist Europe, where noble status was inherited through kinship. Though it might be regarded as an unfair process today, it has been a legal and very important access method throughout world history. The old idea that "phoenixes beget phoenixes and dragons, dragons" reflects how this was once a major and institutionalized method of accessing leadership positions. It was not until the Revolution of 1911 that China legally abandoned this system, which had lasted for more than four thousand years since the Xia Dynasty—far longer than most other countries. In the following years, though the "ghost" of this inheritance still haunted China occasionally, it lost its judicial foundation

1 Frank Parkin. *Class Inequality & Political Order.* London: Mac Gibbon & Kee, 1971. 13.
2 Gerhard Emmanuel Lenski. *Power and Privilege: A Theory of Social Stratification.* Hangzhou: Zhejiang People's Publishing House, 1988. 2.

as a system. However, even today in some monarchical states, social stratification follows the conventional practice of status inheritance.

Second, "status attainment" may also be based on networks of kinship, relatives, and friends. Though there is a tendency to avoid mentioning the influence of networks of private relations, it plays a significant role in "status attainment," as becomes clear if we take the sources of elite groups in the world into consideration. Networks of kin, relatives, and friends have been well-developed in Chinese society since ancient times. A prominent contribution of Confucianism is the codification of the rules of interpersonal relations. A set of criteria was formulated for the relations between ruler and subject, father and son, husband and wife, elder brother and younger brother, and friend and friend; these rules became the core content of social operation. Since a network of private relations is prone to form factions and gangs, it has both positive and negative features. China's more than thirty years of reform have been a grand exercise in, and transformation of, social relations. One of the major reasons behind the reform is an attempt to change Chinese society from one operated by networks of private relations to a society managed by law. The paradox networks of private relations are hard for people to shake off. Quite a few explorations have therefore been made to search for an equilibrium point between a society run by private relation networks and a society ruled by law.

Third, recruitment systems can be based on recommendation. It was said that before the Xia Dynasty led by Da Yu, all the rulers of states came into power through recommendation. In ancient China, there once existed a system for the recommendation of "filial and honest persons" (孝廉) and of scholars. During the Cultural Revolution, university enrollment also adopted a recommendation system: "recommendation by the masses" ("群众推荐"). The recommendation system is still active in today's social life. For instance, Xiao Shenyang, an actor, recently won a "May 1st Labor Merit Certificate" after being recommended by the All-China federation of Trade Unions, which is a rather prestigious award. Though it aroused controversy, it proved that the recommendation system still exists in the Chinese society.

Fourth, recruitment systems can be based on selection. Recruitment systems based on recommendation are similar to those based on selection, but the latter carries more features of the political system of a state. Recruitment depends on the number of votes cast in the selection process. At present, more and more countries are adopting "recruitment based on selection" as their primary point of access to political positions. China has conducted some new experiments in selection systems in recent years—such as village elections, town and township governor elections, and the competitive election for members of the Central Committee, alternate members of the Central Committee, and members of the Central Commission for Discipline Inspection of the CPC. It could certainly be said that the twenty-first century will be an important era for China to practice and experiment with different selection systems.

Fifth, recruitment systems may be based on examination. The Chinese Imperial Examination System was the first "recruitment system based on examination" of its kind. Today, the systems of civil servant and public servant selection used throughout the world have all adopted the process of examination to select government officers. The fundamental principle of examination is fair competition; it is recruitment based on the results of an examination, which is the most widely accepted mechanism of selection for status attainment. Owing to this principle, the University Entrance Examination in China—though flawed in a number of ways—features more traits of fairness than some other recruitment mechanisms. It is commendable, especially at a time when networks of private relations and favors reign supreme. Now, the scope of the examination system has greatly expanded. A variety of examinations—such as examinations for lawyers, accountants, appraisers, and all kinds of technicians—have come into being.

Sixth, recruitment systems can be based on diplomas and degrees. This is an extension of the examination system, which further standardizes "recruitment."

Seventh, market competition or economic competition (measured by success or failure in business management and market transactions) can form the basis of a recruitment system. Since the 1990s, China has openly advocated developing a market economy—a mechanism that has provided more opportunities for individual development. Studies have shown that some people who are far distant from political power succeed in market competition and become rich.

Eighth, recruitment systems can be based on achievement and appraisal of one's performance. A promotion or a raise for most ordinary employees, like clerks and wage earners, depends on their hard work. If one not only fails to achieve but also causes great damage by continuously making mistakes in one's work, one should not get promoted but accept punishment and be demoted. That would be a loss of status. Also, to appraise those who make no mistakes but "pine away for years," or in their own words, "work hard for many years even if have not performed meritorious deeds" can be placed in this category.

By analyzing the above eight mechanisms of "recruitment" or "status attainment" one can see that the first two ("status attainment" based on consanguinity or rank and "status attainment" based on networks of kinship, relatives, and friends) constitute "unfair competition" and reek of unfairness in modern Gesellschaft. The third through eighth bear the characteristics of competitive mechanisms, complying with the ideal of fair status attainment in a modern society. In the long run of historical transformation, access to status attainment or recruitment increasingly depends on competition in a modern society.

If one recalls the two different attitudes towards fairness mentioned previously ("equity based on status differentiation elimination" and "equity based on equal opportunity") and compares them with the attitudes before and after the economic reforms in China, it is not difficult to see that the first was the dominant attitude before the economic reforms. At this point, the dominant attitude was that equity could only be achieved once differentiation had been eliminated. After the economic reforms, along with the popularization of Deng Xiaoping's idea of "letting some people grow rich first," the dominant ideology shifted focus to the fairness of "status attainment" and "recruitment processes." This was not simply denying social stratification, but advocating "status attainment" through competition, providing more competitive opportunities for upward mobility for all social strata. In the last thirty-plus years, Chinese society has witnessed a great transformation of social stratification structures and the concept of fairness.

3. The Significance of Studying Social Structure and Social Stratification

Though social structures and social stratification seem on the surface to be quite academic subjects, they are actually practical topics closely connected with the lives of ordinary people.

To begin with, social stratification affects the lives of everybody in a society. For one thing, the current global economic downturn is something not only Chinese people have to face, but also the rest of the world, because the global economy is inseparably related to everyone in the world. For example, what was the cause of the present economic slump? Has it something to do with social stratification? The answer is definitely yes. It is now known that one of the main reasons for the economic downturn in China, which began at the end of 2008, was "scant demand"—there was a surplus of commodities on the market that could not be sold out for the following reasons. First, orders from abroad fell drastically while domestic production continued to produce a surplus. The central government proposed to expand domestic demand and called on people to buy these overstocked products. Why didn't people buy them? Was it because they had too much at home? Fifty-six percent of the Chinese population lived in the countryside; it was not because they had too much at home, but too little. They did not buy because they earned very little and had no spare money to afford extra purchases. Thus, we faced an acute contradiction: large amounts of products that could not be sold and a large group of farmers who were in dire need of those products, but could not afford them. From the perspective of social stratification and social structure, the problem of overproduction and under-consumption was caused by the oversized proportion of the low-income group in China's social structure. The low-income group did not buy those products, not because those in that group were not in need of them, but because they lacked the consumptive capacity and were short of money. Therefore, the key to expanding domestic demand was the adjustment of social structure. Only when the social structure changed and the majority of the Chinese population had the required consumptive capacity could the problem of overstocked commodities eventually be solved.

Next, the major institutional factor that has influenced Chinese people's lives for over thirty years has been the policy of reform and opening up. This reform has a close connection with social structure and social stratification. It should be made clear what these reforms were. These reforms were changes in the systems and regulations of the social lives of Chinese people. Before the reforms, for example, there were only people's communes in the countryside. After the reform, "the household responsibility system" ("包产到户") was put into practice. Essentially, the rules of life and labor implemented for farmers to survive had changed. Authority over production management shifted from the leaders of the communes and brigades to each household and every farmer. Therefore, the relations between farmers and leaders, between households, between country people and city residents all changed, substantially influencing the social relations between individuals and groups. This, in turn, affected social structure and social stratification. Taking into consideration the relationships between social groups—especially the political, economic, and social status differences between them—the most prominent changes in the last thirty years have been the alterations made to the social stratification structure. Or, in sociological terms, the major change since reform could be summed up as a change in social stratification. The changes in the Chinese social structure were not only the result of the transformation of political, economic, and social relations, as well as the ideological paradigm, but also the dynamic power driving further political, economic, social, and cultural transformation. The social stratification structure, therefore, was the core of the issue under discussion. It should be the main concern of investigations into Chinese social stratification structures aiming to explore the changes of individuals and groups. Reforms, transformations, revolutions, and improvements in any society will eventually lead to changes among individuals and groups. Analysis of social stratification structures focuses more on the interest structures and status structures among individuals. Simply put, the economy, politics, material interests and social resources, which were the main focus of the reforms in China, were placed in an important position. The external manifestations of the reforms were rapid economic growth, changes in the appearance of cities and the countryside, and the expansion of material wealth. The total number of buildings built in China during the reforms was more than the total sum of the past few centuries. The profound internal changes of the reforms were the changes in social relationships, lifestyles, life goals, and life attitudes of individuals. Because of the changes among individuals, new demands and targets were created to propel the reforms forward. Accordingly, changes in social stratification structures could be interpreted as continued renovations of the reform.

Finally, social structure and social stratification are the loci of social contradictions and problems. A recent survey shows that the widening gap between rich and poor is attracting wide concern. The rich-poor division, or polarization, is becoming an increasingly prominent issue. On the one hand, the owners of coal mines are swimming in money, and on the other, their miners work under very tough conditions for extremely low wages. The idea that "inequality rather than want is the cause of trouble" ("不患寡而患不均") is deeply rooted in Chinese society, and the rich-poor division naturally causes dissatisfaction. An increasingly common folk chant says: "In grand restaurants, grafters are found; shoot first, investigate later; ensure justice is served" ("大饭店往里看,里面都是贪污犯,先枪毙后审判,基本没有冤假案"). It is not true that all those in the expensive restaurants are embezzlers guilty of corruption, but the fact that such a folk chant is going around indicates the existence of dissatisfaction. Research has demonstrated that there is a close relationship between most of China's social contradictions and conflicts and the division of wealth, interests, group interests, and social stratification. The case of farming land requisition led to the problem of "landless peasants" ("失地农民"). In the cities, "housing demolition" ("拆迁") conducted to facilitate real estate development caused conflicts. There have also been movements for the protection of the rights of real estate developers, property owners, and tenants, as well as contradictions between urban administrators and migrant populations. All these conflicts and contradictions are directly linked with social stratification. Therefore, social stratification is the dominant research field in China today for improving social safety, social harmony, and social stability.

Chapter 2

The Characteristics of Chinese Social Structure

Social stratification is a universal phenomenon; regardless of whether we examine ancient Chinese society or modern Chinese society, there exist social strata. What then are the characteristics of Chinese social stratification in terms of China's long history? What are they in comparison with those of other countries and nations? The following nine points are significant features deserving of mention.

1. Immense Social Difference

China has been a country of immense social difference throughout its history and this has been consistently true to the present day. Some people believe that it was the economic reforms which began in 1979 that set the Chinese society on the road to social differentiation. Actually, social difference has never been small in Chinese society. Of course, it is natural for people to hold different attitudes regarding social difference. As defined in the first chapter, social stratification refers to the phenomenon that social strata are generated according to the possession of different quantities of social resources. Social resources refer to any of the full range of socially defined resources including political, economic, and cultural resources. Accordingly, the tremendous social differences in Chinese society are manifested in various ways. For example, there was a rapidly growing income differential and a widening gap between the rich and the poor in the 1990s. However, does this mean that social difference was small in the 1980s and earlier? I wrote a paper analyzing and comparing "economic stratification" and "political stratification."[3] As was argued in that paper, social difference during the Cultural Revolution was not small. Although economic differentiation then was not great, political differentiation was significant. Some people were labeled as belonging to one of the "Revolutionary Five Groups" ("红五类") while others were classified as belonging to the "Five Reactionary Gangs" ("黑五类") which faced political discrimination. Could we say that this

3 Li Qiang, "Economic Stratification and Political Stratification." *Sociological Studies* 4 (1997): 34–43.

was a small social difference? The fierce political struggle at that time was reflected in significant social difference. Thus, if one wishes to take on the difficult task of spanning the tremendous gap of difference, it is necessary to be aware of the fact that it has been formed by thousands of years of economic, political, cultural and conventional inertia.

It is natural for people to ask why there is such great difference. It is because China has been a centralized society since its unification in the Qin Dynasty, and wealth and resources have been highly concentrated ever since. Subsequent dynasties further strengthened this centralized system. Mao Zedong once said that "all dynasties afterwards followed the administrative system of the Qin Dynasty." In other words, societies after the Qin copied the centralized political system of the Qin. It was natural that central zones and marginal ones formed in such a centralized system. All social resources were, first of all, concentrated towards the central zones before diffusing out towards the marginal ones. That undoubtedly led to significant differences in resource allocation between the central zones and the marginal ones. I will use the terms "central group" or "nuclear group" to refer to those people in the central zones, and "marginal group"[4] to refer to those in the marginal zones, though this is simply my own form of expression.

Reviewing the odyssey of over thirty years of reform, we know that the gap in resource allocation between the central zones and marginal zones was once narrowed in the 1980s, but in the 1990s and in the twenty-first century, all kinds of statistics provide evidence that the gap is now widening instead of narrowing.

2. Urban-Rural Separation

The immense difference mentioned above has manifested itself in differentials between town and country. The Chinese society has been one of great social difference since ancient times with an imbalanced distribution of resources between town and country. Though people might think it was the city-countryside household registration system established in the 1950s that led to the gap between city and countryside, the gap actually has a rather long history. In ancient

4 Li Qiang. *Social Stratification in Social Transformation in China.* Shenyang: Liaoning Education Press, 2004. 23–27.

Chinese society, there was a division between the so-called "urbanites" ("国人") and "hicks" ("野人") The word *urbanite* refers to those who lived in the cities, while the word *hick* referred to those who lived outside the cities or in the countryside. China is a country that, at one time, had the most walled cities in the world, which shows that people in cities and towns were physically separated from those in the countryside. In the past, all the cities closed their city gates at night. Once the gates were closed, mobility between city and countryside was entirely blocked. When comparing the social structure of American society and Chinese society, Fei Zhengqing argued that two societies existed in China. One was the society of farmers—the majority of the population—who performed unchanging agricultural work in villages and on farms, sheltered by trees and woods on the original land inherited from their ancestors; the other was the society of the upper classes, composed of landlords, scholars, businessmen, and officials—powerful and influential urban households in cities and towns—which possessed a certain degree of social mobility.[5] Certainly, the strict household registration system established in the mid- and late 1950s further consolidated this social separation. During the twenty years between 1959 and 1979, social mobility between city and countryside was essentially halted. Although the recent economic reforms have witnessed more free social mobility between city and countryside, it still (from the perspective of long-term historical development) requires a great deal of effort to narrow the gap between the two.

Social problems in today's Chinese society, almost without exception, have a lot to do with the two societies of city and countryside. In the matter of social stratification, the most prominent is the hierarchical stratification between urban and rural areas.

3. Pyramid-Shaped Social Structure

The third feature of social stratification in the Chinese society is that of a pyramid-shaped social structuration. It is a normal practice for sociologists to utilize graphs or charts when analyzing social structure. People can be arranged in ranking order from low to high—for instance, according to income distribution, the group with the

5 Fei Zhengqing. *America and China*. Beijing: World Affairs Press, 2000. 20.

highest income being placed at the top and the group with the lowest income at the bottom, forming a social hierarchy chart. Chinese society has had a pyramid-shaped or cone-shaped social structure since ancient times, with a small group of people at the top of the ladder and very large groups of people occupying the middle and bottom sections of the ladder. The traditional Chinese society was a royal or imperial society; the majority of the population was made up of peasants while the minority was made up of the royal family, lords, and high officials. Though there were gentry at the mid- to high-level, they were few in number. In his early criticism of the Imperial Examination System, Kang Youwei asserted that few people entered their names for imperial examination and only one in a thousand or even ten thousand could pass. Since 1949, especially since the economic reforms, the Chinese social structure has undergone some obvious changes but, generally speaking, the pyramid shape remains. This small top and large base structure can be seen in the separation of city from countryside. By 2008, the urban population amounted to 45.68% of the total Chinese population, while the rural population made up 54.32%, proving that a modern social structure had not yet been formed in China. Only if the proportion of the population engaged in farming decreases greatly and the size of the middle group constantly increases can a modern social structure take shape.

In chapter 11, the data of the Fifth Population Census will be analyzed by means of the International Socio-Economic Index of Occupational Status (ISEI) in an attempt to uncover the true structure of Chinese society. This analysis will show that the problem is even more serious than is indicated by the pyramid-shaped structure; an accurate model of China's social structure actually forms an inverted T-shape. That is to say, the majority of the Chinese population—the farming group—has stayed at the bottom as the long horizontal line of the inverted T-shape, while the urban population has formed the vertical column.

4. Status Society

Traditionally, great importance had been attached to status in Chinese society. Statuses and ranks have been strictly stratified, which can most easily be seen in the social division of officials and civilians, as well as in different rankings of officials. This situation has remained unchanged to this day. Nowadays, the seating arrangements of ceremony attendants, the order of speech deliveries, screen time on TV, and name lists in newspapers are all strictly arranged according to hierarchy. This is nothing new. Chinese society has always been extremely careful about status, rankings, and hierarchies; even those taken to the heather in the novel, *The Water Margin*, had a solemn ceremonial seating arrangement.

Before the economic reforms, the linking of welfare with the household registration system and unit system reinforced the status system. For instance, the registered status of urban citizens, the registered status of farmers, the status of cadres, and the status of workers were, for the greater part, lifelong statuses that could not be changed. Another prominent feature of the status system during that period was the importance of political statuses, such as poor and low–middle class peasant, rich peasant, clerk, revolutionary cadre, etc. Since the economic reforms, great progress has been made in weakening the power of status. The former rigid status system, such as the regulation that "farmers should not move into cities" has had much less power. What we cannot ignore is that the status system was built over thousands of years and could hardly be abolished within a short time. Since the economic reforms, a new problem has emerged. Newly formed statuses are often unable to find their proper positions in the old social hierarchical system. The new rich classes complain about their status being ignored or disrespected in formal social interactions. In official documents of recent years, the term "new social class" has been brought into use. The term refers to the newly instituted classes, including private enterprise owners and freelance professionals. Numerous regulations and rules have been officially remolded to track the new statuses' movement into the original status system so that the new classes could adapt to the existing social status system.

5. Official-Civilian Society

Chinese society has always been an "official-civilian" society with relation to the features of its status system, within which officials and civilians formed two distinctive groups of different social statuses, a tendency best manifested in the imperial dynasties following the Han Dynasty. According to studies into class structure in the

Han Dynasty, officials not only stood higher on the social ladder but also possessed quite a lot of social wealth. Additionally, they were stipulated to be a privileged group. In the rules and regulations of the period, they were entitled to special material means of subsistence according to their corresponding statuses.[6] Officials became the dominant group and thereby controlled social resources. When the Imperial Examination System was established in the Sui Dynasty, access to officialdom—including official reproduction and promotion—became a possibility. There existed other means of access to officialdom, such as the Supervisor Recommendation System, Recommendations, the Military Exploit System, and the Donation System. Despite these other means, the Imperial Examination System remained the primary mechanism for official generation.

The official system became the center of Chinese society, attracting elites of all ranks because official positions meant power, money, prestige, and status. Conversely, the business class (although in possession of assets) did not have a correspondingly high social status, as Confucianism looked down upon businessmen, preventing them from holding their own with the official classes at the center of society. Furthermore, the statuses of businessmen would not be stable unless they established good relationships with the necessary officials—they might occasionally be harassed by the government if they lacked important ties with officials. In imperial society, the confiscation of a rich businessmen's property was a frequent occurrence. The official system was much more developed in China than in other countries. Officialdom and political power controlled the most important resources of the entire society and the division of statuses was chiefly a matter of political power. Society's assets became largely subordinated to political power. This probably had much to do with the centralized political system that had been maintained since the Qin Dynasty. Although three social groups existed—officials, landlords, and farmers—officials and landlords were always intermingled. Members of landlord families were the main candidates for the Imperial Examination and formed the bulk of the gentry class. Intimate relationships between landlord and official families were often created by wedlock. Therefore, the essential division was mainly between officials and civilians.

6 Qu Tongzu. *Social Structure in Han Dynasty.* Shanghai: Shanghai People's Publishing House and Century Publishing Group, 2007. 100.

6. The Great Role of Elites

"Elite" ("精英") is a sociological term referring to the most outstanding group of people in a society. What, then, is meant by the most outstanding group? From the perspective of social stratification, there are two interpretations. First, "elite" may refer to those who stand at the top of the ladder in a social stratification system. For instance, political elite refers to political leaders; economic elites are top-level entrepreneurs. Second, it can also refer to the most talented people. Both interpretations may refer to the same group of people, as well as to different groups, such as the governing or non-governing elites, the elites in political power or elites out of political power, and the circulation of elites. None of these will be the main topic here, since elite theory has established its own system including the theories of G. Mosca, V. Pareto, R. Michels, C. W. Mills, Harold D. Lasswell, A. Giddens, and Joseph Schumpter. I have given lectures on elite theory and will try to offer more details in my future works, but here I will make only a simple analysis of the character of Chinese elites in social stratification.

It is generally believed that elite groups are small in size in all societies. Elites are few in number but their decisions have enormous consequences, especially those of political elites, who have been able to significantly influence the world situation throughout history. The most prominent feature of Chinese elites, especially political elites, is their great role in Chinese society. This may have been caused by the powerful administrative functions of the government developed during more than two thousand years of centralism and the deep-rooted elite culture, which developed after the Qin Dynasty. In the centralized system, the most important commands were transmitted down from a central decision-making arena. All inferior officials were appointed by higher authorities and were required to be obedient to their superiors. The superiors held absolute authority over their inferiors. Chinese politics has primarily been an elite politics, whether one is considering it from the perspective of its five thousand years of civilization or its three thousand years of recorded history. Political activities are manipulated by supreme authorities. Elite politics and democratic politics are in contradiction with one another. In elite politics, decision-making power is bestowed on competitive elites, while in democratic politics, it is placed in the hands of the public.

The prominent role of elites in Chinese society made the selection of elites extraordinarily important. How to select elites, what the makeup of elite groups would be, whether elite groups would be solidaristic, and the sources from which elites would be drawn all became key concerns. Basically, four challenges confront Chinese elite groups: the configuration of elites, the circulation of elites, the interchange of elites, and elite solidarity.

"The configuration of elites" refers to the proportion of different types of elites in society. Three basic elite groups can be found in all societies[7]: political, economic, and technical elite groups. The three groups should be proportionally configured; none of them expand in an unlimited fashion. Before the economic reforms, the political elite group in China was oversized, while the economic and technical groups were somewhat undersized. Particularly during the Cultural Revolution, China swarmed with political elites while economic and technical elites were little known. One of the important measures taken since the economic reforms was to give prominence to the essential roles both economic and technical elites play. In the initial stages of the reforms, Deng Xiaoping convened the National Conference on Science and the National Conference on Education to raise the status of technical elites. Increased emphasis was also placed on the training of entrepreneurs—the economic elite group. For rather a long time, there had not been a clear line between the functions of government and enterprise; therefore, economic elites were forced to hide themselves among the political elites, as political elites had completely replaced economic elites. Since the mid-1990s, positive effort has been made by the government to promote the separation of the functions of government from those of enterprise. Some economic elites have begun to leave the shelter of government. Furthermore, economic diversification and the institution of a market economy over the past thirty years has created a group of economic elites, such as the chairmen of the boards of directors of state-owned enterprises, managers, private entrepreneurs, township entrepreneurs, enterprise contractors, and senior employees in foreign-funded enterprises. These measures, including the streamlining of government agencies, also forced some political elites to transfer over to the economic elite

7 For Mills, C. Wright, there is a fourth elite group, that is, the military elite group. In China, it is hard to separate political elites from the military ones. Therefore the I speak of three elite groups instead.

group. The expansion of the economic elite group has undoubtedly been significant in the revitalization of the Chinese economy.

However, not everyone could successfully transfer over to the economic elite group. Some former officials, not having a specialization in management, drove their enterprises to bankruptcy. They were necessarily eliminated from the elite group. Here, we encounter the notion of the circulation of elites.

The circulation of elites refers to the succession of elites. G. Mosca, an Italian sociologist, argues that an elite group can maintain its vitality only if it absorbs elites from the lower groups,[8] which confronts us with the question of how to deal with relations between existing elite groups and future elite groups. The existing elites are prone to bring their kin into their own group, expecting their progeny to remain elites. But, from the perspective of social equity, it is undesirable to monopolize elite status in a modern society. The governing group must therefore adopt fair procedures to ensure both the smoothness of elite circulation and easy access to elite groups for future generations. These procedures may include giving examinations, awarding diplomas, open recruitment, voting by experts, selection, public bidding, and competitive bidding. The examination, public bidding, competitive bidding, and selection are fair in form and their evaluation indexes are objective. On account of their advantages with respect to culture and social relations, the ratio of the offspring of elites entering into elite groups will be slightly higher, as is the case in every country around the world.

During the ten years of the Cultural Revolution, the "National Higher Education Entrance Examination" ("高考") was abolished. As a result, traditional cadre circulation, as well as economic and technical elite circulation, was brought to a halt. The circulation of elites then adopted a "hyper-normal" approach. Some people began to obtain their access to the political elite group by becoming "Rebels" ("造反派"). That was why Deng Xiaoping asserted the need to revive the Entrance Examination as soon as the Cultural Revolution ended, so as to set elite circulation back on track. Through more than thirty years of effort, the circulation of Chinese technical elites has become a relatively full-fledged system. In these thirty-odd years, technical elite circulation has made great progress. The academic degree system has

8 G. Mosca. *The Ruling Class*. Beijing: Yilin Press, 2002. 173–74.

been restored and the evaluation and assessment systems for many kinds of technical titles have been established. But currently, the ubiquity of fake diplomas, the sale of diplomas, those studying simply for the sake of having a doctoral degree, "professors," and "tutors for Ph.D. candidates" reveal the many different problems that still exist with the circulation of technical elites.

With the establishment of a market system, economic elite circulation has been brought into being. Success or failure in running an enterprise is becoming the sole criterion for evaluating economic elites. A group of competitive entrepreneurs is becoming established. According to the 2007 China Entrepreneur Survey System, only 20.2% of respondents have an educational background of technical secondary school, high school or less; 35.6% have a junior college educational background; 26.6% have an undergraduate degree, and those with doctoral and postgraduate degrees and diplomas accounted for 17.6%. Among those, 31.3% majored in economics; 48.7% in management; 24.5% in science, engineering, agriculture, and medicine; 7% majored in art, history, philosophy, and law; and 13% majored in other disciplines.[9] It is obvious that the educational standard of China's economic elites has had a perceptible improvement. China's entry into the WTO further optimized the competitive and circulatory mechanism for economic elites, generating world-class economic elites and eliminating clumsy and low-level managerial staff from the ranks of entrepreneurs and elites.

In recent years, reforms have been made to the promotion system of cadres in China. Factors such as diplomas and schooling have been stressed and differential voting has been frequently adopted, notably improving the circulation system of the political elite group in China.

"The interchange of elites" refers to mobility among political, economic, and technical elites. A typical example of elite interchange can be found in the aforementioned reforms of separating the functions of government from those of enterprise and the streamlining of government agencies, where many political cadres became entrepreneurs. As has been discussed, elites are groups of outstanding people in a society. Therefore, it is reasonable for elite groups to undergo interchange. Practices in modern countries exhibit that elite interchange is common worldwide.

9 China Entrepreneur Survey Team. *Special Survey Report on the Growth and Development of Chinese Entrepreneurs 2007.* Beijing: China Entrepreneur Survey System, 2007. 3.

Before the economic reforms in China, it was not common for the upper classes to interchange. This was caused by both an absence of class identification and the fact that the rapid expansion of political elites and the extinction of economic elites prevented the formation of a base for elite interchange. Although a few technical elites became political elites, examples of political elites becoming technical elites were rare. However, it was an international practice for political elites to become technical elites. Henry Kissinger taking up teaching at a university after leaving his governmental position is a good example. In China, the transformation of political and technical elites into economic elites was referred as *xiahai* (下海) (plunging into the sea of business), which was not yet quite accepted by the public. Although elite interchange has benefits for class identification and social solidarity, this does not necessarily mean that all elites were interchangeable; certain preconditions are necessary for such interchange.

In China, the notion of interlocking elites was advanced by Professor Sun Liping, Professor Shen Yuan, and me in a paper in 1998.[10] The theory of interlocking elites originated with H. D. Lasswell and was developed by C. W. Mills in his book, *The Power Elite*—both used the term negatively. To Mills, the interlocking of elites in American society built a coalition of elites who were out of touch with the "average Joe."[11]

On the Chinese scene, the interlocking of elites has had both positive and negative aspects because the opposite of interlocking elites is the splitting apart of elites, which once led to catastrophe in China. When the People's Republic of China was newly established, a generation of elites—whose achievements were best displayed in the *Common Program of the CPPCC* (Chinese People's Political Consultative Conference)—was interlocked. Unfortunately, the anti-rightist campaign of 1957 resulted in the disintegration of this elite group and caused conflicts within the institution, eventually leaving China's elites in a position of long-term separation and conflict. China then fell into a series of long and troubled political struggles. It was not until the end of the Cultural Revolution that Deng Xiaoping called the National Conference on Science and National Congress of Education, a step that could be understood as an important measure to

10 Sun Liping, et al. "The Current Directions and Underlying Problems in the Transformation of Chinese Social Structure." *Strategy and Management* 5 (1998): 1–17.
11 Mills, C. Wright. *The Power Elite*. New York: Oxford UP, 1956. 7–8.

reverse the schisms between China's elites. In the mid- and late 1990s, the interlocking of economic, political, and technical elites was eventually achieved through building mutual trust and all-around cooperation. Such relations obviously maintained the stability and aided the development of Chinese society. In 2002, the Constitution and Program of the Chinese Communist Party was revised at the Sixteenth National Congress of the Communist Party of China. "New classes" were allowed to apply for party membership, marking the cooperation of the elites both within and outside the dominant institutions. Generally speaking, the interlocking of elites was conducive to social development and stability in China, but one should not ignore Mills' warning that the interlocking of elites can lead to a tendency to neglect the interests of the lower social classes.

7. The Important Function of the Grassroots

Stressing the great role of elites does not necessarily mean that one can neglect the important functions of the grassroots of a society, especially in a society like China's. As was mentioned above, the social structure of China was pyramid-shaped for a long time. The size of the lowest and most basic social stratum was huge. Provided that attention was paid to this group and their interests, no serious social problems would occur. Through the long and winding history of feudal rule in China, the low and basic stratum was made up of farmers. In times of intensified social contradictions, it can clearly be seen that China's many farmers were able to significantly influence the stability of society. Chairman Mao Zedong concluded that there had been more than one hundred large- and small-scale peasant uprisings in China since the late Qin Dynasty. The scale and frequency of these uprisings is unparalleled anywhere else in the world.[12] It is generally admitted by historians that there seems to be an inherent law of periodicity in Chinese history: whenever a new dynasty is set up, there is always a period of rapid growth until a peak is reached, after which time, owing to the centralization of wealth and the aggregation of the poor at the lower strata, social contradictions become intensified. Violent peasant uprisings may then be organized at the bottom of society and a new dynasty may arise. For thousands

12 Mao Zedong. *Selected Works of Mao Zedong*. Vol. 1. Beijing: People's Publishing House, 1967. 588.

of years, there were unceasing uprisings by farmers or the low, lower, and lowest social strata. That is why the governing authorities always emphasize social stability. A time of prosperity is called *dazhi* (大治) (great order)—for example, Wen Jing's or Zhe Guan's time—while a time of trouble is called *luanshi* (乱世) (troubled times), which actually refers to the troubles and resulting political rectification. Thus, the rise and fall of a dynasty results from social division and the intensification of contradictions. Successive Chinese rulers usually adopted lenient policies, which Professor Jian Bozan called "concessive policies" (让步政策), to narrow social divides, moderate social contradictions, and ease social tension. By such means, social stability would usually be regained.

By observing the slogans of peasant uprisings since the late Tang Dynasty, it is simple to discover that their criticism was directed increasingly explicitly at the lack of equity in property and wealth distribution. When unequal distribution of wealth threatened the survival of society's lower and lowest groups and the vital needs of farmers were not met, society entered a period of "high tension"—a tension of social structural. At such times, Chinese society became a powder keg awaiting the lighting of a fuse.

8. Mechanisms of Social Mobility

Social mobility refers to the vertical change of one's position on the ladder of society—from low to high or from high to low. Sound mechanisms of social mobility can usually be found in a fair and just country, where opportunities are offered for people in lower social strata to move to higher social strata. From ancient times, Chinese people have created and maintained good mechanisms of social mobility, which were naturally of a Chinese character. Although Chinese society was strictly stratified and social difference was immense, an Imperial Examination System was established as early as 587 AD. Considered globally, this was quite advanced at that time. In European and other countries, the aristocracy prevailed and titles were handed down based on family and familial lines of descent. There was no equivalent system to select talented people through examination in those countries. Besides, the Imperial Examination System was one that could convey talented people to top and upper social strata, enabling them to become governing officers or important social administrators through examination. Of course, people might well wonder whether it was the descendants of the rich or the offspring of the poor that tended to pass the examinations. According to extant historical data, 50% of those who passed the highest level of the Imperial Examinations were born to families of low social class from which nobody within three generations had passed any type of Imperial Examination. We should therefore admit that it was a comparatively effective mechanism of mobility. Later, a breakthrough was made by European countries in constructing mechanisms of mobility when civil service systems were acknowledged and market mechanisms developed. Since the market was open to all individuals, the Western mechanism gradually surpassed the Chinese one. This reminds us that mechanisms of mobility should be improved constantly.

As far as social mobility is concerned, modern Chinese society has a considerably high degree of social mobility. Sociologists divide society into two types: one type allows little or no possibilities for individual mobility; the other provides significant possibilities for individual mobility. For example, the aristocracy was a structure with little social mobility. Titles were inherited in such a way as to block the upward mobility of people from other strata to the top stratum. In the sixty years following the establishment of the People's Republic of China, China has had a comparatively high degree of social mobility, although the causes of social mobility have varied greatly. In the first thirty years, politics was the main reason for individual mobility; in the following thirty years, mobility came more from economics than politics. Based on my calculations, the rate of intergenerational mobility in the 1990s in China was similar to that of America. Both countries had high levels of social mobility, evidencing the vitality of the Chinese society.

9. The Significant Role of Family in Spanning Social Differences

The final feature is the significant role of Chinese family and clan structures in penetrating classes and maintaining social solidarity. The previous five features mentioned stress stratification, but social amalgamation takes place in all stratified systems. The eighth feature is the high level of social mobility through which social amalgamation is realized. In this section, emphasis will be placed on the role of kinship in social amalgamation. Family and clan have played essential roles in Chinese society. A "clan" refers to a group of people descended from the same ancestors or the same father, or who share the same surname. The size of a clan can increase generation after generation, for ten generations or even for one hundred years. Therefore, clans are usually huge in size. A "family" refers to a set of people related by blood and includes a few generations; this can also be referred to as an extended family, including the immediate family and related kinfolk such as uncles, aunts, nieces, and nephews. Such a set is not usually small. In a family or clan there will always be rich members and poor ones. So, clans and families are composed of people with different economic and class statuses. Though they differ in status, the blood bond and family culture—including such features as genealogy and family trees, as well as a sense of belonging—tend to maintain the internal unity of a family or clan.

As has been discussed above, Chinese society was stratified by politics before the economic reforms, but after the economic reforms it gradually became a society stratified by economics. Due to the tremendous changes in social structure, mixed classes exist in the same family. In a family, some brothers and sisters become rich while others remain poor. Some become bosses while some are unemployed. Unlike in Western countries, mutual support among family members is common in China. In such a way, kinship performs the function of transferring capital, narrowing class divisions within a family. In recent years, Chinese cities have welcomed the only-child generation. Family sizes have tended to be small. Only children have no siblings, thus weakening the family function of narrowing the class gap.

This chapter has attempted to summarize the general characteristics of Chinese social structure, which have left an obvious impression on the sixty years of history of the People's Republic of China, and which have developed many new features.

Chapter 3

Measurement of Social Stratification

The measurement of social stratification is the measurement of people's social status. Though the study of social status has a long history in the field of sociology, this chapter's purpose is not to probe into that history, but to provide a brief introduction to the measurement of social status. In order to measure social status, one must, first of all, be aware that multiple types of social status exist: economic, political, power, and prestige. Different methods of measurement are utilized to measure these various types of social status. As economic status is the most important of all these types, a great part of this chapter forms an introduction to the measurement of this type of status—that is, the measurement of people's income, property, and other forms of material wealth. An introduction will also be made in section 3 to the measurement of prestige status, since this measurement has the most sociological characteristics.

There are two approaches to the measurement of status: quantitative and qualitative. This chapter will give an introduction to quantitative measurement. First, however, I would like to acquaint the reader with the difficulties in taking measurements of income and economic status in China.

1. Difficulties in the Study of Measurement of Income and Economic Status

I have been conducting sociological research on people's incomes in China for years. Research on "income"—an economic subject—from the perspective of sociology naturally has its own special purpose. My primary concern is with differences in income because I regard income as an index reflecting the differences in status between people, so one can study the stratification of social status by studying this index. But why am I particularly concerned with this index?

People's economic status can be studied in terms of property and income, which are obviously related to each other. To put it in economic terms, property is monetary stock while income is a monetary flow. Income is the total amount of money gained during a specific period of time—for example, a year. A person's property is the amount that they accumulate from their yearly surplus, which naturally includes inheritance and similar windfalls. Compared to income, the composition of property wealth is far more complex. Some forms of property are impossible to convert into money, which makes analyzing property statistically even more challenging. A person, if they do not have a very clear picture of their "family possessions," can hardly estimate how much property their own family possesses. Given this situation, a simpler way of studying economic status is to study income, especially recent income, as people tend to be able to recall such details more accurately. Another reason for taking income as an indicator of economic status is that, in China, income is a very frequent topic in people's conversations. Just as with talking about the weather, talking about income is common, even in conversations between strangers on a train. Up until now, the contents of employee pay slips in most Chinese work places have been public information and have not normally been considered private. Therefore, questions about a person's income will not generally appear offensive.

As income has been set as a major indicator of social status in our research, it is worthwhile to mention the difficulties in conducting research on people's incomes in China. In rural areas, surveys on household income are not a matter that can be resolved with one or two questions. When employing a general question like, "What is your family's yearly income?", the surveyor is not likely to get a satisfactory answer from rural residents. This is because the composition of a rural resident's income is rather complicated. Without careful calculation, even the rural resident himself cannot possibly figure out what his own income is. Normally, a rural household's income consists of their income from crop production, which includes the part that is consumed by the family themselves and is, therefore, often ignored. It must additionally include income from livestock breeding and subsidiary production, such as raising poultry, breeding fish, and planting trees; pig, chicken, and duck breeding should also not be forgotten. Usually, the eggs consumed by the breeders are not included in income assessments. Finally, rural household incomes should not exclude gains from nonagricultural undertakings. These can be family-run handicraft businesses, small-scale manufacturing, transport services, restaurants, and shops. Income from part-time jobs and money sent back from family members who work as migrant workers in other places must also be taken into account as contributing to rural household income. To calculate income from such a variety of sources is a rather complicated task, even if it is assumed that all the respondents are telling the truth; if they are not, then things become even trickier.

Surveying urban residents' incomes is an equally challenging task. To start with, the composition of urban employees' salaries is growing increasingly complicated in China. An individual's salary is usually made up of their base pay, fluctuating salary, position-related pay, pay-scale salary, job subsidies, housing allowance, travel allowance, and telephone allowance, among others. Likewise, employees' bonuses are intricately categorized into team bonuses, department bonuses, monthly bonuses, quarterly bonuses, and year-end bonuses. It is often the case that the employees themselves cannot tell what constitutes their pay. These complications in calculating salary aside, an urban resident's income refers to not only the salary he receives from his job, but also his non-occupation-related income. Although income derived from one's occupation is comparatively simple to calculate because there are payrolls and bonus stubs to record it, assessing a resident's non-occupation-related income is much harder. This can be pay received from a part-time job, profits from the stock market, or gratuities given by friends and relatives for helping them run businesses, which are very often ignored and excluded from the resident's income in surveys. The survey is even more difficult when the person's non-occupation-related income exceeds his salary, which is the case for some urban residents.

Another intricacy of urban residents' incomes is that formal income and actual income are, more often than not, significantly different. Some urban governmental officials, for instance—judging from survey responses alone—have comparatively low monetary incomes, but for these individuals welfare or nonmonetary income is an important factor. For one thing, they have significant privileges and allowances when it comes to housing. It is even more complicated to calculate the benefits gained by some governmental officials from the special resources afforded by their power. If we converted all this into monetary terms, their incomes would be significantly higher. Based on their survey forms, an employee of a foreign-funded enterprise may seem to earn much more than a governmental official. However, if housing privileges and allowances—as well as other material benefits—are included, the actual incomes of the two are approximately the same, given that they have similar qualifications, work experience, educational backgrounds, and capabilities. That means an invisible market mechanism is actually working to regulate people's income levels in China.

Income can also be classified as on-payroll or off-payroll, as well as legal or illegal. It would be even more difficult to survey illegal incomes, such as incomes made from the gray and black markets.

Taking all these intricacies into consideration, I feel that it is necessary that some index be designed to test the credibility of reported incomes. Three methods have been attempted. The first is to verify income by studying consumption. This involves asking about expenses after inquiring about a family's income so that responses to each section can be compared. If consumption is significantly greater than income, it reveals an omission in reported income. The second way is to inquire about both the total income and its constituent parts—such as income from salary, part-time jobs, private businesses, stock investment, renting out real estate, and so on. This allows the researcher to check whether the constituent parts match the total. The third approach is to observe the housing conditions of the survey respondents and study the relationship between income and housing at different levels. Generally speaking, the income level of a family is reflected in their housing conditions. Families living in simple, poorly furnished buildings usually belong to low-income groups, while those housed in better buildings are at least fairly well-off. Those who can afford to live in upscale residential neighborhoods, expensive condominiums, or luxury villas are, without exception, from high-income groups. That is to say, there actually exist designators and visible features from which we can estimate a resident's or a family's income level. Simply put, research on economic status needs thorough investigation instead of relying solely on official statistics.

2. Specific Methods of Measuring Wealth Disparity and Status Difference

There are a lot of methods for measuring the disparities in property, income, and status; a few examples are the Inequality Index, the Equal Division Method, and the Gini Coefficient. An introduction to each of these devices is given below.

2.1 The Inequality Index

The Inequality Index is a device used to measure disparities in social stratification and social structure. To be more specific, the Inequality Index is used to indicate social inequality by determining the percentage of the highest-income earners and that of the lowest-income earners in the total population. This method inevitably involves defining the high- and low-income groups and calculating their incomes. Conventionally, those who are under the poverty line or with incomes lower than half of the average income are defined as being in the low-income group. Those with an income double or greater than the average are defined as being part of the high-income group. That is to say, once specific standards are set, the Inequality Index can be calculated and changes in the inequality levels of a specific country or area can be compared.

Table 3-1 demonstrates the changes in the Inequality Index for Chinese rural farmers since the reform and opening of the Chinese economy. The low-income group is defined as those farming families whose per capita annual net income is 50% lower than the average national income; the high-income group is made up of those with a per capita income of double or higher than the average.

Table 3-1 Inequality Index for Chinese Farmers in Rural Areas (1980-2005)

Year	Percentage of the Lowest-Income Group in the Rural Population (%)	Percentage of the Highest-Income Group in the Rural Population (%)	Inequality Index
1980	9.38	5.99	15.37
1985	12.22	13.26	25.48
1990	13.76	8.69	22.45
1995	7.97	12.27	20.24
2000	18.42	9.81	28.23
2005	19.97	13.16	33.13

The Inequality Index measures the relative level of inequality without taking people's absolute income level into account. This means that the Inequality Index measures only the relative inequality levels among different social groups as well as changes in the proportion of each group in the population. During the period between 1985 and 2005—although the absolute income level of Chinese rural farmers improved a lot—the level of inequality between the wealthy class and the low-income class rose as well.

2.2 The Equal Division Method

The Equal Division Method was first proposed by the British scholar F. W. Paish in an essay entitled "The Real Incidence of Personal Taxation."[13] The Equal Division Method divides the population into five equal-sized sections according to the size of their per capita income and measures the proportion each section forms of the total income of the population. A balanced distribution of wealth would be one in which one-fifth of the population possesses one-fifth of the total income of the population. Naturally, the actual proportions of the population and income are always unbalanced, thus showing the level of wealth disparity.

Table 3-2 demonstrates the income distribution of several countries measured by the Equal Division Method.

Table 3-2 Income Distribution in Several Countries Measured with the Equal Division Method

Population Divided into Five Sections	Brazil (%)	China (%)	The Russian Federation (%)	Indonesia (%)	Philippines (%)	The U.S. (%)	Sweden (%)	Japan (%)
The Lowest 1/5	2.6	4.7	6.1	8.4	5.4	5.4	9.1	10.6
The Second Lowest 1/5	6.2	9.0	10.5	11.9	8.8	10.7	14.0	14.2
The Middle 1/5	10.7	14.2	14.9	15.4	13.1	15.7	17.6	17.6
The Second Highest 1/5	18.4	22.1	21.8	21.0	20.5	22.4	22.7	22.0
The Highest 1/5	62.2	50.0	46.6	43.3	52.3	45.8	36.6	35.7

Source: World Bank. *World Development Index 2006*. Beijing: China Financial and Economic Publishing House, 2006. 76–78.

13 F. W. Paish. "The Real Incidence of Personal Taxation," *Lloyds Bank Review*. London: Lloyds Banking Group, 1957. 43.

As shown in the above table, the gap between the rich and the poor in Brazil is the widest, with the lowest fifth of the population possessing only 2.6% of the total income while the highest fifth possesses 62.2%. In contrast, countries like Sweden and Japan have comparatively narrow gaps between the rich and the poor, with the lowest fifths of the population possessing 9.1% and 10.6% of the total income respectively, while the highest fifths possess 36.6% and 35.7%. The three largest countries—China, the Russian Federation, and the U.S.—lie in the middle. However, compared with the other two, China's wealth disparity is more substantial. The proportion of the lowest fifth of the populations of China, the U.S., and Russia possess respectively 4.7%, 5.4%, and 6.1% of the total income, which shows that the lowest fifth in China has a comparatively low percentage of the national income among these three countries. The proportions of the highest fifth in the three countries possess 50.0%, 45.8%, and 46.6% respectively, demonstrating that the highest fifth of the Chinese population takes a higher percentage of the national income than those of the U.S. or the Russian Federation.

2.3 The Gini Coefficient

The Gini Coefficient is an index measuring the level and distribution of inequality that was developed by the Italian economist Corrado Gini and is based on the Lorenz Curve. Compared with other measurements, the Gini Coefficient is a more thorough and accurate indicator of inequality in the distribution of wealth and income, and is therefore commonly adopted in the study of social stratification. The Gini Coefficient is a value between zero and one. A Gini Coefficient of zero indicates absolute equality—that is, even distribution of income or wealth to every member of a society. However, a Gini Coefficient value of one indicates absolute inequality, which means income or wealth is centralized in the hands of one person without anything being distributed to anyone else in society. Obviously, neither of these two extremes can exist in any society, so the Gini Coefficient of any country or society is always a value greater than zero and smaller than one. A value closer to one indicates greater inequality in the distribution of wealth. Listed below are the Gini Coefficients of several countries (see table 3-3).

Table 3-3 Gini Coefficients of Several Countries

Country	Brazil	China	Russian Federation	Indonesia	Philippines	The U.S.	Sweden	Japan
Gini Coefficient	0.580	0.447	0.399	0.343	0.461	0.408	0.250	0.249

Source: World Bank. *World Development Index 2006*. Beijing: China Financial and Economic Publishing House, 2006. 76–78.

As shown in the above table, Brazil has the greatest Gini Coefficient (0.580) and Japan the smallest (0.249). China has a Gini Coefficient greater than the Russian Federation, Indonesia, the U.S., Sweden, and Japan, but smaller than the Philippines and Brazil. Using this method is a very convenient way to compare the levels of wealth disparity in different countries or societies. Although the Gini Coefficient uses the same method of measurement as the Equal Division Method in principle, the Gini Coefficient is comparatively simpler and more intelligible when making comparisons. It is, therefore, the most popular approach to measuring and comparing levels of inequality. At present, some countries have very great Gini Coefficients. According to the *World Development Index 2006* released by the World Bank, the countries with the ten greatest Gini Coefficients are: Namibia, ranked first with a Gini Coefficient of 0.74; followed by Lesotho, 0.632; Botswana, 0.630; Sierra Leone, 0.629; The Central African Republic, 0.613; Swaziland, 0.609; Bolivia, 0.601; Haiti, 0.592; Columbia, 0.586; and Brazil ranked tenth with a value of 0.580. In contrast, the top seven countries with the smallest Gini Coefficients are: Denmark, 0.247; Japan, 0.249; Sweden, 0.250; The Czech Republic, 0.254; Norway, 0.258; Finland, 0.269; and Germany, 0.283.

Differences in social status are not only reflected in one's economic situation, but also in one's occupation and education. Therefore, several other methods of measuring inequality of social status will be introduced as follows.

2.4 Socioeconomic Score (SES)

A common approach to the measurement of social status is the Socioeconomic Score, or SES, a method first proposed by the American sociologist and social statistician O. D. Duncan.[14] An SES is obtained by examining and assessing individuals' incomes—as well as educational and occupational status by scoring their income, educational, and occupational statuses respectively—and then calculating a final score. The rating of income status and educational status is comparatively easy to undertake; income can be rated according to the monetary value of one's income, and education can be scored according to the level of schooling completed or education received. It is, however, more difficult to rate occupational status. The main factor in deciding a person's occupational status is occupational prestige, including their power status and income status. Prestige status, power status, and income status are interrelated. Considering the situation in China, I have made some adaptations to commonly used socioeconomic scoring and tried to use them in measuring China's metropolitan residents' socioeconomic status, as shown below (see table 3-4).

Table 3-4 Socioeconomic Score of Metropolitan Residents in China

Education, income and occupation	Categories /Items	Score
I. Education	Graduate Schooling and Above	7
	University	6
	Junior College	5
	Senior High School, Vocational School, Secondary School	4
	Junior High School	3
	Elementary School	2
	Illiterate/Barely Literate	1

14 O. D. Duncan. "A Socioeconomic Index for All Occupations and Properties and Characteristics of the Socioeconomic Index." *Occupations and Social Status.* Ed. A. J. Reiss. Glencoe, Ill: Free Press, n.d. 109-161.

Education, income and occupation	Categories /Items	Score
II. Household Members' Per Capita Income (RMB: yuan)	6001 and above	7
	4001–6000	6
	3001–4000	5
	2001–3000	4
	1001–2000	3
	501–1000	2
	500 and below	1
III. Occupation	Senior Managers and Professionals	7
	Middle Managers and Professionals	6
	Junior Managers and Professionals	5
	Clerks	4
	Skilled Workers	3
	Manual Workers	2
	Temporary Workers, Migrant Workers from the Countryside, and the Unemployed	1

In the above table, household per capita income is taken as the index for income status, because an individual's income is always affected by their family members' income or expenses. Here "family members" means those living together and economically not separated under common economic accounting. According to the table, the criterion for scoring the comprehensive value of people's socioeconomic status (which was divided into seven categories) is as follows:

Top level: total 21 points (full mark);
Upper level: 18–20 points;
Middle upper level: 15–17 points;
Middle level: 13–14 points;
Middle lower level: 9–12 points;
Lower level: 6–8 points;
Bottom level: 3–5 points.

2.5 International Socio-Economic Index of Occupational Status (ISEI)

Sociologists H. B. Ganzeboom, Paul M. de Graaf, and Donald J. Treiman made an attempt to establish an internationally applicable index to measure people's social and economic status, which later became known as the International Socio-Economic Index of Occupational Status, or ISEI. The index was based on three indicators: income, education, and occupational prestige. The aforementioned sociologists collected thirty-one sets of data from sixteen countries. The data was collected between 1968 and 1982. Among the sixteen countries in question were included both developed countries like the U.S. and underdeveloped countries like India. The data geographically covered East Asia and South America, and was thus highly representative. They selected 271 occupations from the International Standard Classification of Occupations formerly constructed by Professor Treiman and—after rating income, education, and occupational prestige using their data—they arrived at an International Socioeconomic Index of Occupational Status for the 271 occupations. With the ISEI, it is now possible to know a person's status index from their occupation and to compare different people's statuses. Table 3-5 is a list of the ISEI of seven categories of occupations, each of which is composed of the average of all the occupations in each category. Table 3-6 is a list of partially detailed occupations in category five: Service Personnel and their ISEI.

Table 3-5 ISEI of Seven Categories of Occupations

Occupational Categories	ISEI
Professional, Technical, and Related Staff	67
Administrative Staff and Enterprise Managers	67
Clerical Staff	49
Sales Personnel	51
Services Personnel	38
Agricultural, Animal Husbandry, and Forestry Personnel	25
Production, Transport Workers, and Manual Laborers	34

Table 3-6 ISEI of Partial Occupations in Category Five: Services Personnel

ISEI Score	Major Professional Group Code	Minor Professional Group Code	Professional Subclass Code	Professional Sub-Subclass Code	Occupation
40	5000				Service & Shop Marketing Staff
38		5100			Personal Service & Security Personnel
34			5110		Travel/Courier Service & Related Personnel
34				5111	Tourist Service Personnel & Train Crew
34				5112	Train Conductors
34				5113	Travel & Museum Guides
32			5120		Janitors & Restaurant Service Personnel
30				5121	Building Management Staff/Housekeepers
30				5122	Cooks
24				5123	Waiters, Waitresses, & Bartenders
25			5130		Private Maids & Personal Care Servants
25				5131	Nurses & Baby-Nurses
25				5132	Personal Care Staff in Institutions
25				5133	Personal Care Staff at Home
25				5139	Other Care Staff
30			5140		Other Personal Care Staff
29				5141	Hairdressers, Barbers, & Beauticians
19				5142	Domestic Staff (Valets & Maidservants)
54				5143	Undertakers, Embalmers, Funeral Servants
19				5149	Other Personal Care Staff, etc.

ISEI Score	Major Professional Group Code	Minor Professional Group Code	Professional Subclass Code	Professional Sub-Subclass Code	Occupation
43			5150		Astrologers, Fortune-Tellers, etc.
43				5151	Astrologers & Related Personnel
43				5152	Fortune-Tellers, Palmists, & Related Personnel
47			5160		Protective Services Personnel
42				5161	Firefighters
50				5162	Police Officers
40				5163	Prison Guards
40				5164	Armed Forces, Soldiers
40				5169	Other Protective Services Personnel
43		5200			Salespersons, Models, & Demonstrators
43			5210		Fashion & Other Models
43			5220		Shop Salespersons & Models
37			5230		Stall & Market Salespeople

With a scale like this, it is possible to assign scores to all occupations and, eventually, quantitatively evaluate the social status of a specific group or the whole population of a society. Needless to say, adaptations might sometimes be needed when applying such a scale according to the situations of different countries.

2.6 Self-Evaluation and the Evaluation of Others

The topic of this section is the subjective evaluation of economic status. Subjective evaluation can be divided into two kinds: self-evaluation and evaluation by others. More accurately, evaluation of the living standard, income level, and wealth status of individuals or groups can be made from two angles: one is the evaluation of a person's living standard by that person himself and his group, the other is evaluation by other people. These two evaluations usually lead to divergent results, as the evaluators have different experiences and standards of assessment. For instance, when a survey is conducted in a poverty-stricken area, surveyors may believe that some families should be considered poor households, while members of those families tend to believe that their living standards are at the upper-middle level and would not categorize themselves as poor. This is what happened when a survey was carried out in areas inhabited by various ethnic groups in Yunnan province. People from outside counted most of the residents of the village as belonging to the poor, but the villagers themselves thought they were quite well-off. To demonstrate the differences between the two evaluations, I shall now list the results of a door-to-door questionnaire survey conducted several years ago by my students among urban residents of Beijing. A random sampling of 630 families from thirty neighborhood communities in eight districts in Beijing was surveyed, among which 602 proved valid. The survey respondents were asked: "What level do you think your household income is at with respect to Beijing City?" The surveyors were also asked to independently make the same assessment of the respondents' household incomes with respect to Beijing City. Answers to these two questions are arranged below in table 3-7.

Table 3-7 Self-Evaluation and Others' Evaluation of Family Income Level

		Surveyors' Evaluation (Number of Samples)						Total
		High	Upper-Middle	Middle	Lower-Middle	Low	Missing	
Self-Evaluation	Upper-Middle	4	12	4	1			21
	Middle	6	74	131	32	5	1	249
	Lower-Middle		14	114	96	10		234
	Low		4	14	30	50		98
Total		10	104	263	159	65	1	602

The data in the table is composed of the correspondents' and the surveyors' separate evaluations. In general, self-evaluation was lower than the evaluation of the surveyors. Among the self-evaluations, none of those surveyed categorized themselves as being part of the high-level group, whereas the surveyors regarded ten as being so. Of the 234 respondents who counted themselves as belonging to the lower-middle level income group, fourteen were assigned to the upper-middle group by the surveyors, 114 to the middle group, and ten to the low group. Of the ninety-eight respondents who counted themselves as belonging to the low-level income group, four were placed in the upper-middle group by the surveyors, fourteen to the middle group, and thirty to the lower-middle group. It is obvious that self-evaluations are significantly different from the other evaluations. What, then, is the point of conducting both forms of evaluation? In China's present situation, people's self-evaluations of their income status and wealth status are very important. Disparity is mainly embodied as the difference in objective living conditions in a society where some people are still struggling to survive; however, when survival is no longer a problem for most members of a society, subjective attitudes towards wealth disparity (especially self-evaluation of property) become significant. If most members of a society think their living standards are at middle or upper-middle level, a stable mentality is formed in that society; if most members think that their living standard is low, the society will be unstable and is likely to develop a series of social problems, or even turbulence. Subjective self-evaluation is usually unstable since it reflects the respondents' mentality rather than their reality; it is often incorrect and therefore needs to be adjusted by others' evaluations.

3. Measurement of Occupational Prestige

As mentioned above, the subjective evaluation of social status is an important tool for measuring social status. The measurement of social status that has the most particularly sociological characteristics is the measurement of occupational prestige. There are three dimensions to the evaluation of social status in sociology: wealth status (economic status), power status, and prestige status. Prestige status is a subjectively assigned status. Technically defined, prestige status is measured based on the results of others' evaluations. To be more concrete, it is the societal acknowledgment of an occupation or a status. The identification of wealth status and power status is comparatively easier because both of them have unambiguous and objective indicators. The measurement of prestige status is more challenging because it involves subjective evaluation, which is unavoidably more arbitrary. The measurement of prestige status is usually achieved through prestige surveys, and the most common focus of this type of survey is measuring occupational prestige.

Largely due to the influence of "leftist" ideas, no study has yet been made in China on prestige. Actually, differences in occupational prestige are objective phenomena that exist universally in every society. For instance, during one of my prestige surveys, a respondent asserted that all occupations are of equal status and should not be graded differently. I asked him, "What occupation would you wish your own child to choose in the future?" He said he certainly would like his child to have a better occupation. I then asked him to grade different occupations by number, imagining that his own child were to choose one of the occupations listed. In the end, different scores were assigned to different occupations. Therefore, even in a society such as China's, there are differences in occupational status that appear when individuals choose occupations.

Compared with research conducted in the international sociological community, there has been a dearth of occupational prestige research in China. First, empirical studies and prestige surveys are rarely conducted in China. In the currently available literature, there are four important studies—all of which were carried out before 1990. The earliest occupational prestige survey in China was conducted in Beijing by Professor Lin Nan and Mr. Xie Wen in 1983. The survey measured the prestige of fifty occupations. The results of this survey showed that doctors rank the highest in their occupational status, followed by engineers. The first few top places in the status list were all intellectual occupations. The most impressive thing about the results is that cadres at the bureau and departmental level (who were thought to enjoy a high status) ranked below high school teachers. The main problem with the survey is the rough classification of occupations using general titles such as "university teachers," "natural scientists," "social scientists," and so on, which reflects the underdeveloped classification of occupations in China at that time. Afterwards, in July 1985, another occupational prestige survey covering ten occupations was conducted in China by the editors of the journal *Chinese Youths*. The survey, surprisingly, collected more than seventy thousand responses.[15] However, the survey adopted

15 "Analysis of the Results of a Questionnaire Survey: Through 70,000 Questionnaire Responses." *China Youths* 9 (1985): 5–9.

a nonstandard classification of occupations. College students and postgraduates were considered occupations, and cadres and technical professionals were put into the same category. The results showed that college students and postgraduates ranked the highest while peasants ranked the lowest. Because of this irregular classification, subsequent analyses of the survey's results were inevitably affected. In the same year another survey, led by Hong Biao, was conducted in Beijing among 572 residents covering ninety occupations. The results showed that professors, economists, painters, writers, and lawyers were ranked as the top five occupations in terms of prestige status. After that, in 1989, a prestige survey on rural occupational status was carried out among China's rural residents by Dai Jianzhong, which revealed rural residents' evaluations of occupational status. The survey classified rural occupations into seventeen categories. The results showed that governmental officials, soldiers, and police officers were the top three occupations in terms of prestige—a completely different result from that provided by urban residents. This survey also had some deficiencies, both in its classification of occupations and in its methodology.

Fortunately, the occupational prestige survey carried out in Beijing and Guangzhou in 1990 by Jiang Laiwen from Peking University was of a much higher academic standard, both in its method of measurement and in its classification of occupations. Three conclusions were drawn from this survey: respondents from different subgroups had basically the same perceptions of occupational prestige; occupational prestige varies by area; and, compared with the occupational prestige survey made in Beijing in 1985, the occupational prestige gap between occupations was significantly narrower. The survey also studied the key variables for Chinese evaluation of occupational prestige, including professional knowledge and technology, amount of power, working environment, irreplaceability, income, and ethics, among others.[16]

From 1997 to 1998, when I was giving lectures on "social stratification and mobility" to undergraduate students at Renmin University of China, I had my students from the Sociology Department conduct two occupational prestige surveys in Beijing. The data listed in this chapter is from one of those two surveys.[17] Quota sampling was adopted and

16 Jiang Laiwen, et al. "A Study on Occupational Prestige in Beijing and Guangzhou." *Sociology and Social Survey* 4 (1991): 11–15.

17 The two surveys acquired very similar results but, due to some defects in the verification of the data in the second survey, this chapter quoted the data from the first survey.

the proportion of each quota was based on the distribution of people's occupations in Beijing. The survey was made of 468 respondents: 55.7% were male and 44.3% female; 8.9% were between eighteen and twenty-nine years old, 21% between thirty and thirty-nine, 22.2% between forty and forty-nine, 17.7% between fifty and fifty-nine, and 30.1% above sixty. The questionnaire covered one hundred occupations common in Beijing. The respondents were asked to grade each of the following occupations based on their own comprehensive evaluation of their prestige. The score values were then collated according to the North-Hatt Occupational Prestige Scale. The results of this survey are shown in table 3-8.

Table 3-8 Occupational Prestige Survey in Beijing in 1997

Ranking	Occupation	Prestige Score	Ranking	Occupation	Prestige Score
1	Scientists	88.95	51	State Organizations, Office Clerks	57.16
2	University Professors	86.37	52	Postmen	57.03
3	Engineers	82.97	53	Soldiers	56.97
4	Physicists	81.94	54	Property Developers	56.38
5	Doctors	79.98	55	Tour Guides	56.07
6	Economists	79.23	56	Publishers	56.03
7	Sociologists	78.57	57	Hairdressers/ Beauticians	55.47
8	Judges	78.25	58	Librarians	54.96
9	Pilots	76.84	59	Labor Union in Units, President	54.68
10	Procurators	76.82	60	Small Enterprise, Managers	54.55
11	Lawyers	76.58	61	Insurance Salespeople	54.07
12	Architects	74.96	62	Employees, Foreign Companies	53.38
13	Senior Officers	74.95	63	Bus Drivers	51.94

Ranking	Occupation	Prestige Score	Ranking	Occupation	Prestige Score
14	University Faculty	74.94	64	Workshop Directors	51.94
15	Governors of Banks	74.47	65	Farmers	51.57
16	Translators & Interpreters	73.40	66	Fashion Models	51.39
17	Musicians	72.74	67	Garage Workers	51.11
18	Writers	72.27	68	Train Stewards/ Stewardesses	51.09
19	Painters	71.45	69	Electricians	50.34
20	Coaches	71.33	70	Sanitation Workers	50.17
21	Journalists	71.22	71	Steelworkers	49.85
22	Editors	71.00	72	Veterinarians	49.79
23	TV Program Anchors	70.64	73	Tailors	49.51
24	Radio Announcers	70.39	74	Taxi Drivers	49.04
25	Athletes	69.27	75	Neighborhood Committee Chiefs	48.91
26	State Organizations, Bureau Directors	69.19	76	Purchasing Agents	48.57
27	Presidents of Companies	68.71	77	Paging Service Operators	48.44
28	Film Directors	68.44	78	Housing Bureau, Clerks	48.43
29	Secondary & Primary School Teachers	68.31	79	Carpenters	48.14
30	Intermediate Military Officers	68.03	80	Textile Workers	48.12
31	Fashion Designers	67.39	81	Ticket Sellers	47.93
32	State Organizations, Division Chiefs	65.64	82	Stock Speculators	47.37
33	Customs Officers	64.98	83	Miners	47.34
34	Geological Prospectors	64.23	84	Private Business Owners	47.15
35	Company Managers	64.04	85	Pop Singers	47.06
36	Kindergarten Teachers	64.03	86	Construction Workers	46.88
37	Nurses	63.93	87	Pressmen	46.30

Ranking	Occupation	Prestige Score	Ranking	Occupation	Prestige Score
38	Air Stewardess	63.76	88	Small Shop Owners	46.03
39	Large Enterprises, Directors	63.55	89	Cooks, Small Restaurants	45.75
40	Firefighters	63.25	90	Shop Assistants	45.60
41	Accountants	63.10	91	Township Enterprises, Workers	43.64
42	Public Security Officers	62.13	92	Migrants Farmer-Businessmen in Cities	42.21
43	Senior Employees, Foreign Companies	61.91	93	Security Personnel in Units	42.20
44	Agro-Technicians	61.85	94	Migrant Workers in Cities	41.71
45	Chefs, Big Restaurants	59.49	95	Porters	41.22
46	Bank Clerks	58.95	96	Nannies	38.80
47	Traffic Police	58.50	97	Labor Contractors	37.63
48	Actors /Actresses	58.48	98	Salvage Collectors	36.61
49	Computer Dealers	57.67	99	Rickshaw Pullers	36.51
50	Industrial and Commercial Tax Collectors	57.50	100	Gatekeepers	36.37

People often wonder whether scores assigned to occupational prestige are really reasonable. As we know, a prestige survey measures the opinions of every individual, and each individual has opinions different from those of others. Judging from the data collected in the survey, for instance, both the highest and the lowest scores were assigned to each occupation by different respondents. The calculation was made by working out the cumulative scores assigned by all the respondents, which is exactly what an opinion survey should do—everyone's opinion should matter and affect the final results, but only very slightly. The final scores are cumulative totals, thereby representing everybody's opinions. Occupational prestige evaluations are neither a reasonable nor unreasonable matter—as long as the design, methodology, and analysis of the survey are carefully and seriously conducted, the evaluation will be of value.

The fact that occupations such as scientists, university professors, engineers, physicists, and doctors rank foremost in the table reflects that factors such as knowledge, technology, and education are the most important in people's evaluations of occupational prestige. Occupations involving those elements tend to get high scores from respondents, a trend that is universal to respondents of any social class.

Interestingly, governmental officials' prestige is not high. In international occupational prestige evaluations, those occupations symbolizing justice often get quite high scores. For example, in a lot of countries, judges of the supreme judicial courts usually rank first and governmental officials are also often highly rated. However, in Beijing's occupational prestige survey, governmental officials got comparatively low scores. Traditionally, official rank was considered to be the sole criterion of one's worth. Officials were highly respected, but the survey here shows the decline of that tradition. Governmental officials are generally nowadays censured and do not get high scores. Some respondents said that they did not have good impressions of governmental officials. Industrial and commercial tax collectors did not get high scores either, largely due to their commonplace abuse of power and racketeering. That is to say, widespread corruption and a deterioration of social morality had caused governmental officials to be held in low esteem. Further analysis on this point will be made as follows.

Farmers have long been stereotyped as having very little prestige, but in our survey their scores are not low. Farmers' prestige scores were, surprisingly enough, higher than even fashion models. That shows that when people evaluate an occupation, they attach importance to its contribution to society. Agriculture-oriented thought has traditionally been prevalent in China because of the idea that we would have nothing to eat without farmers. The importance of social contribution is also reflected in the scores given to soldiers, which were higher than those for occupations like property developers and publishers.

In general, people tend to set their occupational status first and then evaluate others' occupations by taking their own as reference point. This means that their opinions about their own occupation will affect their evaluation of other people's occupations. In our survey, we found that most people have a high estimation of their own occupations. For example, respondents working in governmental offices assigned higher scores to directors, section chiefs, and clerks in state organizations; migrant workers in cities also assigned high scores to themselves. Of course, there were also a few people who, despairingly, gave very low scores to their own and similar occupations.

Personal preference and personal experiences can also affect people's evaluations of an occupation. For example, respondents who had encountered doctors that expected them to pay bribes gave low scores to doctors, and those who had watched television reports about corrupt governmental officials assigned low scores to governmental officials. Some respondents gave low scores to bus conductors, simply because they had been treated rudely by some of them. Drivers did not normally evaluate traffic police highly because traffic police frequently caused problems for them. One respondent who worked in an ineffective factory assigned low scores to managers of enterprise simply because he thought that his boss's corruption, extravagance, and incompetence were the causes of his low income. Beijing residents gave low scores to migrant workers in their city, but not to farmers working in their hometowns, because they blamed these workers for traffic problems, rising crime rates, and the unemployment of urban residents. Generally speaking, those occupations with which people had regular contact tended to be evaluated based on the respondents' own experiences and impressions. Since every respondent evaluated occupational prestige according to his own experiences, the results acquired from these people can be said to be fair because everyone recorded different appraisals according to their own experiences.

Additionally, concepts conveyed through media such as newspapers, radio, and television had an impact on people's evaluations. For example, sanitation workers—who had often been portrayed by the mass media as dedicated workers with beautiful souls—ranked the highest among nontechnical workers, even higher than steelworkers and carpenters. Also, the frequent repetition of such slogans as "science and technology are the primary productive forces" and "China can be revitalized through science and education," led to science- and education-related occupations being highly estimated.

Compared with previous surveys, this survey recorded great changes in some occupational statuses. In 1990, writers ranked first on the prestige scale in Beijing's occupational prestige evaluation, but their rating fell to 18th in the present survey. Painters plunged from 7th place to 19th, fashion designers from 11th to 31st, pop singers from 47th to 85th, and fashion models from 48th to 66th. There were, of course, some occupations that retained their positions. For example, both in the 1990 survey and in this one, professors, engineers, physicists, and doctors were among the top five.

In May and June 2009, when I was giving lectures on "social stratification and social mobility," I had my students conduct another occupational prestige survey in Beijing.[18] Quota sampling was adopted and the rate of quota was based on the distribution of Beijing residents' occupations in 2008. This time, they acquired 430 valid samples. Of them, 56.5% were male and 43.5% female; 60.5% were under thirty years old, 30.5% between thirty and forty-nine, and 9% between fifty and seventy. The questionnaire included ninety-nine occupations with a similar classification of occupations to that used in the 1997 study. Listed below (in table 3-9) are the averages of the scores assigned by the respondents to those occupations, according to their own opinions.

Table 3-9 Occupational Prestige Survey in Beijing in 2009

Ranking	Occupation	Prestige Score	Ranking	Occupation	Prestige Score
1	Scientists	83.46	51	State Organizations, Clerks	56.85
2	Physicists	79.09	52	Labor Union in Units, President	56.76
3	Pilots	76.92	53	Actors & Actresses	56.57
4	University Professors	76.88	54	Commercial and Industrial Tax Collectors	56.50
5	Senior Military Officers	76.58	55	Private Business Owners	56.47
6	Company Presidents	75.30	56	Librarians	55.86
7	Sociologists	75.20	57	Fashion Models	55.54
8	Economists	75.18	58	Nurses	55.00
9	Musicians	74.12	59	Pop Singers	55.00

18 I would like to express my gratitude to those students who attended the course entitled "Social Stratification and Mobility" at Tsinghua University in 2009 and took part in the survey. Special thanks are given to Liu Haiyang and Li Ying, who did a lot of data processing work for the project.

Ranking	Occupation	Prestige Score	Ranking	Occupation	Prestige Score
10	Doctors	73.66	60	Housing Bureau, Clerks	54.87
11	Judges	73.54	61	Soldiers	54.86
12	Governors of Banks	72.76	62	Veterinarians	53.95
13	Prosecutors	72.53	63	Neighborhood Committee, Chiefs	53.40
14	State Organizations, Bureau Directors	71.69	64	Tour Guides	53.22
15	Engineers	71.47	65	Stock Speculators	52.91
16	Customs Officers	71.17	66	Postmen	52.61
17	Lawyers	70.67	67	Workshop Directors	52.24
18	University Faculty	70.56	68	Computer Dealers	51.14
19	Architects	70.05	69	Purchasing Agents	51.12
20	Directors of Large Enterprises	69.35	70	Electricians	50.79
21	Senior Employers, Foreign Companies	68.85	71	Small Shop Owners	50.49
22	TV Program Anchors	68.21	72	Taxi Drivers	50.19
23	State Organizations, Division Chiefs	67.55	73	Bus Drivers	49.72
24	Writers	67.41	74	Beauticians/ Hairdressers	49.39
25	Film Directors	66.16	75	Steelworkers	49.34
26	Painters	66.15	76	Migrant Businessmen in Cities	49.26
27	Translators & Interpreters	66.14	77	Sanitation Workers	49.23
28	Fashion Designers	66.01	78	Carpenters	48.39
29	Company Managers	66.00	79	Train Stewards/ Stewardesses	48.02
30	Editors	65.67	80	Cooks at Small Restaurants	47.91
31	Journalists	65.10	81	Farmers	47.74
32	Secondary & Primary School Teachers	64.93	82	Tailors	47.70
33	Geological Prospectors	64.26	83	Garage Workers	47.53
34	Athletes	64.07	84	Labor Contractors	47.47

Ranking	Occupation	Prestige Score	Ranking	Occupation	Prestige Score
35	Radio Announcers	64.03	85	Township Enterprises, Workers	47.21
36	Intermediate Military Officers	63.78	86	Printers	47.00
37	Coaches	63.60	87	Textile Workers	46.96
38	Air Stewardesses	62.28	88	Ticket Sellers	46.41
39	Public Security Officers	62.24	89	Miners	46.22
40	Chefs of Grand Hotels	62.14	90	Migrant Workers in Cities	45.63
41	Directors of Small Enterprises	62.10	91	Insurance Agents	44.59
42	Agro-Technicians	61.81	92	Construction Workers	44.51
43	Kindergarten Teachers	61.30	93	Paging Service, Operators	42.44
44	Firefighters	60.58	94	Nannies	41.14
45	Publishers	59.23	95	Porters	41.05
46	Property Developers	59.09	96	Gatekeepers	40.70
47	Employees of Foreign Companies	58.30	97	Security Personnel in Units	39.84
48	Traffic Police	57.83	98	Rickshaw Pullers	39.58
49	Bank Clerks	57.51	99	Salvage Collectors	38.84
50	Accountants	57.16			

Compared with the 1997 survey in Beijing, the 2009 survey saw seven changes in occupational prestige.

First, there were changes in the top ten occupations. As is shown in the above table, the first ten occupations with the highest scores among the ninety-nine listed occupations are scientists (83.46), physicists (79.09), pilots (76.92), university professors (76.88), senior military officers (76.58), company presidents (75.30), sociologists (75.20), economists (75.18), musicians (74.12), and doctors (73.66). In 1997, the top ten occupations were scientists, university professors, engineers, physicists, doctors, economists, sociologists, judges, pilots, and prosecutors. On the whole, the top ten occupations in the 2009 survey were those of an intellectually high level and those involving a high-level of professional techniques and skills (such as piloting aircraft, medical treatment, engineering and playing music), reflecting respondents' high estimation of knowledge and professional skill. Scientists and sociologists ranked first and seventh respectively, retaining stable prestige over the years. Compared with the 1997 survey, changes can be observed in the 2009 survey. As far as ranking was concerned, legal professionals such as judges and prosecutors—who had high prestige ten years ago—were no longer among the top ten. This is unusual, as judges and prosecutors were consistently ranked as the top two professions in international prestige surveys. Doctors (who ranked fifth in 1997) fell to tenth place in 2009, but physicists and economists ranked slightly higher in 2009 than before. Pilots rose from ninth to third place, and three new occupations joined the top ten: musicians, senior military officers, and company presidents. They rose from seventeenth, thirteenth, and twenty-seventh respectively in 1997 to ninth, fifth, and sixth respectively in 2009. Among these, the change in the prestige of company presidents was the most significant, suggesting that with the development of a market economy in China, enterprise management personnel became more highly esteemed by the people. Their score was even higher than that of administrative governmental personnel.

It is also important to examine the range of changes to occupational prestige. Further comparison showed that some occupations had a high range of change in 2009. Compared with the rankings from 1997, the ten occupations that saw the greatest changes in prestige in 2009 were: private business owners (rising by twenty-nine places), pop singers (by twenty-six places), senior employees of foreign companies (by twenty-two places), company presidents (by twenty-one places), directors of large enterprises (by nineteen places), directors of small enterprises (by nineteen places), clerks of housing bureaus (by eighteen places), customs officers (by seventeen places), stock speculators (by seventeen places), and small shop owners (by seventeen places). These are mostly senior or middle management personnel or staff in non-state-owned enterprises, as well as a few civil servants.

The ten occupations that experienced the greatest drop in rank in 2009 were: insurance agents (by thirty places), nurses (by twenty-one places), computer dealers (by nineteen places), operators of paging services (by seventeen places), beauticians and hairdressers (by seventeen places), coaches (by seventeen places), garage workers (by sixteen places), farmers (by sixteen places), postmen (by fourteen places), and engineers (by twelve places). Among these, insurance agents fell to 91st place. The surprisingly low evaluation of insurance agents might be the result of people's resentment of their constant promotion of their products, leading to a low occupation prestige rating. These enormous changes in occupational prestige evaluation were manifestations of the changes wrought in people's values by social transformation.

Changes also took place in the prestige scores of educational and research occupations. These occupations include scientists, university professors, physicists, economists, sociologists, university faculty, primary and secondary school teachers, and kindergarten teachers. Among these professions, scientists and sociologists maintained a stable level of prestige, ranking first and seventh respectively in both surveys. The prestige of university professors decreased slightly; lecturing faculty members were also graded lower; economists also fell by one place; and elementary school, secondary school, and kindergarten teachers all suffered losses of prestige in the 2009 survey.

Despite this, educational and research occupations had comparatively high status, with five out of the eight occupations placed in the top ten. This is evidence of people's high evaluation of these occupations, especially those concerned with research such that they require professional knowledge. However, four teaching occupations—including university professors, university lecturers, primary and secondary school teachers, and kindergarten teachers—all fell in the rankings.

There was also a rise in the prestige of economic and finance-based occupations. These occupations included company presidents, governors of banks, directors of large enterprises, company managers, directors of small enterprises, and stock speculators.

Among these occupations, company presidents made it to the top ten, while governors of banks and directors of large enterprises were among the top twenty; stock speculators made especially significant increases in their status, moving from 82nd place to 65th. The economic and finance-based occupations experienced a general improvement in prestige status. Company presidents and directors of both large and small enterprises gained nearly twenty places, indicating that China's market economy had changed people's criteria for evaluating these occupations.

Changes can also be observed in the prestige of governmental officials and civil servants. These include bureau directors in state organizations, customs officers, division chiefs and clerks in state organizations, industrial and commercial tax collectors, and clerks in housing bureaus.

The 2009 survey showed that people evaluated governmental officials more positively than before. Bureau directors and division chiefs in state organizations reached higher levels (moving up twelve and nine places) to reach a rank in the top half of the list—14th and 23rd places respectively. Customs officers and clerks in housing bureaus also rose in rank significantly—the former rising from 33rd place to 16th and the latter from 78th to 60th. However, unlike governmental officials, the prestige of clerks in state organizations remained unchanged, ranked at 51st place as before. This enhancement in the prestige of governmental officials and civil servants is closely related to the phenomenon of university graduates competing fiercely for civil service positions. This might also have something to do with the revival of the traditional "mandarinate-oriented" ("官本位") idea of official rank being considered the sole criterion of one's worth, which faded away in the early stages of China's economic reform but has undergone a resurgence in recent years.

The prestige of farmers also underwent significant changes. This category includes three groups: migrant businessmen, migrant workers in cities, and farmers in the countryside. Among these, the prestige of migrant businessmen increased greatly, with their rank rising from 92nd place in 1997 to 76th place in 2009—even higher than ticket sellers and insurance salespeople, and approaching the prestige of skilled workers like steelworkers and carpenters. Migrant workers in cities slightly enhanced their prestige as well, rising from 94th place in 1997 to 90th in 2009. However, farmers in the countryside underwent a significant depreciation in prestige status, dropping from 65th place in 1997 to 81st place in 2009. Apparently, despite farmers' great contributions to the development of cities and the great attention paid to them by Chinese society in recent years, farming is not ranked particularly highly in terms of prestige. Even farmers based in rural areas received poor evaluations, probably due to their low efficiency and low incomes.

There were also changes in the prestige of property developers. Despite the rapid development of the real estate business in the last twenty years, the prestige of property developers remained intermediate, ranked at 54th place in 1997 and 46th in 2009, eight levels higher but without as large a change as might be expected for such a lucrative profession. Rising housing prices have led many to criticize property developers for land hoarding, artificially inflating the housing market, and fraudulent transactions, all of which have led to popular dissatisfaction with property developers—a sentiment that has transferred to public evaluations of the occupation. Occupational prestige thus plays a role in balancing and rectifying the indiscretions of those with money and power. Occupational prestige evaluation is, therefore, a special kind of public opinion.[19]

Apart from what has been discussed above, we can also study the difference in evaluation by different social groups. Table 3-10 and table 3-11 below show the differences in evaluation by groups with different educational backgrounds and different occupations.

Table 3-10 Correlation Coefficients of Prestige Evaluations and People with Different Educational Backgrounds

	Illiterate	Elementary School	Secondary School	High School	Junior College	University	Graduate Schooling and Above
Illiterate	1.000						
Elementary School	0.7680	1.000					
Secondary School	0.8319	0.8236	1.000				
High School	0.8751	0.8358	0.9421	1.000			
Junior College	0.8879	0.8155	0.9301	0.9532	1.000		
University	0.8740	0.8119	0.9370	0.9637	0.9748	1.000	
Graduate Schooling and Above	0.8990	0.8195	0.9298	0.9525	0.9745	0.9869	1.000

Table 3-10 shows that each respondent's educational background is highly relevant to how they evaluate different professions, especially evaluations made by people with high school educations and above. As the correlation coefficients here are all greater than 0.95, these data clearly illustrate that people with higher educational backgrounds have similar criteria for evaluation. It is also worth noting that illiterate respondents and primary school graduates had comparatively smaller correlations with the other five groups, showing that people

19 Li Qiang, "Conflicted Occupational Prestige Evaluations in Social Transformations in China." *Sociological Studies* 4 (2004): 100-111.

with different educational backgrounds evaluate occupational prestige differently, and that the difference was more obvious in 2009 than in 1997.

In the 2009 survey, the respondents' occupations were categorized into ten classes: senior and middle management, senior and mid-level technicians, common office clerks, manual workers, private business owners, commercial and service business personnel, students, retirees, farmers, and domestic workers. The correlation coefficients of prestige evaluations and the ten groups are listed in table 3-11.

Table 3-11 Correlation Coefficient of Prestige Evaluations and Different Occupational Groups

	Senior Management	Senior Technicians	Office Clerks	Manual Workers	Private Business Owners	Service Staff	Students	Retirees	Farmers	Domestic Workers
Senior Management	1.000									
Senior Technicians	0.9743	1.000								
Office Clerks	0.9826	0.9858	1.000							
Manual Workers	0.9718	0.9684	0.9710	1.000						
Private Business Owners	0.9266	0.9267	0.9420	0.9098	1.000					
Service Staff	0.9368	0.9492	0.9532	0.9289	0.9142	1.000				
Students	0.9035	0.8959	0.8986	0.8975	0.8181	0.8716	1.000			
Retirees	0.8926	0.8955	0.9045	0.8841	0.8491	0.8627	0.8013	1.000		
Farmers	0.9178	0.9321	0.9386	0.9232	0.8919	0.9395	0.8593	0.8610	1.000	
Domestic Workers	0.7893	0.7720	0.7875	0.7921	0.7644	0.7545	9.7648	0.7044	0.7624	1.000

Table 3-11 shows that evaluations made by people with different occupations were highly relevant to this topic—with the exception of domestic workers, whose correlation with other groups was low (around 0.7). Commercial and service personnel, whose correlations with other occupations were low in 1997, had greater correlation coefficients in 2009 (over 0.9). Generally speaking, the correlativity among different occupational groups was higher in 2009 than ten years before, but it was lower compared with those from other countries and different from the results acquired by Li Chunlin in 2001[20] (for all the classes except white-collar workers and self-employed workers, the correlativity was greater than 0.98). The 2009 survey shows that differences endure among different occupational groups in terms of their prestige evaluations, which differs from the structural functionalist perspective of Treiman, so "conflicting" evaluations remain prevalent in Chinese society.

20 Li Chunlin, "Prestige Stratification of Contemporary Chinese Society." *Sociological Studies* 2 (2005): 74–102.

Chapter 4

The Characteristics of Chinese Social Stratification Structure Before the Reforms and Opening Up

The previous three chapters have mainly introduced social stratification and the overall situation of China's social stratification from the perspective of general theories and methods of social stratification. The following chapters will analyze more specifically how Chinese social stratification has changed since the founding of the People's Republic of China sixty years ago. This will be illustrated with reference to two historic periods. The first period lies before the reforms and opening up, which this chapter will focus on. The other period begins after the reforms and opening up and will be covered in the next chapter.

China's social stratification structure prior to the reforms of the 1980s was a new, gradually formed type of social structure that came into being after the 1949 revolution. This structure was completely different from that which existed before the revolution, since this social stratification structure was formed through the shock of a major revolution. I believe that it had the following four characteristics.

1. Non-Property Based Social Stratification Structure

Although there are more than ten kinds of criterion for examining social stratification,[21] economic and property stratification are still the main yardsticks of social stratification. In China's thousands of years of recorded history, stratification by property, land, and economy has consistently been the main standard for explicit social status differentiation. However, before the reforms, one of the most prominent features of Chinese social stratification was that it did not primarily employ economic indicators, property, or income to distinguish social status. Why was this so? This was because the 1949 revolution overthrew the original economic form of ownership, influencing every region, every village, every settlement, and every family. The breadth and depth of these changes in economic stratification were unique in the entirety of Chinese history. The 1949 revolution completely shattered the

21 Li Qiang. "A Brief Analysis on Ten Criteria of Social Stratification." *Academia Bimestris* 4 (2006): 40-46.

traditional Chinese class system, a feat that was accomplished in two major steps. The first step was the land reform movement that commenced immediately after the revolution—the land, production tools, and property of landlords were seized in the name of the revolution and divided equally amongst the peasants. Since then, in the rural areas of China, the class stratification system could not be understood merely by evaluating individual possession of land and property. The second step was the movement towards the socialist transformation of capitalist industry and commerce, which began in 1956. By purchasing property at low interest rates, the government transformed the national Chinese bourgeoisie and urban private enterprises first by establishing "public-private joint ownership," then later gradually transforming them into state- and collective-owned enterprises. Finally, asset owners were entirely deprived of ownership rights during the gradual process of revolution. At the same time, private housing was also transformed. Therefore, after the socialist transformation commenced in 1956, class in a real economic sense ceased to exist in urban and rural areas, from the point of view of economic and property status.

While economic stratification was shattered, political stratification was gradually introduced into the Chinese social stratification system. In a theoretical sense, applying political and ideological standards to social stratification can be traced to Mao Zedong. In his early writings, Mao put forward the concept that political attitude was an important standard of social stratification. For example, in his article, "Analysis on Classes in the Chinese Society," he stratified the national bourgeoisie not only according to economic criteria, but also according to political criteria. He assigned them into the following categories: unenlightened gentry, enlightened gentry, the national bourgeoisie right-wingers, and the national bourgeoisie left-wingers, etc. In addition to classifying the petty-bourgeoisie economically, he politically subdivided this group into the petty bourgeoisie right-wringers, petty bourgeoisie centrists, and petty bourgeoisie left-wingers.[22] After 1949, China's social stratification policy largely reflected the political stratification thought of Mao Zedong. Political stratification had a great impact on China from the mid-1950s until the late 1970s. Political stratification dominated China's social stratification system as the main means of stratification.

22 *Selected Works of Mao Zedong*. Vol.1. 1–11.

China's political stratification, evolving from a theory into a social policy and social practice, originated from the classification of class status in the rural land reform movement in the late 1940s and early 1950s, and later the large-scale rectification movement ("整风运动") and the consolidation of party organizations movement in the cities. It continued to classify the class status of urban employees by referring to the standards of rural areas. In the following political movements, some new class statuses were created, some class statuses became "political labels," such as "rightists" ("右派"), "those with serious historical bad records" ("有严重历史问题"), and so on. The classification of rural class status during the land reform and classification of status in urban areas in the early 1950s was initially based on economic stratification because at that time classification was based on economic relations and the resource ownership relationship. However, until the late 1950s, this kind of social stratification no longer held a meaning in terms of economic stratification; by that time, landlords no longer possessed the land they had owned before 1949. At that time, capitalists no longer possessed the enterprises they had run before the "socialist transformation of industry and commerce" ("工商业社会主义改造") in 1956. As for assigning "political labels," this was obviously a purely subjective, ideological stratification. Thus, social stratification at this time was actually a political stratification.

So, what were the strata according to political stratification at that time? Political stratification was comprised of several elements, but the most important was the identification of the so-called "class" status. In this context, "class" meant class in the political sense and could be divided into two types: individual class status and family origin status.

According to the principle of the political stratification, class based on either individual class status or family origin status could be stratified into three basic levels: good, bad, and middle status (or good, bad, or middle family origin status). At that time, "good" class status was reserved for workers, poor peasants, lower-middle class peasants, revolutionary cadres, revolutionary soldiers, and family members of revolutionary martyrs, etc. The "bad" class status included landlords (including "bullies," warlords, old society bureaucrats, bankrupt landlords, administrators of communal lands, etc.), rich peasants, capitalists, counterrevolutionaries, "bad men," and "rightists," etc. The middle-class status included middle-class peasants (including well-to-do middle-class peasants), intellectuals, freelancers (including clerks), religious professionals, small handicraftsmen, petty dealers, and peddlers, etc. During the Cultural Revolution, the terms "Revolutionary Five Groups" and "Five Reactionary Gangs" referred to groups with good and bad class status respectively, which turned political stratification into the grounds for serious political discrimination.

In addition, people's political status was classified according to whether they were Communist party members, ordinary people, or those with bad historical records. People were also classified as being activists, stragglers, or general actors according to their political performance. With respect to complexity of social relationships, people were divided into the following groups: people with complex social relationships, people with simple social relationships, and people with general social relationships. According to the standard of that time, the more complex the social relationships that someone had—overseas relations were considered complex, for example—the less politically reliable they were. The more simple a person's social relationships were, the more politically reliable they were considered.

Political stratification was not directly linked with economic interests at this point. During the period in which the political and legal order could be maintained, the status obtained from political stratification was unrelated to that obtained from economic stratification. For example, the political status of capitalists was very low, but their economic status and standard of living could be relatively high. Another example: after being assigned the "rightist" label, many people's political status plummeted, but their economic positions and living standards remained basically unchanged. Of course, in the turbulent era of the Cultural Revolution, political and legal order was erratically maintained—illegal searches and the confiscation of property were common—but that is another matter.

When the social order was normalized, what was the social significance of political stratification, given that political stratification and economic stratification were not directly linked? According to the available evidence, political stratification was mainly embodied in political struggles, ideology, public opinion, and social psychology. Those with bad class status became the targets of attack during various political movements. They were always treated as negative examples that must serve as object lessons. In the newspapers, on the radio, and through various other kinds of mass media, they were always the object of political criticism. Their social reputation was very low and they were looked down upon. In daily life, they were a group lacking in public trust. Whenever problems occurred, they were always the first to be blamed. Thus, as time went by, the psychology of this group was significantly affected, resulting in serious inferiority complexes and even leading to the development of personality disorders among those in this group.

Additionally, political stratification also profoundly affected a person's life chances, which were particularly reflected in the influence of family origin status. From the 1950s to the late 1970s, the influence of family origin status on the development of young people was very profound. First, family origin status affected a person's chances of receiving higher education; having a good or bad family origin status was an important standard for university admission. Within universities, some important disciplines—particularly those considered likely to involve expertise with cutting-edge technology—only admitted those students who had good family origin status. The rare opportunities to study abroad were definitely only for those who had good family origin status. Second, family origin status affected a person's employment opportunities. Under the planned economy, there was little scope for people to choose their own careers; instead, young people's jobs were generally assigned by administrative organizations at all levels. These organizations attached great importance to family origin status when arranging employment. At that time, due to the political atmosphere, many professions were considered politically sensitive—such as professions relating to aerospace, national defense, the military industry, sophisticated technology, security, administration, leadership, intelligence, and customs inspections, etc.—which were reserved for those with good family origin status. There should not have been any political sensitivity at the grassroots level, but leaders tended to assign important jobs to people with good family origin status, according to the political criteria of the day. Again, the impact of family origin status could be felt in all aspects of social life at that time. For example, at that time, many people were keen on joining the army or joining the Communist Party and other career paths in which family origin status often played a decisive role. Another example: young people often took family origin status to be an important factor when choosing a spouse.

The aforementioned individual class status and family origin status are similar to a caste system, which is basically a closed social stratification system. This means that once class status and family origin status are identified, they cannot be changed. The following three sections of this chapter will expound in detail on the stratification system of social status.

2. Relatively "Equalized" Income Distribution Policy

Corresponding to the non-property-based social stratification system—from 1949 to the beginning of Deng Xiaoping's reforms in 1979—China pursued an income distribution policy characterized by "equalization" ("均等化"). Here the term "equalization" will be employed rather than "equalitarianism," because equalitarianism was theoretically antithetical to the government before the reforms.

In terms of the degree of inequality present in urban and rural areas, cities were much more equalized at this point. According to a report written by the World Bank Economic Study Group in 1980, the Gini coefficient for the per capita income of urban households in China before the reforms was only 0.16, the lowest Gini coefficient ever recorded at that point. In countries with sophisticated welfare systems and a high degree of equalization—such as Austria (0.231), Denmark (0.247), Sweden (0.25), Norway (0.252), Finland (0.256), and Germany (0.281)—the Gini coefficient was above 0.2. Why was the Gini coefficient for the per capita income of urban households so low in China? The report written by the World Bank Economic Study Group believed that there were three reasons for this. First, before the reforms, Chinese residents were essentially unable to draw income from private property, such as rent, dividends, and profit, etc.[23] Of course, before the socialist transformation of capitalist industry and commerce, China's private businessmen still accrued income of this type. Additionally, from 1956 until September 1966, they were still able to obtain fixed-rate incomes from private shares. However, after September 1966, property-based income was entirely unavailable. Secondly, before the reforms, urban residents had no income from self-employment; so-called self-employment activities range from the activities of small tradespeople and peddlers, and

23 World Bank Economic Survey Group. *China: Development of the Socialist Economy.* Beijing: China Financial and Economical Publishing House, 1983. 59.

independent income-generating activities depending on professional skills. These kinds of activities generate income diversity and inequality in many countries. Thirdly, income was relatively equal. Although there was a graded salary system, the salaries of managerial, professional, and technical personnel were actually not high compared with other developing countries. Not only that, after the Cultural Revolution began in 1966, even the bonuses reflecting differences in labor were cancelled as the equalization of urban employees' incomes reached a new extreme.

In addition to the above, the planned and administrative allocation of resources was also an important cause of the high degree of equalization in the cities. Of course, the planned allocation of resources was chiefly because of the extreme shortage of resources. This was particularly true during the three years of famine from 1959 to1962; there was a shortage of almost all items including food, edible oils, eggs, meat, sugar, cloth, and cotton. Only by highly organized administrative means could the available materials be distributed through the use of ration coupons, allowing people to eke out a living. But, these methods of planned and administrated resource allocation continued and lasted for a long time after the end of the famine.

Another reason leading to the high degree of equalization of urban residents' income was the equal distribution of housing resources. After the socialist transformation in 1956, most of China's urban housing became publicly owned and was distributed or rented out by the state. Since housing was not primarily purchased by means of a market, the real estate market disappeared. If people were in want of housing, their only access to it was through "housing distribution." There were, of course, rules and regulations governing housing distribution. At that time, each work unit[24] had "housing allocation regulations." I once researched the management of housing distribution in different work units,[25] finding that the basic principle of distribution was equalization. Only those without housing or those with poor quality housing were able to apply for housing. There were also strict age limits and marriage restrictions for housing applications. Only those who had reached a certain age—such as twenty-eight years old or thirty years old—and

24 All public and collectively owned entities in China before the reforms were referred to as "work units" ("单位"), including government organizations, schools, hospitals, companies, enterprises, grocers, etc.
25 Li Qiang. *Social Stratification and the Gap between the Rich and Poor.* Xiamen: Lujiang Publishing House, 2000. 218-31.

those who were married were able to apply for housing. This was a simple egalitarianism based on natural causes. Again, household size was also an important factor in the allocation of housing. Naturally, only the officially registered urban population in the "residence booklet" was considered for housing (after the mid-1970s, it was also stressed that the family planning policy could not be violated); thus, the more members a family had, the more housing space it could be assigned.

In short, the economic equalization of urban residents' living conditions was closely related to a whole set of equalization policies.

This change was not replicated in the countryside, however. The most prominent achievement in rural equalization resulted from the revolution in 1949—namely, rural land reform. Land reform in the early 1950s completely changed the status quo in which the landlords (a very small proportion of the population) possessed a large quantity of land and the majority of poor peasants had none. Therefore, the enormous gap between rich and poor created by the unequal possession of land and other means of production in rural areas disappeared in the 1950s. However, the available data indicates that before the reforms, the degree of inequality in rural China was still much greater than that in the city. According to the World Bank Economic Study Group's report, the Gini coefficient for the per capita income of rural households was 0.31. As mentioned above, in the same year, the Gini coefficient for the per capita income of urban households was only 0.16. Why was the degree of inequality in rural China so much greater than that in the city? First of all, this was caused by the huge regional differences that exist in rural China. Irrespective of whether the data under examination is for income level or level of consumption, China's coastal rural areas were much better off than those in the countryside of Northwest, Southwest, and other inland areas. After China's rural collectivization, farmers in each village became relatively equal economically, but significant economic disparities endured between different villages. In particular, the differences between different regions were enormous. Additionally, even after the collectivization, farmers still had some opportunities to diversify their sources of income. When the famines began in 1959, private family plots in rural areas, free markets, small contracting, and the household responsibility system became very common and greatly increased the likelihood of farmers developing other sources of income outside of the collective economy. In fact, even during the Cultural Revolution—when economic and political control was the strictest—it was hard to stop farmers from climbing up mountains to pick different types of mushrooms or from keeping bees, chickens, ducks, pigs, or conducting other economic and business activities. So, before the reforms, the Chinese government's effort to achieve the economic equalization encountered a lot of resistance in rural areas. The government's ability to regulate income distribution was far less well-established in the countryside than that in the city. Therefore, the income gap between farmers created by individual differences and differences in business activities was much wider than in the city.

Of course, compared with countries of similar developmental level before the reforms, income distribution inequality in China's rural areas was still very low. Table 4-1 compares the indicators of per capita household income in rural China with that of some countries of a similar developmental level.[26]

Table 4-1 International Comparison of Rural Household Per Capita Income

Country, Time of the Survey (year)	Proportion of Each Income Group in Total Income (%)			Gini Coefficient
	The Poorest 40%	The Richest 20%	The Richest 10%	
China , 1979	20.1	39.4	22.8	0.31
Bangladesh, 1966–1967	19.9	41.7	26.1	0.33
India, 1975–1976	20.2	42.4	27.6	0.34
Pakistan, 1970–1971	21.9	38.8	24.0	0.30
Sri Lanka, 1969–1970	18.6	42.5	26.4	0.35
Indonesia, 1976	16.4	46.0	32.0	0.40
Malaysia, 1970	12.2	54.8	39.3	0.50
Philippines, 1970	17.3	46.7	31.7	0.39
Thailand, 1975	17.8	46.5	31.3	0.39

26 World Bank Economic Survey Group. *China: Development of the Socialist Economy.* Beijing: China Financial and Economical Publishing House, 1983. 61.

It can be seen from table 4-1 that, before the reforms, the degree of inequality in China's rural areas (Gini coefficient: 0.31) was lower than that of rural areas in other countries of a similar developmental level. Rural China's richest 10% possessed 22.8% of the total wealth, while in other countries of similar developmental level this figure was as high as 39.3%, indicating a greater degree of polarization than in China.

In addition to the equalization of the economy and the distribution of income in urban and rural areas, a number of other social policies particularly reflect the features of equalization. Here only three aspects will be mentioned. After 1950, China's urban and rural employment rate increased significantly—especially the employment rate of women. A wide range of employment opportunities caused an increase in the number of people with incomes, which, to some extent, shaped the general low-wage system and was one of the reasons for the equalization of income at that time. A high employment rate also led to an equalizing trend in people's social status. The equalization of education was also significant. In 1949, China's primary school enrollment rate was 20% and the illiteracy rate was 80%, but by 1965 the country's primary school enrollment rate had reached 89%—much higher than that of other low-income countries in the world at that time. The final critical factor was the equalization of basic medical and health services. In the 1960s and 1970s, the public health system in urban areas covered almost all employed laborers, while the cooperative medical system in rural areas covered about 90% of the rural population.

In short, after summarizing the state of income distribution before the reforms, particularly the income distribution policies, it would still be realistic to say that it was centered on "equalization."

Of course, "equalization" did not mean that people's finances were uniform, but the so-called "equalized" income distribution policy meant that the government's policy aimed to narrow the gap and achieve equalization. In fact, the economic gap was not small. In the terms of that time, there were three major differences: the difference between workers and peasants; the difference between urban and rural areas; the difference between mental workers and manual workers. However, this summarization of three differences is actually not accurate. The differences between workers and peasants or the difference between urban and rural areas were part of the same issue, so including both is a repetition. In fact, there were four prominent differences: regional differences, differences between urban and rural areas, differences between mental workers and manual workers, and differences in the rank-based wage system of cadres. Further explanation of these differences will be made below.

Since these differences existed before the reforms, why then do we refer to an equalized income distribution policy? Difference is always relative; if compared with the conditions before 1949, the gap appears greatly reduced. When compared with countries of a similar developmental level, China still appears to have a low level of social inequality. Moreover, the CPC's national policy consistently attempted to reduce economic disparities and achieve equalization.

3. The Stratification System of Social Status

The first feature mentioned above was the breaking up of the property stratification system, and the second feature was the equalized income distribution policy. According to Martin King Whyte, an expert on China, these two aspects are both forms of de-stratification.[27] From the 1950s to the 1970s, China remained in a state of ceaseless political movements. In wealth and income distribution, there was a tendency towards de-stratification. Since de-stratification was underway, was China becoming an "equal" society? Things were not quite that simple. Property ownership and income were just two aspects of the differentiation between the different groups in China; there were many others besides. According to Frank Parkin's theory of social closure,[28] all kinds of social groups use certain criteria to limit access to scarce resources and opportunities to one's own group. To effect this, different social or natural attributes are chosen as the justification for excluding others from using them. These attributes included: ethnicity, language, social origin, region, religion, and so on. Just as Parkin said, when property ownership was no longer a condition for social closure, other conditions immediately emerged to replace it. These conditions included household registration, family origin status, the time at which one signed up for work, rank, and work unit ownership. By the mid-1950s, this type of non-property-based social stratification had become a relatively stable institutional system that

27 Yanjie, ed. *Market Transformation and Social Stratification: China under Analysis of American Sociologists.* Beijing: Sanlian Bookstore, 2002. 33.
28 Parkin, Frank. *Marxism and Class Theory: A Bourgeois Critique.* New York, Columbia UP, 1979. 11–13.

endured until 1979. This social stratification system can be termed a "status system."

In sociology, the first person to study status and status groups was the German sociologist Max Weber.[29] The complexity of his conceptualization of status lay in how it emphasized not only the role of subjective social status assessment, but also that status was based on objective lifestyle, education, occupation, and family origin. Status systems, according to Weber, are sets of institutional systems approved by laws, regulations, and norms. In the Chinese case, the status groups that emerged were slightly different from those hypothesized by Weber. As mentioned above, in the mid-1950s, status was inextricably linked to politics, producing status groups such as poor peasants, lower-middle peasants, landlords, rich peasants, employees, and revolutionary cadres.

This status system reflected the fact that Chinese society was still very traditionally structured at that point. For example, farmers accounted for more than 80% of the total population. However, it also reflected that after the class system was broken up, property became less important in determining status, so other characteristics of status stratification became more prominent. Such a system centered on status, specifically manifested in the following four aspects: household registration status, worker and cadre status, cadres' rank status, and work unit status.

From the mid-1950s until 1979 (and even today), China has implemented a strict household registration system. According to this system, once a person is registered as living in one place, then it becomes very difficult to move to another place. The main function of the household registration system is to separate people into two categories: urban and rural registered residents. All rural residents are referred to as farmers, although many of them are not engaged in agricultural production. It is extremely difficult for Chinese farmers to change their residential status and it is extremely difficult for them to enter into another status group.

The formation of the Chinese urban and rural household registration system was a historical process. During the period immediately following the founding of New China (the People's Republic of China) in 1949, the central government did not strictly control movement between urban and rural areas. According to the then-prevailing law, citizens had

29 Max Weber. *Economy and Society*. Vol.1. Beijing: Commercial Press, 1977. 333–39.

the freedom to move from one residence to another. However, with the upsurge of industrialization in the 1950s, farmers began to crowd into the cities in huge numbers, leading to shortages of food, non-staple foods, transport, housing, and urban services, etc. In response, the central government twice implemented more stringent limits on farmers moving into cities. These new measures were first implemented in the mid-1950s when the central government specifically instructed farmers not to crowd into cities. In June of 1955, a strict household registration system was established, marking the beginning of China's household registration system. The second time was during the Great Leap Forward in 1959 and 1960 because at that time China faced severe financial difficulties and a catastrophic shortage of food and daily necessities. The government began to implement a policy that almost completely prohibited farmers from moving into cities. According to the prevailing regulations, each year the country allowed only 1.5% of agricultural residents to acquire urban residence registration booklets. This included rural youths being admitted to university and cadres being promoted and invited to move into the city with their families. It was impossible for common farmers to become urban residents. This policy was maintained until 1979. The strict household registration system formed a dual social structure characterized by the division between urban and rural areas. Differences between urban and rural areas became the most basic form of social stratification. For example, by comparing the consumption levels between urban and rural residence (table 4-2), it can be found that the consumption level of China's urban residents was much higher than that of rural residents.

Table 4-2 Comparison of Consumption Levels of Urban and Rural Residents before the Reform[30]

Year	Consumption Level (RMB)		Urban and Rural Residents' Consumption Ratio (Supposing the Agricultural Consumption Level is 1)
	Agricultural Residents	Nonagricultural Residents	
1952	62	148	1: 2.39
1953	69	181	1: 2.62
1954	70	183	1: 2.61
1955	76	188	1: 2.47
1956	78	197	1: 2.53
1957	79	205	1: 2.60
1958	83	195	1: 2.35
1959	65	206	1: 3.17
1960	68	214	1: 3.15
1961	82	225	1: 2.74
1962	88	226	1: 2.57
1963	89	222	1: 2.49
1964	95	234	1: 2.46
1965	100	237	1: 2.37
1966	106	244	1: 2.30
1967	110	251	1: 2.28
1968	106	250	1: 2.36
1969	108	255	1: 2.36
1970	114	260	1: 2.28
1971	116	267	1: 2.30
1972	116	295	1: 2.54
1973	123	306	1: 2.49
1974	123	313	1: 2.54
1975	124	324	1: 2.61
1976	125	340	1: 2.72
1977	124	360	1: 2.90
1978	132	383	1: 2.90

It can be seen from the table above that, before the reforms, the consumption ratio for urban and rural residents fluctuated between 2.28 and 3.17, a marked difference. Moreover, at that time urban residents also enjoyed various forms of welfare not available in rural areas, such

30 State Statistic Bureau. *Forty Years' Striving 1949-1989*. Beijing: China State Statistics Publishing House, 1989. 456.

as food subsidies, housing subsidies, medical insurance, and work-unit-provided children's education. Urban and rural residents could therefore be said to live in two different worlds, forming two different lifestyles. At that time, it was almost impossible for farmers to change their lifestyle and enter into the other world. There were also various limitations placed on urban and rural areas. For example, the system of food coupons and non-staple food coupons was a very important limitation. There were still means by which farmers could change their status, but these were quite few. For young rural residents, being accepted by a university was the main approach. If one passed the college entrance examination and was admitted to an institution of higher education, then he could become an urban resident, which farmers referred to as "jumping over the dragon gate" ("跳龙门"). However, due to the generally poor quality of education in rural areas, very few people were able to "jump over the dragon gate." As farmers accounted for the majority of the population, this system that encouraged great differences between urban and rural areas naturally produced the characteristics of a pyramid-shaped social stratification structure. The household registration system's urban-rural segregation firmly constrained the farmers to a certain area and allowed them no freedom of mobility, and so also constrained the productivity and enthusiasm for social activities of the largest group in the Chinese population. This was clearly not beneficial for China's economic development, social modernization, and social progress. This was also one of the important reasons for the slow pace of economic development seen during that period of time.

Before the reforms of 1979, urban employees could basically be divided into two social status groups: cadres and workers. The term cadre ("干部") can be understood in a broad and narrow sense. Broadly speaking, all administrative staff were referred to as cadres in China. This appellation originally came from the fact that people initially called Communist Party officials coming from liberated areas "cadres." Later, however, the term was generalized and used to refer to all managerial staff and leaders, including top officials like the chairman of the country and the premier, as well as those low-ranking officials like rural village chiefs and team leaders. Even monitors and group leaders in primary and secondary schools were referred to as class cadres.

Cadres, as a status group, were not the same as cadres referred to in this broad sense. In the narrow sense, cadres referred to a social status.

Attaining this status depended on being registered in the national cadre's quota system by the personnel department in accordance with the relevant regulations. Being registered in the national cadre's quota system depended on a "file identity management system." Every urban employee in China had his own file, which was maintained by his work unit. These files recorded the person's life experiences, family background, and relatives, etc. There are essentially two categories of file status: cadre and worker. The distinction between these two statuses was not only a method of file management, but also reflected significant differences in material interests. These two statuses differed a great deal in wage level, working conditions, travel subsidies, housing, medical care, pensions, and so on. Generally speaking, the treatment received by registered cadres was superior to that of ordinary workers. Thus, the vast majority of employees sought to be included in the national cadre's quota system. However, for the vast majority of those registered as workers, it was rather difficult to achieve cadre status. It was no less difficult than making the change from being a registered rural resident to a registered urban resident.

Before the reforms, what kind of people could become cadres and be registered in the national cadre's quota system? First, education was the most important factor. All those who received full-time higher education—including three years of college education or a four- to eight-year university education, full-time vocational education, or secondary technical education—could obtain cadre status after being assigned to work in work units according to the government's assignment plan. Those who graduated from part-time vocational schools, correspondence colleges, part-time colleges, radio and television universities, or cadre training classes could not be immediately registered as cadres, even though they held junior college diplomas. Those assigned to cadre posts with cadre quotas assigned by the national personnel department could be registered as cadres. However, the quotas were often very limited and there were very stringent requirements regulating who attained the rank of cadre in this way. For example, a college degree or further education was often required. The aforementioned graduates of part-time vocational colleges and part-time universities therefore had to wait for an opening in the cadre quota to be registered as cadres. During the Cultural Revolution, a lot of cadres were promoted from amongst the workers, peasants, and soldiers.

According to the statistics, they accounted for about two million people. Although they had obtained the position of cadre, they retained their original status, a practice commonly known as "allowing a worker to act as a cadre" ("以工代干"). These workers who were allowed to act as cadres could not be registered as cadres in the national cadre's quota system, but retained their status as workers. Until after the mid-1980s, when the state allocated more cadre quotas to grassroots work units, the serious issue of "allowing a worker to act as a cadre" was gradually resolved. Again, when demobilized soldiers left the army to become civilians, they could be registered as cadres. Demobilized soldiers who had held the rank of company commander or above while in the armed forces became cadres. Of course, after they took office in local work units, they were guaranteed to be registered as cadres in the national cadre's quota system. When common soldiers completed their terms of military service, they were usually registered as workers or returned to their hometowns to become farmers.

Employees in cities who did not meet these criteria were generally registered as workers. This division between cadre status and worker status was distinct. Each employee was aware of his status. In terms of numbers and proportions, people with cadre status accounted for between one-sixth and one-seventh of the urban working population, while the rest were of worker status. This distinction between cadre status and worker status maintained the social stratification and social order in a society where property stratification had been eliminated. However, because the channels of social mobility were too narrow and workers made up the majority of urban employees, their vitality and enthusiasm for work were greatly constrained.

Under the hierarchical system in place before the reforms and opening up, stratification by cadre rank was central to the system in which official rank was considered to be the sole criterion of one's worth. The hierarchical stratification of cadres and official ranks was taken as the baseline of social stratification, and from this starting point the stratification system for the entirety of Chinese society was derived. Why did cadre and official ranks become the systematic base of Chinese social stratification? This naturally had something to do with Chinese tradition. However, judging from the developments following 1949, the system in which official rank was considered to be the sole criterion of one's worth was also formed on the basis of the following two factors. First, in July 1955, a set of countrywide unified rank and salary standards for governmental officials was established. Secondly, in 1956, the salary system of state-owned enterprises, public institutions, and state organizations was reformed. According to the new regulations, cadres were divided into thirty ranks. All these ranks were matched with a corresponding salary. These specific salary standards were maintained for almost thirty years, although several adjustments were made later.

The cadre stratification system in place before the reforms and opening up bore the following characteristics. Cadre rank stratification formed the base of other social stratification systems. In 1956, when the State Council introduced cadre ranks and corresponding salary grades, other agencies, organizations, and social departments were stratified accordingly. For example, the salary of teaching staff, engineers, doctors, laboratory personnel, editing and publishing staff, library staff, and general production workers were all graded according to this model. Thus, a variety of social stratifications for other major social groups were derived from the baseline of cadre stratification. The planned economy was also an important foundation of the official rank-based system at that time. China implemented a planned economic system from the 1950s to the 1980s. The right to manage the whole society's production, distribution, exchange, and so on was mostly concentrated in the hands of government departments at all levels. The government managed not only economic activities, but also the various activities of society, ideology, culture, etc. This top-down, hierarchical power structure became the mainstream of social life in China. People's positions in this power structure became the most important factor in determining their positions on the social ladder. Additionally, the economic system in place before the reforms and the official rank-based system were mutually supporting. By the end of 1956, China's urban and rural socialist transformation was complete. Various economic sectors had been amalgamated into a collectively owned, state-controlled economy; economic distribution and personal income were also simplified. Wages became the main source of income for the vast majority of urban workers. Therefore, wage stratification was equivalent to the then-prevailing income and property-based stratification. Economic status was derived from one's wage grade and everyone's economic status was exactly the same as one's wage grade. Thus, the cadre-salary stratification system became the basis of the economic stratification of the whole of the Chinese society. Finally, a welfare system providing benefits, treatment, services, and so on was established and pegged to cadre-salary grades. For example, on one occasion, China's Ministry of Finance produced specific stipulations regarding the standards of housing, transport, vehicles, entourages, secretarial services, medical care, direct supply of goods and materials to fixed locations, furniture, household appliances, rent, utility bills, and entertainment for cadres of different ranks in state organizations, enterprises, and public institutions. Pegging this system to the different levels of cadre status helped form the core of social status stratification from the 1950s to 1980s.

In short, the cadre rank-based stratification system became the basis for all social stratification after its formation in the mid-1950s and endured as China's most important social stratification system. Even the turmoil of the Cultural Revolution could not shake this system. For example, even cadres who were "sent down" ("下放") to the countryside during the Cultural Revolution were considered to be high status individuals. Common people still judged the status of these cadres according to their registered ranks. After the Cultural Revolution, these cadres—having been deprived of authority during these years—quickly regained their ranks, demonstrating the solidity of the official rank-based system.

Before the reforms, urban employment in China was mostly of the so-called "iron rice bowl"[31] ("铁饭碗") system, so it was a relatively common phenomenon for a person to work in the same work unit all his life and there was little mobility between work units. The work unit, therefore, became the center of most people's activities for their whole lives and the relationships between work units and employees tended to be quite close. The unit provided its employees with a full complement of services and security, leading to employees becoming subordinate to and dependent on their work units.[32] Due to the close links between a work unit and all aspects of its employees' lives, each work unit represented a microcosm of Chinese society. Different work units constituted distinct communities and differed from each other a great deal. Different status groups were actually formed in this way.

31 "Iron rice bowl" ("铁饭碗") is a vivid expression meaning life-long employment.
32 Andrew G.Walder. *Communist Neo-Traditionalism: Work and Authority in Chinese Industry*. Berkley: U of California P, 1986.

One's living standard, and indeed all aspects of one's life, was closely related to the work unit; even social status was often related to one's position in the work unit. The higher the work unit's position and grade, the higher the employee's status, and vice versa. Thus, China's urban employees had a "work unit status."[33] The differences in resources, status, prestige, and so on between different work units were bestowed upon each employee in the unit. An important feature of the work unit status system was that there were great differences between different work units, but the differences between the members of the same unit were quite minimal. Egalitarianism prevailed within individual work units. In addition, work unit employees enjoyed life-long employment and inter-unit mobility was very limited, resulting in a situation in which the quality and efficiency of a worker's activities made no difference to their prospects. This system thus dampened individual motivation, placing it in conflict with the competition mechanisms of a market economy.

4. Monolithic Social Aggregation

What is the meaning of monolithic social aggregation? There have been many differences and controversies when it comes to patterns of analysis of social stratification. One school advocates the approach of analyzing huge social groups; another advocates the approach of analyzing social stratification broken down into small groups.[34] The object of analysis for the former school of thought is a huge collection of social groups, which we may refer to as a "monolithic social aggregation." This would include, for example, large classes and strata: the working class, the peasant class, the bourgeoisie, the landlord class, and intellectuals, etc. These all belong to a "monolithic social aggregation." The latter holds that these monolithic social aggregations should be broken down into small groups for analysis, such as small occupational groups: carpenters, bricklayers, engineers, doctors, managers, professors, and so on. The former argues that only "monolithic social aggregations" can explain the significant conflicts of interests present in society and the earthshaking class struggle that results. Marx is a typical representative of this school. For example, in

33 Lu Feng. "Work Unit: A Special Form of Social Organization." *Social Science in China* 1 (1989): 71-88.
34 Li Qiang. *Ten Lessons on Social Stratification*. Beijing: Social Sciences Academic Press (China), 2008. 135–36.

his book, *Class Struggles in France, 1848 to 1850,* Marx analyzes the intense political and economic conflicts occurring in French society. Marx's approach is to analyze "monolithic social aggregations," and Marx identifies a number of these groups in French society, including: the financial bourgeoisie, the industrial bourgeoisie, the petty bourgeoisie, the peasant class, the proletariat, and the lumpenproletariat, etc. Mao Zedong's article, "Analysis of the Different Classes in Chinese Society," is a continuation of Marx's analysis of "monolithic social aggregations." Since we are quite familiar with Mao's analyses, I will not discuss this paper in detail.

The other school of thought believes that in social life and social practice, the real social interaction between individuals occurs within small groups and that the only tangible organization of individuals is within small occupational groups. While people cannot normally observe the effects of "monolithic social aggregations" in the conflicts of interest taking place in their societies, a variety of small organizations play a significant role. The main theoretical representative of this school is the American sociologists David B. Grusky and Jesper B. Sorensen, although there are others.

The key difference between the two schools is not the approaches they take to analysis, but the objects of their analyses—social groups as entities whether they are "monolithic social aggregations" or broken-down small groups. In Marx's era, class was truly an integral social aggregation. That Marx and Engels organized the First International and the Second International is proof of this. Mao Zedong, mobilizing the broad mass of farmers, launched three revolutionary uprisings from 1924 to 1949. The scale of these uprisings had rarely been seen before in history. These large-scale conflicts were also embodiments of "monolithic social aggregations." Because huge social classes have been decreasingly evident since the 1970s in Western societies, the appearance of a model focused on analyzing small groups is consistent with the status quo of many Western societies.

Prior to the reforms, how were Chinese social groups constituted? I believe they were "monolithic social aggregations." Why was China's pre-reform social stratification characterized by "monolithic social aggregation?" Firstly, "monolithic social aggregations" were closely linked with the Chinese revolution before the reforms. As mentioned above, during the first half of the twentieth century, Mao Zedong instigated three revolutionary uprisings in China. The antagonistic relationships between classes were quite obvious in these rebellions. The opposing forces formed rival camps, such as the opposition between landowners and peasants concerning land issues. The establishment of the People's Republic of China was the result of these social conflicts, so these rival classes and the relationships between them endured for a long period of time after the founding of New China. Land reform after the founding of New China and the socialist transformation of industry in the mid-1950s were also continuations of the conflicts deriving from "monolithic social aggregations." As mentioned above, after the end of class stratification based on property, efforts were made in China to form a new status system, leading to the development of huge new status groups. Boundaries between status groups were very obvious and mobility between status groups was very limited, further strengthening the trend towards monolithic social aggregation. China's four main social aggregations at that time were: workers, peasants, cadres, and intellectuals. Although there were differences between these four social aggregations—such as differences between workers in state-owned enterprises and those in collective enterprises—the overall system was unified. Economically focused enterprises and units being relatively few in number, combined with a unified management model and a unified income distribution system, reinforced these social aggregations. In addition, political movements were also an important factor in strengthening these huge social aggregations.

"Monolithic social aggregation" was also closely related with the prevailing ideology of class struggle and the policies it produced. Before the reforms and opening up, the dominant ideology was that of class and class struggle; during the Cultural Revolution, the conflicts between huge monolithic social aggregations were pushed to an extreme. From the 1950s to the 1970s, one after another, various political movements and political practices strengthened this type of social aggregation. These included the classification of class status, the socialist transformation movement, the anti-rightist struggle, the Four Clean-Ups Movement ("四清运动"), and so on. Additionally, before the reforms, China's "monolithic social aggregation" was also closely related to the patterns of social resource distribution and social mobilization during that period. Prior to the reforms, the planned economy pursued the centralized allocation of resources by the state. The state adopted a unified and centralized system of resource allocation management, and employed a variety of methods to fully intervene in income distribution. Social groups and class strata formed in this way, therefore, developed consistent characteristics. In other words, the national unified policy shaped the trend of "monolithic social aggregation."

Finally, what were the social consequences of "monolithic social aggregation" before the reforms? The "monolithic social aggregation" and the period's political movements had a causal relation. As mentioned above, the ongoing political movements contributed to the development of "monolithic social aggregation" with strong political characteristics; at the same time, "monolithic social aggregation" motivated the formation of a variety of conflicting political movements. The Cultural Revolution was a notable example of this trend. The so-called factional struggles that took place during the Cultural Revolution were the most common manifestations of the causal relationship between "monolithic social aggregation" and the outbreak of major social conflicts. During the Cultural Revolution, people in almost all work units—including enterprises, factories, mines, governmental organizations, schools, and hospitals—were split into two factions. The two factions scattered throughout various work units then formed a larger aggregation through alliance. The conflict between these two factions escalated and even caused a large-scale armed conflict between 1967 and 1968.

Therefore, my conclusion is that the "monolithic social aggregation" of social stratification prior to the reforms was directly linked with political movements, political instability, and social conflict. From this perspective, when guiding policy, we should avoid the formation of "monolithic social aggregation" in order to achieve social harmony and social stability. This is a summary of the experiences and lessons of social stratification before the reforms.

Chapter 5

Changes in Social Stratification Structure Since China's Reforms and Opening Up

After 1979, significant adjustments were made to China's politics and economy and a fresh new set of reforms and opening-up policies were adopted. For example, the focus of China's economic policies was shifted from public ownership to various other forms of ownership; from a planned economy to a new system of socialist market economy with gradually increasing market regulation; from a people's commune system to a household contract responsibility system in rural areas. The transformation of China's political policies was another example in which the legal system was improved, the relationship between different social groups was adjusted, and the position of intellectuals was enhanced. In the 1990s, the reform of state-owned enterprises and the work unit system caused changes in housing systems, medical treatment, pensions, and employment in urban areas. These changes significantly affected different social groups and their relationships, as well as altered China's social stratification structure. In this chapter, five aspects of these changes will be analyzed.

1. The Transformation of the Status System

If the main changes in social stratification structure during more than thirty years of reform can be summarized in a single sentence, it may be said that these changes were characterized by a conversion from differentiating social status on the basis of "social status indicators" to a reliance upon "non-status indicators." As stated in the previous chapter, before the reforms, social stratification was not based on property. Social status was not decided by how much property one owned or how large one's income was, but by special status indicators. The differentiation between political and household registration status and between the statuses of workers and cadres has already been analyzed. From the perspective of social stratification, the most significant changes in the status system took place alongside the gradual collapse and decline of the original status system after the reforms, which mainly manifested in the following five areas.

An important change was the end of tight restrictions on the residential registration status of farmers. The restrictions on the household registration status of farmers have been gradually loosened since the 1980s. Farmers were allowed to move into towns and small cities, and open stores or service-oriented businesses to meet urban citizens' needs. At first, the more relaxed policy included significant limitations, allowing people to "leave farming but not their hometowns" ("离土不离乡"), meaning that farmers were allowed to leave agricultural work and start other professions, but could not leave the region of their hometown and move to big cities. However, once the open policy was initiated, it proved very difficult to prevent farmers from leaving both their lands and their hometowns. By 2008, the total number of farmers working and doing business in towns and cities astonishingly reached over 150 million. A number of "farmer entrepreneurs" emerged, with some even becoming millionaires and multimillionaires. Though they were still called "farmer entrepreneurs," their social status was no longer controlled by the household registration system. During the reform of the household registration system, marked by the Chinese State Department's abolishing *The Regulations to Collect and Send Back Urban Vagrants and Beggars* ("城市流浪乞讨人员收容遣送办法"), more positive policies were adopted to urbanize the rural population, with the ultimate goal of achieving status equality between the rural and urban populations.

Changes also took place in the stratification system based on work unit and cadre ranks. Prior to the reforms, the system of work unit and cadre ranks was the product of a single economic element. The reforms diversified China's economy, allowing economic trends such as the private economy, the foreign capital economy, and joint ventures to develop. Additionally, the reforms diversified people's incomes to include salary, income from shares of stock, income from securities, income from real estate, and so on. Therefore, the importance of official salary grades paled in comparison to the potential of property-based income; as a result, this sort of hierarchy was unable to maintain the so-called "standard" system. Meanwhile, the effects of the market economy were also being felt. In a planned economy, governmental departments held most of the power in managing every aspect of life; thus, officials at all levels of the government naturally became the core group of the Chinese society. In recent years, the planned economy has given way to a market economy. The vigorous growth of China's market economy has weakened the grasp of governmental officials on economic power. With the gradual devolution of various government powers, officials no longer play a core role in social economic operations as they did in the past; instead, they have to share power with the market. In addition, the development of the market economy has also directly changed the official rank-based system. In the past, every operating work unit in China was assigned an official rank—including enterprises at the section level, enterprises at the department level, and enterprises at the bureau level. Under conditions of market competition, however, an enterprise's position can only be evaluated by its assets, output, and profits. Thus, an enterprise definitely cannot be granted a high level rank in advance; this can only be earned according to its economic prosperity. Due to the effects of market competition, the official ranks assigned by the government gradually lost all meaning. Some enterprises with low official ranks become highly competitive, holding assets worth hundreds of millions of yuan, rendering the original rank system obsolete.

Of course, during more than thirty years of reform, the changes in the cadre or official rank-based system have not been one-dimensional. In some ways, official ranks and statuses were reinforced. For instance, since the 1990s, along with great increases in government fiscal revenue, official ranks and statuses have tended to be reinforced in social stratification. However, as far as the overall relationship between government and the market is concerned, the marketization of resources and pluralized social stratification has overridden the importance of a single official rank-based system.

How should these changes in the official rank-based system be evaluated? To begin with, they are definitely a form of social progress. Although the former social stratification system of the administrative rank hierarchical system did play an important role in maintaining stability and social order, the fact that this system greatly constrained economic development cannot be ignored. All power—even the management of the economy—was controlled by government administrators, which discouraged people from using their initiative and contributed to China's economic backwardness. When administrative ranks were the only source of status, people yearned to become cadres, which was not good for socioeconomic development. People surged forward to squeeze into cadre contingents with the result that the number of cadres grew, governmental organizations became heavily overstaffed, and this bureaucratic bloating became a chronic illness. This bloat was a heavy burden on Chinese society as it cost billions in administrative expenses every year, an expense that was hard to bear for an underdeveloped country with a low per capita income and a huge, impoverished population. Overstaffing also greatly decreased the government's administrative efficiency. Overstaffing, responsibility shirking, endless disputes over trifles, slack discipline, and demoralization were all common throughout the Chinese government. Therefore, the aforementioned changes in the stratification system were helpful in simplifying the government and increasing social efficiency as well in dissuading people from pursuing the sole aim of cadre membership.

The "file status system" ("档案身份制度") was also dismantled. Before the reforms, the vast majority of people worked in the same enterprises all their lives and it was difficult to transfer from one enterprise to another. One of the reasons for this difficulty was a special administration system that can also be called the "file status system." "File status" was an important foundation allowing "a work unit to possess talented staff." After the market reforms, the free mobility of labor was a great blow to enterprises' possession of talented staff. Thereafter, it was not uncommon for a large number of people to work for an enterprise without the enterprise holding their files. With the emergence of a variety of economic work units, file status became less important, freeing people from their file status. "Talent exchange centers" were also created, the function of which was to keep files and bridge the gap between different systems. Therefore, the constraints of file status on employees in cities and towns were loosened.

Social mobility brought vitality to the Chinese society. It is estimated that one-third of America's GNP results from social mobility. There are currently no such statistics in China, but it is not difficult to find the contribution made by social mobility by analyzing the reasons for the sustainable economic growth that has taken placed in the last thirty years.

Traditional indicators of innate identity were also replaced. Achieved status obtained through individual effort—such as diplomas, schooling, and technical certificates—have played an increasingly important role in social stratification, closure, and selection. Since the college entrance examination was reintroduced in 1977, education and diplomas have begun to play a more important role in the differentiation of social status. Since the 1980s, the central government has emphasized the importance of diplomas when regulating the standards for promoting cadres. Those who do not have university degrees usually cannot be promoted. After the mid-1980s, the academic degree system was officially restored and a system of bachelor's degrees, master's degrees, and doctoral degrees was established. Since the 1990s, a technical certification system was gradually developed to include certifications for accountants, lawyers, and asset appraisers. In the twenty-first century, China's entry into the WTO led to the technical certification system being brought into line with international standards, creating a new standard for differentiating social status.

The exclusive function of property has also become more prominent. As mentioned above, a necessary prerequisite for establishing China's strict household registration system was the breaking up of the class system and the existing patterns of ownership. The class and ownership systems were replaced by the strict household registration system, which formed the core of the CPC's system for maintaining social order and reallocating resources. After the reforms and opening up, the government came to own much less property and the amount of property controlled by private interests increased significantly. Since the 1990s, private enterprises, joint-equity enterprises, and various enterprises with various kinds of ownership have developed spectacularly. The property possession in the private sector has become increasingly diversified. For example, after the housing reforms, the possession of private houses has become much more common. According to data provided by Song Chunhua, former vice minister of the Ministry of Construction, the proportion of Chinese mainland urban residents who own their own houses has reached 82%, higher than 53.6% in Hong Kong and 69% in the U.S. On March 16, 2007, the National People's Congress passed *The Property Law of the People's Republic of China* ("中华人民共和国物权法") in which it clearly declares that private property and public property should be equally protected. On October 15, 2007, Hu Jintao—on behalf of the Central Committee of the Chinese Communist Party—gave the Party's 17th Congress Report, in which he pointed out that "conditions should be created to allow more common people to receive income from property." With the facilitation of all these conditions, it can be concluded that property will gradually become the main standard for assessing one's economic and social status. Status based on a property ownership system is rising and may come to form the new primary system for maintaining order.

In terms of general worldwide trends of structural change, the early period when status stratification disintegrates and develops into economic stratification is normally the period when social contradictions intensify. In China, a difficult situation developed in which a class struggle endured after the class system had been dismantled and, in theory, economic differentiation no longer existed. When China abandoned status stratification and re-adopted economic stratification, what was needed most was a period of stable development—but this period is one in which social contradictions are most easily intensified. This was the most challenging social stratification issue that China had ever encountered.

Of course, this status system had been in place for several decades in China, so its alteration inevitably encountered all sorts of obstacles. In a country like China with a population of more than 1.3 billion people, any change or reform must be made gradually. This was particularly true for the status system, which exhibited the characteristics of a prominent hysteresis effect. In the early twenty-first century, the hysteresis effect characteristics of changes to the status system triggered and intensified some social contradictions. At present, a prominent contradiction exists in the relationship between the still extant household registration system as a form of "social inertia" and new social groups. At present, three important problems require resolution. The first problem is the dual isolation of urban and rural areas, as China is at a critical point in which the status relationship between urban and rural residents needs significant adjustment. Many countries have encountered the problem of rural populations transferring into cities as part of the process of industrialization. However, in most countries, industrialization and urbanization were almost synchronous—meaning urbanization was popularized while industrialization was realized. This was possible because they were not chained to an existing status system. Although China has rapidly industrialized in the last thirty years, the development of urbanization (especially the reform of the urban household registration system) lags behind. Thus, distant from the main track of social development, an enormous group of farmers remain registered in rural areas while working in urban areas, meaning that the household registration system is a significant problem. The second problem is that of the migrant population (especially white-collar workers and members of the upper middle class among the migrant population), as these groups have vociferously demanded their rights. In recent years, migrant white-collars have repeatedly demanded the cancellation of household registration constraints, and many well-educated migrants and entrepreneurs among the migrant population have demanded official urban household registration status.

The government administration has reacted to this call and the Ministry of Public Security has advocated that experimental reform of household registration should be carried out regionally according to local conditions. It also asserted that the basic principles of the experiment were to meet local needs, benefit local residents, facilitate local government financing, and effectively resolve local problems. For example, some time ago, the cities of Shanghai and Guangzhou promoted a "blueprint for a household registration system," which prescribed that migrants could apply for permanent local residence only if they met certain criteria regarding time spent living and working in the city, after they had obtained their blueprint household registration. The Personnel Department of Beijing also loosened its management of household registration for skilled migrants. Zhejiang province proposed that the barrier between urban and rural household registration was to be broken across the entire province. The division between agricultural and nonagricultural household registration was to be eliminated and a unified household registration system was to be developed. Guangdong province declared that it would strive to end all agricultural household registration when conditions were ripe. The contradiction triggered by the continued influence of household registration status has caused interest in the issue to intensify. Compared with thirty years ago, the range of the present household registration status-related interests has contracted significantly, although some still endure. The most prominent contradiction is that the examination scores required for college enrollment and enrollment ratios vary for students with different household registrations. In the twenty-first century, this kind of interest conflict has become particularly serious. During the annual National People's Congress and Chinese People's Political

Consultative Conference, the problem of differing enrollment score requirements for different household registrations has been officially discussed. During the discussion, many congress members attacked "household registration priority" and "household registration privilege" in university admission, stating that this is the greatest injustice in China's education system. They argued that national colleges and universities should adopt the principle that "everyone is equal before examination scores" and that people with different household registration statuses should be entitled to equal admission rights. Such discussion will be the catalyst for the further erosion of status stratification, I believe.

2. Replacement of Political Stratification with Economic Stratification

An important aspect of China's social stratification is the differentiation of political stratification and economic stratification. As discussed above, China's stratification was dominated by politics before the reforms and opening up. At that point, political status was more important than economic status and there were notable differences between people in terms of their political status. Political status had important functions and political struggle was heated, even leading to serious political prejudice. After the reforms and opening up, the importance of political status was greatly reduced and economic status became increasingly important. Therefore, since China's reforms and opening up, the most significant change in social stratification structure is that China has transformed from a society based mainly on political stratification to one based mainly on economic stratification. In other words, economic inequality has replaced political inequality.

The process of transforming the system started with the new policy of "allowing some people to become rich first" ("允许一部人先富") proposed by the Third Plenary Session of the Eleventh Central Committee of the Communist Party of China. On December 13, 1978, Deng Xiaoping pointed out in a speech (considered to be the keynote speech of the Third Plenary Session) that "in the case of economic policy, I think that some areas, some enterprises, and some workers and farmers should be allowed to earn more income and have better lives based on their harder work and greater achievements."[35] Deng's speech and the

35 Deng Liaoping. *Selected Works of Deng Xiaoping (1975-1982)*. Beijing: People's Publishing House, 1983. 142.

policy of "allowing some people to become rich first" became one of the initial impetuses that started China's reform. They also became an important force for transforming the social stratification system that manifested in, among others, the following three ways.

First, there was a significant change in social stratification and in the basic principle of social equity. As mentioned above, after 1949, the basic guideline of national policy was to seek equality in economy, namely wealth allocation and income distribution. Theoretically speaking, this was the "fair result" ("结果公平") principle. A "fair result" can be said to exist only if people obtain relatively similar shares of economic resources through distribution; if people obtain quite different results or the results show a huge gap, it is unfair no matter what the cause may be. However, it could be argued that the new policy of "allowing some people to become rich first" fell under the principle of "equal opportunity" ("机会公平") or "fair process," ("程序公平") on the basis that an equalized result was not real equity[36] because it is unfair to pay equal sums to those contributing varying amounts. Real equity and justice lay in providing people with equal opportunities to compete and setting up an equal procedure for competition. In this way, although people gained different economic incomes, those who were competitive, worked harder, and contributed more gained more than those who were less competitive, worked less, and contributed less. This was fair because this was in accordance with the principle of "work more, earn more; work less, earn less" ("多劳多得少劳少得"). Therefore, Deng's speech on December 13, 1978, was a significant transformation in equity and justice theory in China that led to significant changes in the social stratification system.

Based on this most basic and fundamental policy of "allowing some people to become rich first," a series of policies were formed; it was because of these policies that China's social stratification structure finally changed. This reveals that, in China, policy variation has a great impact on social stratification structure. I believe that one of China's important characteristics is its "government-guided society," in which powerful institutions play the most important role in the decision making about and management of important issues such as politics, economy, and culture. Therefore, a very special situation has appeared in China: policies influence the whole of Chinese society

36 Li Qiang. *Ten Lectures on Social Stratification*. 267.

and can even change the social stratification structure. I have analyzed how policies have changed the structure of social stratification time and time again in the sixty years since the People's Republic of China was founded.[37] I have also proposed that policies in China affected and changed social stratification at three different levels (for more details, please refer to chapter 6 of this book). The first level is "basic tone theory," which indicates the basic direction of policies. The second level is the macro principle of policy. At this level, under the guidance of basic tone theory, various principles and guidelines are applied to practical problems and proposed as potential policies. The third level is that of specific policy, which specifies operational details. The policy of "allowing some people to become rich first" belongs to the first level, which is the most critical and fundamental part of the policies that impacted the social stratification mechanism. Therefore, its impacts can be seen in all directions.

Finally, the "allowing some people to become rich first" policy is a simple expression that is easy for common people to understand. China's reforms are taking place in every aspect of society and at all levels, including politics, economy, culture, the Party, the military, civilian life, and education. For civilians, however, the most impressive and relevant aspect is the policy of "allowing some people to become rich first" because this policy sharply contrasts with that which prevailed before the reforms. In the past, personal enrichment was criticized, but now personal enrichment is encouraged and becomes something tangible from which civilians can benefit. Thus, people have strived for more income and to have better lives. Therefore, this policy played an incomparably important role in starting the reforms among the masses.

After the 1980s, under the guidance of the "basic-tone theory" of "allowing some people to become rich first," a series of principles and specific policies were unveiled. These were the policies that finally changed the original political stratification system and shaped a new system of economic stratification. Those specific policies covered a broad field of areas, so only some aspects will be used as examples. In the case of rural policy, first came the policy of allowing farmers to cultivate their own lands and conduct sideline economic activities, and encouraging and helping them to increase their personal incomes.

37 Li Qiang. *Ten Lectures on Social Stratification*. 327–345.

Later, policies of fixing farm output quotas for each household and contracting work to households were introduced, leading to the formation of the household-based contract responsibility system with remuneration linked to output. The relatively important documents of the CPC central committee include the 1982 No. 1 document, *The Summary of National Rural Work*, which introduced the policy of contracting work to households. The 1983 No. 1 document also promoted the household contract responsibility system, with remuneration linked to output by encouraging farmers to conduct household business and sign contracts for fruit tree cultivation and the management of forests, wild mountains, and lands. Finally, the 1985 No. 1 document proposed the relaxation of controls over the prices of agricultural products and promoted commodity markets. In the case of urban policy, from 1980 it became popular for enterprises to utilize bonus systems, which started the "breaking the iron rice bowl" experiment ("打破大锅饭") that abandoned equal distribution and encouraged working more to earn more. The piece rate wage, contracting system, and responsibility system became popular among enterprises. In the midst of these reforms, a new system was formed by enterprises in which wages and bonuses were decided by enterprises themselves, with the result that the bonuses awarded to some workers and official staff exceeded their wages. Meanwhile, the *Central Governmental Decision on Economic System Reform*—passed by the Third Plenary Session of the Twelfth Central Committee of the Communist Party of China in 1984—clearly states the necessity of developing individual businesses and private enterprises properly, and admits the legitimacy of employing labor, which was an epochal reform. Since then, China's private and individual economies have developed rapidly. Additionally, the opening up vigorously introduced foreign capital, foreign enterprises, and three types of foreign-funded enterprises (joint ventures, cooperative joint ventures, and wholly foreign-owned enterprises). The change in the ownership of foreign-funded enterprise and the changes undergone by employment patterns were also an important aspect of the transformation of China's social stratification.

After the important transitional period of the 1980s, the system of economic stratification stabilized in the 1990s. Almost all the economic policies introduced were conducive to economic stratification. For example, in the case of ownership, the co-existence of diverse forms of ownership as well as the development of the private economy were encouraged. In the case of income distribution, the co-existence of diverse distribution forms was permitted; incomes from the individual economy, and income from interest, dividends, bonuses, and capital were all made legal. The documents produced by the Chinese Communist Party clearly pointed out that production factors such as labor, capital, technology, and management should participate in the distribution according to their contribution.

In the early 1990s, after Deng Xiaoping's southern tour, the pace of China's reforms quickened with the introduction of the market economy. The most crucial policy variable was the *Report of the 14th Congress of the Communist Party of China* in October 1992; for the first time, there was an unambiguous statement that the aim of the reforms was to build a socialist market economy. In November 1993, the Third Plenary Session of the Fourteenth Central Committee of the Communist Party of China passed *The CPC Central Committee's Decision on Several Issues about Building a Socialist Market Economy System*. The rapid development of the market economy led to great changes in the quantity, content, and means of resource allocation among different social classes, strata, and social groups. Take the allocation of land resources, for example. Since the implementation of the State Council's 55th document, *Provisional Regulations on the Sale and Transfer of Urban State Owned Land Use Rights*, in 1990 and the 22nd document of the Ministry of Construction, *The Planning Management Measures on the Sale and Transfer of Urban State Owned Land Use Rights*, in 1992, the individual and group renting of urban land and territory leasing were legalized. Thereafter, the real estate industry began to develop rapidly and a wealthy group made up of real estate group companies and real estate businesses thrived and grew at a rapid pace. After the marketization of the real estate industry, the allocation of house and land property became fundamentally different from that of the past. The core of urban housing reform was the reform of the land system because only after the land system was reformed could real estate developers possess land and, thereby, build houses. House prices were mainly determined by the price of the land on which they stood. Since the beginning of the twenty-first century, the upward surge in housing prices has resulted largely from rising land prices. Who has profited from high land prices? According to a report produced by Han Jun, minister of the Rural Economy Research Ministry in the Development Research Center of the State Council, 40–50% of the capital generated by these increases in land value was taken by various real estate developing companies, economic development zones, and foreign investment companies.[38] Therefore, the new mechanism of resource allocation is fostering a new stratification of social groups. Land aside, other resources—like those held by state-owned enterprises, some mineral resources, and medical resources—also tended to be allocated in a similar way in the 1990s, when a sharp transition to the market economy took place. Since the twenty-first century began, some important laws and regulations further solidified the economic stratification system. One important example is *The Property Law of the People's Republic of China,* passed in March 2007, which explained in detail the legal protection afforded to state, collective, and individual ownership. Thus, all these laws have become an institutional guarantee for economic stratification.

To sum up, both before and after the reforms, China's social stratification system and its contents were fundamentally different. Before the reforms, political stratification played a prominent role and people's status was not determined by economic factors, but by political factors. After the introduction of the market economy, however, the role of political factors gradually weakened as criteria for stratification. Economic stratification replaced political stratification and became one of the main characteristics of China's post-reform social structure.

3. The Widening Gap between the Rich and Poor

After understanding how economic stratification replaced political stratification, the next task is to understand the extent of economic stratification and how the gap between the rich and poor has changed. A great deal of research has been done on this topic and it has attracted widespread attention. Whether in a scholar's report, data produced by a statistics department, or people's own understandings, there is a consensus that the gap between the rich and poor is comparatively large and has grown over the past thirty years. For data before the reforms, we have adopted materials produced by

38 Wang Ping. "Land Politics: A Comprehensive Analysis on Land System of China." *China Reform Forum* (http://www.chinareform.org.cn/cirdbbs/) → theme forum → rural reform → *China Reform,* August, 2005.

the World Bank. The World Bank's 1979 investigation revealed that the Gini coefficient for the average income of rural and urban households in China was 0.33, which can be considered as the gap between rich and poor at the beginning of the reforms. Later research showed that the Gini coefficient increased continuously after this point. A frequently quoted figure is the Gini coefficient of the average income of rural and urban households in 1988, which was 0.382. This result was obtained by Professor Zhao Renwei and Li Shi of the Chinese Academy of Social Sciences. In 1994, a figure obtained through my own research showed that the Gini coefficient has increased to 0.434. From 1996 to 1997, I organized a national survey that showed that the Gini coefficient had reached 0.4577. By comparing Gini coefficient data from several departments from 2005 to 2008, I found that the Gini coefficient of the average income of rural and urban households was not lower than 0.5. But, it should be emphasized that all the above data were not drawn from official government statistics but from scholars' own investigations. These data were obtained through nationwide sampling. Although the design of the sampling methods used differed, the research methods and survey designs were well thought out. Therefore, these figures can be considered valid and reliable. Of course, unlike the State Bureau of Statistics, which has agencies and survey teams throughout the country, scholarly research is quite limited in terms of capability and funding. Therefore, official data from the State Bureau of Statistics should be considered more authoritative. The recent Gini coefficient published by the State Bureau of Statistics was 0.47, which is a bit lower than the results obtained by the above researchers.

To sum up, although the data from various sources differ somewhat, all available research admits that the present gap between rich and poor is relatively large because all the research shows that the present Gini coefficient is definitely no lower than 0.47. Researchers have commonly accepted that a Gini coefficient between 0.3 and 0.4 shows a moderate level of inequality and a coefficient above 0.4 indicates a high level of inequality. China's inequality is clearly much higher than the "moderate inequality level."

The gap between rich and poor has widened a great deal since the reform. What social role does this play? In my opinion, the gap between rich and poor is too complicated to be evaluated on the basis of size.

A narrow gap is obviously preferable, and the gap between rich and poor was narrower before the reforms, as indicated by the above data. The Gini coefficient of the average income of an urban household was only 0.16 at that time, almost the lowest in the world. However, our investigation also proved that the vast majority of urban and rural residents considered the reforms a correct step and were unwilling to return to the past. Therefore, the present situation is so complicated that it cannot be explained clearly through reliance on figures or Gini coefficients alone.

Generally speaking, the emergence of a gap between rich and poor was synchronous with the introduction of the market economy and its rapid development. At the very beginning of the reforms, Deng Xiaoping proposed the principle of "allowing some people to become rich first," intending to stimulate people's enthusiasm for becoming rich and driving the economic development of the whole nation. Social policy cannot be separated from the social background of the time, and the situation at that time was one of low production and labor efficiency. The economy was backward because the then-prevailing policy was "the same gain for more or less work" and "the same gain for good or bad work" ("干多干少一个样，干好干坏一个样"), which was an unreasonable policy that "encouraged laziness and discouraged diligence" ("奖懒罚勤"). For thirty years, making money and becoming wealthy has been a powerful incentive for improving productivity, hence the rapid economic growth seen since the 1980s. Therefore, the significant social function of the profit motive in stimulating China's economy cannot be ignored when the gap between rich and poor is assessed.

Of course, things are always dialectical. It is a serious problem for a country to lack economic momentum, but it becomes an even more serious problem if people are overly focused on pursuing wealth. In contemporary China, wealth has become an important focus of many people's activities. It was reported that, in a recent online survey completed by young people, the pursuit of "more money" ranked first among their ambitions for 72.68% of all the respondents.[39] Comparatively speaking, at present, the Chinese enthusiasm for money exceeds that in many other countries, manifesting as a public mentality characterized by an eagerness to become wealthy. Does this stimulate the economy? It certainly does—the rapid development

39 "The First Desire of China's Youth: Getting Rich." *New Culture Daily* July 20, 2005.

and expansion of China's economy has proved that. But, the excessive impetus to pursue wealth and the rapid polarization of rich and poor have inevitably produced undesirable consequences.

What are the negative functions of the polarization between rich and poor? Obviously, if the gap between rich and poor is too wide, the imbalance of their interests will lead to social contradictions and conflicts. At present, the imbalance expresses itself in many ways—such as the imbalance between urban and rural development, as well as an imbalance between groups that have benefitted from the reform of state-owned enterprises and those that have suffered (like laid-off workers, and those displaced and resettled by urban housing demolition and renovation and various construction projects). The polarization between rich and the poor is a double-edged sword. On the one hand, it stimulates people's desire for wealth and economic development; on the other, it produces many social problems.

Therefore, nothing is absolutely good or bad, and advantages and disadvantages are relative. What should be borne in mind is whether positive or negative functions played a major role in a specific period. It is not difficult to ascertain that at the beginning of the reforms, the positive function of "allowing some people to become rich first" was more prominent, but in recent years, its negative functions have become increasingly apparent. Therefore, at present, the central government is attempting to produce policies to adjust income distribution and narrow the gap between the rich and poor.

How do sociologists explain the ever-widening gap between rich and poor in China? A popular viewpoint is to explain it from the angle of market transition. This viewpoint holds that before the reforms, the planned economy played a leading role in Chinese society; economic resources were not allocated by the market but by government organizations. For instance, the government distributed houses and adjusted salaries, leading to an income distribution structure characterized by "egalitarianism" and "equalization." After the reforms, with the gradual introduction of the market mechanism, economic resources were allocated mainly by the market. Unfortunately, the market has no sense of fairness and follows the rule of survival of the fittest, so it naturally tends to produce imbalances and causes differentials to grow between those who possess wealth and those who do not. Of course, socialists have also pointed out that China's market system is still at the primary stage and far from complete, and that the market is only one distribution link. In a country with a well-established market system, various other systems work in cooperation with the market economy, such as the social security system and social welfare. These systems work to readjust the distribution of wealth and finally realize balanced development. Therefore, the present problem in China does not lie with the introduction of the market system, but rather with its undeveloped nature.

What social consequences are produced by the increasing gap between the rich and poor? Of course, the polarization of rich and poor will lead to tense social relationships, which I once explained using the concept of "social structural strain" (see chapter 11, part 3). However, social structural strain does not mean that social contradictions and conflicts will unavoidably intensify and explode. The mechanism by which the gap causes social conflicts is complicated—not "direct" but "indirect." That is, the widening gap between the rich and poor will not inevitably cause social conflicts. Even in a situation where the gap is widening, the two groups can co-exist peacefully if there is no conflict of public opinion between these two groups—especially if the wealth differential is not linked to other factors. In America, for instance, where there is a huge gap between the rich and the poor, no social conflicts have arisen directly from wealth inequality. This is also the case in China. Most of the time, the poor and the rich get along peacefully—although peasant uprisings, revolutions, and conflicts are relatively common in Chinese history.

To my mind, "social structural strain" caused by a wide gap between rich and poor is an objective structure that lays an important foundation for social contradictions. Whether or not contradictions emerge or intensify depends on another factor: "justice imbalance" ("公正失衡"). This is a subjective public attitude based on perceptions of fairness. When the gap between the rich and the poor is not linked to "injustice" or "unfairness" and people believe that economic competition is reasonable, social conflicts will not arise. Take Bill Gates, for example—one of the richest men in the world—whose annual operating income is measured in the billions of dollars. Few consider this an injustice. Bill Gates not only gets along well with the poor, but is also complimented by them because he donates more than half of his income to charitable causes every year.

Only when the polarization between rich and poor is associated with "injustice" can social contradictions really become intensified. Therefore, to solve such fundamental problems, it is not enough to simply know the proportions of wealth possessed by the rich and the poor. It is necessary to understand why wealth differentials occur, as well as what ordinary people think of being rich or poor, and whether they consider things to be just or unjust. Under the objective circumstance of structural strain, if many people attribute the gap between rich and poor to "social injustice," then an environment in which social conflicts will take place frequently is created. Many recent incidents reflect the problem of "justice imbalance" to some extent. Consider the case of a series of incidents that occurred some time ago, such as the 2003 "BMW hit and run murder," the 2003 "demolition incident," and the 2003 "immigration project." These cases were seemingly about money and being poor or rich, but actually they concerned the more important matters of what is just or unjust, reasonable or unreasonable. Therefore, China's reforms should pay more attention to social justice. Once serious social injustices emerge and various social groups become dissatisfied, existing social contradictions will be intensified.

In such situations, what counteraction should be taken? As mentioned above, the experience of over thirty years of reform has proved that China's reform was government guided. Governmental policies have had great impact on the nation's economic development and social transformation, and almost all system changes have been conducted under the guidance of the government. Therefore, to alleviate the gap between rich and poor, the most important measure is the adjustment of government policy. As for the issue of income distribution, the basic principle that has prevailed in China for a relatively long time can be summarized as "giving priority to efficiency with due consideration to fairness" ("效率优先，兼顾公平"). This can clearly be observed by scrutinizing the government's documents. The principle of income distribution established by the report of the Fourteenth Congress of the Communist Party of China in October 1992 was "balance efficiency with fairness" ("降谷效率与公平"). In November, 1993, *CPC Central Committee's Decision on Several Issues about Building a Socialist Market Economy System* passed by the Third Session of the Fourteenth Central Committee of the CPC established the idea of "giving priority to efficiency with due consideration to fairness." The

Fifteenth National Congress continued to use the same expression. The report of the Sixteenth National Congress continued to use this expression, but also stressed "focusing on efficiency in primary distribution" ("初次分配注重效率") and "focusing on fairness in redistribution" ("再分配注重公平"). In September 2004, the *Decision on Strengthening the Party's Governing Capacity* produced by the Fourth Plenary Session of the Sixteenth Central Committee used the term "focusing on social fairness" ("注重社会公平"). In October 2005, *The Central Committee's Suggestions for the Formulation of the Eleventh Five-Year Plan for National Economic and Social Development* of the Fourth Plenary Session of the Sixteenth Central Committee discontinued the use of the phrase "give priority to efficiency with due consideration to fairness." Instead, it emphasized "focusing on social fairness" and "alleviating the widening income distribution gap between different areas and between some members of society" ("努力缓解地区之间和部分社会成员之间收入分配差距扩"). Recently, the central government proposed that all Chinese citizens should share in the fruits of reform together. From the above analysis, it can be seen that the central government has made explicit adjustment to its policy of income distribution since 2004. Recently, some important policies made by the central government have further affected these adjustments. For instance, in 2006, the central government made the decision to abolish the agriculture tax within five years. Another example is the decision to establish "the New Rural Cooperative Medical Care Scheme" and to provide more government funds to rural cooperative medical care schemes from 2005 to 2007. These policies have all contributed to the adjustment of income distribution.

As stated above, economic reform and development in China is government guided and the central government has made important adjustments to its income distribution policies and has introduced a series of measures focusing more on fairness, as well as policies to transfer funds to the low-income population through redistribution. Therefore, it can be anticipated that, in the coming five to ten years, the tendency of the gap between rich and poor to widen will be alleviated to some extent and the level of polarization will be adjusted.

Of course, taking the long view on policy orientation, I believe that the following five measures cannot be ignored if the gap is to be actually narrowed. First, it is critical to adjust the relationship between rural and urban areas. Actually, the main gap between rich and poor in China is the gap between rural and urban areas; therefore, a fundamental solution is to solve "problems concerning agriculture, the countryside, and farmers." Second, employment is vital to people's livelihoods. Unemployment and poverty are twinned phenomena, so job creation and unemployment reduction may alleviate poverty. Third, it is necessary to promote the establishment of an effective social security system; this is an important tool in adjusting redistribution and narrowing the gap between rich and poor. Fourth, the gap can be addressed by adjusting taxes and improving the progressive personal income tax policy. Fifth, basic education must be dramatically improved. Previous research has revealed that a country's average educational level tends to be inversely proportional to the gap between rich and poor. Therefore, an increase in the average basic educational level may be capable of narrowing the gap between rich and poor. By improving the education level of the whole nation, China will finally form an "olive-shaped" social structure.

4. Changes in the Relationships between Central and Marginal Social Groups

When we compare the interest relationships of people in the 1980s with those of the post-reform 1990s, it can be found that the interest relationships between two important social groups changed considerably. These two groups are "central social groups" and "marginalized social groups."

Since the Qin Dynasty, China has always been a highly centralized society. Chairman Mao Zedong once said: "All dynasties followed the Qin's political system." By this, Mao meant the political system of centralization. This highly centralized administrative system resulted in the monopolization and centralization of social resources in China. Resources were always allocated by the central government, so those who were nearer to the center formed central groups, which could obtain resources more easily. For instance, governmental administrators, senior professionals, and technicians were near to the center. Compared with rural residents, urban residents were nearer to the center, so they obviously obtained more resources. Big cities were nearer to the center than small cities and towns, so they had better access to resources. The relationship between the large, high-ranking work units and small, low-ranking work units also followed this pattern. In short, by comparing any two strata, it can be found that the stratum that is nearer to the center is able to obtain more resources. Therefore, relatively speaking, I define the groups that are nearer to the center as "central groups" (or "core groups") and groups nearer to the edge as "marginalized groups." For example, Chinese cadres of the highest rank are a typical core group, while farmers classically belong to a marginalized group. This phenomenon was prominent in the system of the planned economy.

During the 1980s, after the introduction of the market system, the relationships between the traditionally central and marginalized groups were significantly altered. For the original centrally planned economic system, the market was an extraneous system in which the principles of free competition and fair exchange could be practiced. This was a huge departure from the original planned economic system in terms of how resources were allocated—from the central government to local governments and from the central area to the marginal areas. If illustrated with a graph, this transition would show two circles of resource allocation. One would be the circle of the planned economy, which consisted of a series of concentric circles representing central and marginalized groups. The other would be the circle of the market economy. In terms of the quantity of resources allocated, after the reforms the circle of the planned economy gradually contracted while the circle of the market economy grew. The relationships between these two circles also changed; the circle of the market economy eroded that of the planned economy from the outside. This corrosion began in the marginalized areas, influencing the mode of production and lifestyle of marginalized groups. For instance, at the beginning of reforms in the late 1970s and the early 1980s, "the household contract responsibility system" was adopted in rural areas. This allowed farmers—a typical socially marginalized group—to freely sell their surplus agricultural products, such as grains and vegetables, in the market. This was a typical example of a marginalized social group entering the market.

Consistent with this principle of two economic circles, I think that in the transformation from a planned economy to a market economy, the process of laborers coming to participate in the market economy, can be divided in two phases. In the first phase, marginalized social groups participated a great deal in the market economy and obtained relatively improved economic returns. The core social groups (cadres and intellectuals), however, seldom participated in the market economy, and thus obtained little profit from it. In the second phase, the market economy began to significantly erode the planned economy, not only in the areas of finance, insurance, and real estate, but also in research, education, culture, and healthcare. All of them fell into the orbit of the market economy. These changes led to the central social groups (cadres, intellectuals, and so on) benefitting from income distribution, while such marginalized social groups as manual laborers were at an obvious disadvantage. As a result of the adjustment of these economic interests, it seems clear that the discontent of mental workers such as cadres and intellectuals was relieved to a large extent while the discontent of laborers and farmers was exacerbated.

The results of central social groups entering the market had both positive and negative aspects. The main positive effect was that the "irrational income differential between mental and manual workers in favor of the latter" ("脑体倒挂"), which had been in place for an extended period of time, was resolved. The factors of high technology, new and high-tech industries, higher-education qualifications, and high-quality personnel came to play an increasingly important role in wealth distribution. Meanwhile, the configuration of elites was also affected[40] in that the number of "political elites" contracted while the number of "economic elites" expanded, which should be considered a positive change.

However, the negative effects of the central group entering the market have been quite serious. With their entry, the chances of power being combined with wealth became much higher than in the first phase. Many central social groups took advantage of their public power to seek personal wealth or promote the private interests of their "clique." Therefore, large amounts of state-owned assets were drained away and various methods were used to transfer public property into the hands of private individuals.

As has been established, "status stratification" before the reforms was unfair because people's social statuses were determined by inherited factors instead of being achieved through their own efforts. The rationale behind China's development of a market economy lay in the belief that the market can create an environment for fair competition

40 Li Qiang. *China's Social Stratification in Social Transformation.* 30-31.

in which people strive for their own success. However, in the second phase of the transformation, this expected fairness of markets did not come into being because public power was frequently used to obtain private benefits from the market. The central groups that took advantage of public power in order to benefit their private interests were actually building a new social status group. This phenomenon eroded the fairness of the new system and is actually a historical retrogression.

Recently, in terms of the relationship between central groups and marginalized groups, central groups have increasingly been characterized as taking more social resources. Whether urban residents are compared with rural residents, or big cities with small towns, or higher authorities with lower authorities, it can be found that the gap between marginalized groups and central groups is widening. This makes a sharp contrast with the situation of the 1980s, when lower level authorities were often wealthier than higher level ones; it was much easier to make money in small towns than in big cities; farmers had greater means to acquire wealth, and those who received lower levels of education made more money than those who had received higher education. Of course, it was unreasonable for marginalized groups to take more social resources, but now the status quo is totally reversed. The interests of marginalized groups have been seriously damaged, farmers' living standards have not improved, and some rural areas have withered—all of which are serious problems. Therefore, at present, more attention should be paid to the damage done to the interests of marginalized social groups, and more importance should be placed on the establishment of a social security system capable of supporting marginalized and disadvantaged groups.

5. The Formation of New Social Strata

Previous sections have mainly discussed systems and mechanisms of social stratification, degrees of stratification, and the changes in groups' relationships. The changes in social stratification, of course, mainly manifested themselves in changes in various strata. Before the reforms, the main characteristic of China's classes was "monolithic social aggregation," with the main four "social aggregations" being: farmers, workers, cadres, and intellectuals. These four aggregations have all changed in the last thirty years of reform. First, fundamental occupational differentiation has appeared among agricultural workers. In 2006, the number of registered rural laborers was 480.9 million, of whom 146.8 million worked in township enterprises and so were actually workers rather than farmers. However, because of the restrictions imposed by the household registration system, they were still identified as farmers. Additionally, a large number of farmers moved to middle-sized and large cities to work; this group contained more than 100 million individuals, although exact statistics have never been established for this group. Also, 21.47 million farmers conducted individual business, and some successful entrepreneurs became bosses or even came to run relatively large companies. Therefore, rural farmers who actually worked in agriculture made up less than half of those registered as farmers. Workers have also changed a lot. Nowadays, the largest workforce of manual laborers in the world is in China. Of course, this is a differentiated group. Generally speaking, this group can be divided into three parts: registered urban workers, workers in rural areas employed by township enterprises (as mentioned above), and rural migrant workers working in middle-sized or large cities. Prior to the reforms, workers were all registered urban residents, and worked in state and collectively owned enterprises. In contemporary China, however, the number of workers registered to urban households who directly conduct production operations has decreased considerably. The occupations of manual laborers have mostly been taken on by migrant farmers, and registered urban workers have begun to mostly occupy "white-collar" nonmanual positions. The cadre group has also undergone important changes. Before the reforms, the term cadre was a general one, so enterprise operators, managers, and many professionals were all considered to be cadre members. Since the reforms, with the promotion of a modern corporate system, enterprise managers have become distinct from civil servants. The management of the civil service has been gradually standardized, and the term "civil servant" is increasingly used to refer specifically to those who exercise state administrative power and perform duties in government departments. The civil service has undergone several reforms geared towards streamlining its organization and a national civil service examination system was established. During the reforms, cadre members and civil servants experienced changing fortunes. After Deng Xiaoping's

1992 southern tour, with the rising tide of the market, a large number of cadre members in government organizations "plunged into the commercial sea" ("下海") and went into business. However, by the late 1990s, many of those who went into business returned to cadre groups, and civil service posts became highly competitive positions for graduates. Finally, the status of intellectuals changed. Before the reforms, during the period of political stratification and intense political infighting, the status of intellectuals declined. Therefore, from the very beginning of the reforms, Deng Xiaoping strongly emphasized the need to raise the status of intellectuals and advocated "the theory of intellectuals being a part of the working class" ("知识分子是工人阶级一部分的理论"). The Organization Ministry of the Central Committee of the CPC therefore promulgated several documents to "implement policies regarding intellectuals" ("落实知识分子政策") with the aim of improving their political and economic treatment. The phenomenon of the "irrational income differential between mental and manual workers in favor of the latter" was extant in the 1980s, but by the end of the 1990s, the economic, political, and social statuses of intellectuals had improved significantly.

Over the past thirty years, the greatest change in classes and strata in China has been the emergence of new social strata. Among the new strata, the most notable groups are private entrepreneurs and the individual owners of small businesses. Of course, there are other new social strata besides these two groups, so why have these two been picked out for special discussion? They are particularly important because both are involved in the same form of ownership: private economic ownership. This contrasts sharply with the traditional public ownership that was the norm before the reforms, so they have been the focus of a great deal of post-reform research. Also, the private and individual sectors of the economy were both the objects of socialist transformation in 1956. After 1956, not only did the private and individual sectors disappear, but also the entire individual and private strata were transformed into common laborers and workers. Therefore, the appearance of these two strata was not simply that of newly developed strata, but the rebirth of formerly extant social groups. Finally, these two strata have developed into populous groups and become a decisive social force in China.

Since the policies of reform and opening up were implemented in 1979, the private and individual sectors have undergone a process of development. In terms of documents issued by the Central Committee of the CPC, the idea that nonpublic economic activities might be allowed to develop was first expressed in *The Central Committee's Resolutions to Some Problems Regarding Accelerating the Development of Agriculture,* passed by the Fourth Session of the Eleventh Central Committee of CPC in September 1979. This document proposed that "private family plots, livestock, sideline production, and rural fair trade are subsidiary and supplementary parts of the socialist economy and should not be criticized as remnants of capitalism."[41] Of course, the private and individual sectors were still not clearly discussed in this document. In August 1980, the Central Committee of the CPC held a conference on national employment in Beijing. The records of this conference advocated that "the individual sector of the economy should be encouraged and supported in its proper development and all law abiding individual laborers should receive the respect of society." It could clearly be seen that the first motivation for encouraging the development of the individual economy was to solve the problem of urban youth unemployment caused by and exacerbated during the Cultural Revolution, as well as to resolve the severe inadequacy of social and economic service networks in cities. Therefore, this initial encouragement of the individual economy was for those in urban areas, rather than people in rural areas. At that point, the rural individual and private sectors were developing very slowly. This was the case because farmers had just begun to take on contracts to cultivate land, people's communes had just disintegrated, and regional surplus labor was still not a prominent problem. Also, the central government was still legislating against the employment of labor at that time. Some rural areas had some private businesses, but those business owners still dared not publicly admit employing workers. Therefore, no exact data on the rural individual and private sectors was included in the official statistics of the time.

In July 1981, the State Council of China issued a document regarding the development of the urban individual economy: *The Regulations on the Individual Economy of Urban Non-Farmers.* The significance of

41 The Literature Research Center of the CPC Central Committee. *Selections from Important Documents Issued since the Third Plenary Session of the Eleventh CPC National Congress.* Vol. 1. Beijing: People's Publishing House, 1982. 173.

this document lay in the fact that it mentioned employment. It decreed that, if necessary and sanctioned by the departments of industry and commerce, one or two workers could be employed by an employer and some individuals with particular special skills could employ up to five apprentices. Although the passages dealing with employment in the individual economy were still worded quite cautiously, they opened the door to employment and proved the legal validity of employment. Since employing a small staff was legalized in urban areas, what, then, would be the case in rural areas? In January 1983, *A Number of Issues on the Present Rural Economic Policies* was published by the CPC Central Committee; this document stated that employment in rural areas should be implemented in accordance with *The Regulations on the Individual Economy of Urban Non-Farmers*. This provided legal grounds for employment in rural areas. Thereafter, the individual and private sectors entered a period of rapid development. A series of policies and regulations enacted thereafter created an increasingly free social environment for the development of the individual and private economy. For example, in the constitutional amendment of the PRC in 1988, the eleventh article pointed out that "the state permits the private economy to exist and develop within the confines of the law. The private economy is a complement to socialist public ownership." *The Property Law of the People's Republic of China*, passed in March 2007, reinforced the protection provided by the state to private property and private ownership. To sum up, under the protection of various policies, the individual and private sectors of the economy developed rapidly.

By the end of 1981, the number of registered individual businesses in both urban and rural areas had reached 1.829 million nationwide, with 2.274 million people employed. By the end of 1986, the total number of individual businesses and private enterprises was 13.425 million, employing a total of 35.456 million people. For data on the development of individual businesses and private enterprises after the 1990s, please refer to the following two tables (see table 5-1 and table 5-2).

Table 5-1 Individual Businesses in 1992–2007[42]

Year	Individual Businesses (Ten Thousands)	Employees (Ten Thousands)	Registered Capital (Billions)
1992	1534	2468	60.1
1993	1767	2939	85
1994	2187	3766	131.9
1995	2529	4614	181.3
1996	2704	5017	216.5
1997	2851	5442	257.4
1998	3120	6114	312
1999	3160	6241	343.9
2000	2571	5070	331.5
2001	2433	4760	343.6
2002	2378	4743	378.2
2003	2353	4737	418.7
2004	2350.8	4587.1	505.63
2005	2466	4901	574.4
2006	2595.6	5159.7	646.87
2007	2621.4	5309	703.47

42 Source: All-China Federation of Industry and Commerce. *The Yearbook of China Private Economy from 2002 to June 2004*. Beijing: China Zhigong Publishing House, 2005. 76.

China's Association of Private Economy Studies. "An Analysis on the Private Economy during the Tenth Five-Year-Plan Period." *The Internal Reference of Private Economy* March 31, 2006.

China's Association of Private Economy Studies. "The Number of Private Enterprises Surmounted 5 Million, Figures from the State Administration for Industry and Commerce." The *Internal Reference Documents for Private Economists* August 17, 2007.

The National Bureau of Statistics of PRC. *China Statistical Yearbook (2006)*. Beijing: China Statistics Press, 2006. 125.

Table 5-2 Private Enterprises in 1992–2007[43]

Year	Enterprises	Employees (Thousands)	Registered Capital (Billions)
1992	139633	2,320	22.1
1993	237919	3,730	68.1
1994	432240	6,480	144.8
1995	654531	9,560	262.2
1996	819252	11,710	375.2
1997	960726	13,490	514
1998	1200978	17,090	719.8
1999	1508857	20,220	1028.7
2000	1761769	24,070	1330.8
2001	2028548	27,140	1821.2
2002	2435282	34,090	2475.6
2003	3005524	42,990	3530.5
2004	3650670	50,170	4793.59
2005	3984200	55,650	5429.5
2006	4981000	65,863	7574.72
2007	5205000	69,275	8300

43 Source: All-China Federation of Industry and Commerce. *The Yearbook of China Private Economy from 2002 to June, 2004.* Beijing: China Zhigong Publishing House, 2005. 75.

China's Association of Private Economy Studies. "An Analysis on the Private Economy during the Tenth Five-Year-Plan Period." *The Internal Reference Documents for Private Economists* March 31, 2006.

China's Association of Private Economy Studies. "The Number of Private Enterprises Surmounted 5 Million, Figures from the State Administration for Industry and Commerce." The *Internal Reference Documents for Private Economists* August 17, 2007.

China's Association of Private Economy Studies. "Chinese Academy of Social Science Reported That There Are Many Defects in Private Enterprises." *Internal Reference Documents for Private Economists* February 17, 2006.

China's Association of Private Economy Studies. "A Comprehensive Report on the Data Analysis of the Seventh Private Enterprises Sample Survey in 2006 (1)."*The Internal Reference Documents for Private Economists* March 16, 2007.

The National Bureau of Statistics of PRC. *China Statistical Yearbook (2006)*. Beijing: China Statistics Press, 2006. 125.

How many private entrepreneurs were there? In total, there are records for 9,486,288 investors for 3,650,670 private enterprises, with an average of 2.59 owners for each private enterprise. In June 2006, the total number of investors was 11,841,493 for 4,648,297 private enterprises with an average of 2.55 owners for each private enterprise. Calculated according to the ratio of 2.55 owners per private enterprise in 2006, the total number of private entrepreneurs in 2007 was around 13,272,750. These 13 million private entrepreneurs were a new force in Chinese society. By observing the rapid growth of private enterprises from 1992 to 2007, it can be anticipated that this group will continue to expand. If every individual business had one owner, then in 2007, China was home to more than 26 million businessmen. Added to the number of private entrepreneurs in 2007, the total number of private owners reached nearly 40 million. If all the family members of private owners can be considered to be of the same class or stratum, then the population of this class is huge. In short, the emergence and development of such a huge group is the most prominent manifestation of social structural change since the reforms.

In view of the significant changes in social structure and the new social stratum, the CPC also made important adjustments to its administrative policies to adapt to the changing social structure. In November 2002, the Sixteenth National Congress of CPC amended its Constitution, changing the first article regarding the qualifications for party membership from "Chinese workers, peasants, soldiers, intellectuals, or other revolutionaries over eighteen years old" to "Chinese workers, peasants, soldiers, intellectuals, or other outstanding individuals in other social classes over eighteen years old." This important modification helped the ruling party solve the long-standing problem of whether or not the class of private business owners could be admitted into the CPC. The modification also indicated that the CPC, as the ruling party from then on, would absorb the economic elites who emerged from this new type of economy. This was significant for the solidarity of economic elites.

According to the seventh sampling survey of China's private enterprises conducted in 2006, CPC members accounted for 32.2% of all private entrepreneurs in private enterprises registered after 2001.[44]

44 The Research Team about China's Private Enterprises. "A Comprehensive Report on the Data Analysis of the Seventh Private Enterprises Sample Survey in 2006." *The Internal Reference Documents for Private Economists. China's Association of Private Economy Studies* 2 (March 2007).

This is an astonishingly high proportion. If one observes the social status and function of private entrepreneurs, it can be seen that private entrepreneurs stand out conspicuously in both economics and politics. It is estimated that by 2005, there were more than 9,000 private entrepreneurs elected as deputies to the NPC (The National People's Congress) at or above the county level and more than 30,000 private entrepreneurs were made members of the Committee of the Political Consultative Conference at or above the county level. Additionally, 233 private entrepreneurs were assigned to the executive committee of the All-China Federation of Industry and Commerce (accounting for 56% of all executive committee members), and eight of these were made Chairmen of the All-China Federation of Industry and Commerce.[45]

From the perspective of education, since the end of the 1990s and the beginning of the twenty-first century, the composition of China's group of private entrepreneurs underwent significant changes. This stands in sharp contrast to the late 1980s and early 1990s, when the general status of private entrepreneurs was relatively poor. According to a nationwide sample survey conducted by the "China's Private Entrepreneurs Stratum" research team in 1993,[46] the educational level of private entrepreneurs at that time was: 5.5% university graduates or above, among whom only 0.6% were postgraduates; 11.7% had three-year college educations; 35.9% had graduated from secondary technical schools, vocational high schools, and common high schools; no more than 36.1% had graduated from junior middle schools; 9.9% had only graduated from elementary schools; and the remaining 1.0% were illiterate. This poor level of education had changed completely by the twenty-first century. According to a nationwide sample survey of 2,998 valid samples of private entrepreneurs prepared by the All-China Federation of Industry and Commerce in 2004,[47] the educational level of private entrepreneurs in China was much higher than that in 1993. University graduates or above accounted for 20.7%, among whom 5.7% were postgraduates; those with three-year college educations

45 China's Association of Private Economy Studies. "Chinese Academy of Social Science Reported That There Are Many Defects in Private Enterprises." *The Internal Reference Documents for Private Economists.* February 17, 2006.
46 Zhang Xuwu, et.al. *China Private Economy Yearbook (1978-1993).* Hong Kong: Hong Kong Economic Reporter, 1994. 117.
47 All-China Federation of Industry and Commerce. *China Private Economy Yearbook (2002 to June, 2004).* Beijing: China Zhigong Publishing House, 2005. 30.

amounted for 31.1%; those who graduated from secondary technical schools, vocational high schools, and common high schools accounted for 33.6%; those who graduated from junior middle schools made up 12.9%; and those who graduated from elementary schools accounted for only 1.7%. Their educational qualifications were much better than the average Chinese worker. Therefore, private entrepreneurs became a "high quality group" during this period.

The social origins of private entrepreneurs also changed significantly. An investigation in 1993 revealed that in the late 1980s and early 1990s, a large proportion of private entrepreneurs were originally farmers and workers. It indicated that 53.5% of rural private entrepreneurs used to be farmers, 25.2% of urban private entrepreneurs were workers, and 17.2% began as farmers, together accounting for 42.4%. Since 2000, from which occupations or strata have private entrepreneurs mainly come? Based on surveys conducted in 2004 and 2006, the proportion of China's private entrepreneurs who used to be cadres or engaged in management and professional technical work has greatly increased. The 2004 survey[48] revealed that, before they started their own businesses, 15.6% had been leaders of government departments, enterprises or public institutions; 10.4% had been general officers; 14.2% were leaseholders and contractors of state-owned or collective enterprises; 2.8% were village cadres; 1.4% had been county and department level cadres; 13.0% were professionals; and 10.0% were supply and sales staff. All together, this proportion had risen to 68%. The survey conducted in 2006[49] revealed that the proportion of those who had been departmental leaders or officers, professionals, supply and sales staff, and private and individual businessmen remained steady at 67.4%. This change in occupational origin reflects a significant change in the composition of the private entrepreneurial group after thirty years of reform.

The compositional changes in this group, and the stratum of private industry and commerce (as well as their improving quality), are of great significance. This stratum, as an independent stratum of actors, broke

48 All-China Federation of Industry and Commerce. *China Private Economy Yearbook (2002 to June, 2004).* Beijing: China Zhigong Publishing House, 2005. 30.

49 The Research Team of China's Private Enterprises. "A Comprehensive Report on the Data Analysis of the Seventh Private Enterprises Sample Survey in 2006 (2)." *The Internal Reference Documents for Private Economists* March 23, 2007. The Research Association of China's Private Economy.

through the constraints of the past status system and work unit system and became one of the most energetic groups in the market economy. Its improved qualities and status reveal that it will play a more positive role in the future construction and development of socialism.

Of course, other new strata besides private entrepreneurs and businessmen have developed; they are categorized into six groups in the report of the Sixteenth CPC National Congress. They are entrepreneurs and technicians in private technological enterprises, administrative and technical staff employed in foreign companies, self-employed businessmen, private entrepreneurs, employees in intermediary organizations, and freelancers. Therefore, apart from private businessmen and private entrepreneurs, there are still four other new strata detailed in CPC policy. As a matter of fact—in terms of their social status and social function—there is no difference between "entrepreneurs and technicians in private technological enterprises" and "administrative and technical staff employed in foreign companies," and they are similar to other types of administrators and technicians, and so cannot be identified as an independent social stratum. Additionally, "employees in intermediary organizations" and "freelancers" are both minority groups. Therefore, since the reforms, the most prominent and distinctive new stratum has been that of owners of the private economy. It is also an experiment of the Chinese socialist revolution and construction to acknowledge the social status of private entrepreneurs and owners in the private economy. The future development of this stratum must be monitored.

Chapter 6

The Great Influence of Policy Variables on Social Stratification Structure

Analyzing Chinese social stratification since the foundation of the People's Republic of China and stratification structure since the reform process began in 1978 reveals the great influence of the variables of state policy on Chinese social stratification. The influence of the variables of state policy on social stratification structure is a key characteristic of contemporary Chinese society. Many previous studies have mentioned this characteristic,[50] but their examinations of the specific mechanisms by which state policy influences social stratification were inadequate. In this chapter, the concept of the "policy packet" will be used to suggest that the specific mechanism by which state policy influences social classification has three levels of operation: that of developing the "basic tone theory;" that of turning the "basic tone theory" into "principles of broad policy;" and that of applying these principles to "specific, enforceable policies." Also, the concept of a "government-led society" will be put forth in this chapter. Of course, this is just a theoretical hypothesis and further verification is needed.

1. The Characteristics of a Government-Led Society and Social Stratification Under the Influence of Policy

One of the most important characteristics of Chinese society is that it is a government-led society. The term government-led society refers to a society in which central and local institutions of governmental authority play the most important role in managing and making decisions on important political, economic, ideological, and cultural issues. The function of government guidance was not only influential in the era of the planned economy, but also after the market-oriented transformation. By observing the last thirty-plus years of reform, it becomes clear that the role of the government is still the most

50 Martin King Whyte and William Parish. *Urban Life in Contemporary China*. Chicago: U of Chicago P, 1984; Xueguang Zhou. *The State and Life Chances in Urban China: Redistribution and Stratification 1949-1994*. Cambridge: Cambridge UP, 2004.

important. In China, the processes of launching reforms, putting forward and implementing important reform measures, adjusting the ideological line, and making reform plans are all carried out according to government instructions. The government-guided nature of its society is key to understanding China. Since it is a government-led society, the influence of the government's policy on Chinese society is great. In fact, in countries with normal and stable systems, the influence of policy variables on the social system and social structure is quite limited. Policy cannot easily change the social structure because a stable system and stable structure are the result of social change occurring over a long period of time. Once a society's systems and structures are stabilized, it is not easy at all to change them. In a society that has long advocated the rule of law, the effects of short-term policies will be limited because the law is quite stable.

However, this is not the case in China. Since the establishment of the People's Republic of China, China has remained in a continuous state of revolution, reform, and experimentation. China's system has changed tremendously many times in this period, so the system cannot be said to be stable. This instability means that policy has played an important role. The so-called reforms have been an effort to correct constantly the system through policy variables.

Of course, I must state my position here. It is necessary to understand the realities of contemporary China's situation, but it is also necessary to question whether the current reality is ideal. The government-guided pattern is not the most ideal model; however, it is the reality. In the economic field, the pace of reform is relatively quick; the role of the market and the role of government have almost reached a point of equal importance. Many resources are allocated by the market. However, if we combine the political, economic, social, ideological, and cultural fields and view them comprehensively, China can be seen to be a typical government-led society.

Regarding this issue of what constitutes a reasonable and ideal model, is a government-led society reasonable? No government is without the ability to influence four major aspects of society; it is a question of the extent of its regulation. The difference is the degree to which the government intervenes. Of course, the government-led type of society involves intense intervention and too much government intervention has a negative effect. So, for a reasonable and ideal model, there must be some boundaries to government influence, intervention, or regulation.

I have just mentioned boundaries for government intervention in the economic field and the market, of which there are four. First, government intervention, regulation, and control should not be direct intervention in the operation of specific enterprises, but should instead be legislation, regulation, and adjustments of the market in accordance with the principle of management. Second, government intervention should not be simply administrative, but should make use of economic levers, such as adjustment of central bank interest rates. Third, the relationship between the state and the market and enterprises should not be one-way, but one of mutual restriction. Legislation, laws, and regulations should not only restrict the actions of enterprises and the market, but also those of the government. Fourth, legislation and regulations should be universally enforced. If all parties—including enterprises—are involved in the process of legislation, then the laws and regulations will be acceptable to all. If actions are taken within these four boundaries, then they will be comparatively fair.

Taking social reality into consideration, an important characteristic of Chinese society is that policy can change social structure. However, not all policies can change the structure. An examination of the actual process of policy-changing structure reveals two key trends. First, important changes are always connected with major historical events. I once led a study into the influence of important social events on Chinese people's life trajectories,[51] such as land reform and the socialist transformation of the 1950s, the Cultural Revolution, the reform and opening-up process, Deng Xiaoping's southern tour, etc. Second, when significant events take place, they result not in the implementation of one policy, but "policy packets"—a series of policies concerned with politics, economic, culture, and ideology. A "policy packet" is a series of interrelated polices with a clear, guiding direction.

In this section, a brief analysis will be made regarding the way that three key historical events and corresponding policy variables have influenced Chinese social structure. These events were: the socialist

51 Li Qiang, et al. *Course of Life: Important Social Events and Chinese People's Life Trajectory.* Hangzhou: Zhejiang People's Publishing House, 1999.

transformation of 1956, the Cultural Revolution after 1966, and the opening and reform process that began in 1978. However, these events will not be assessed as having been equally important. In addition to analyzing the background of the policies, I will take the status change of intellectuals as an example of how policy influences social stratification when analyzing the changes since the reform and opening up in 1978.

2. The Influence of Three Significant Historical Events and Policy Variables on Social Stratification Structure

Below, I will analyze the influence of Chinese policies on social stratification during the socialist transformation in 1956, Cultural Revolution in 1966, and the reform and opening up in 1978.

Policies developed for and during the socialist transformation in 1956 included the establishment of rural cooperatives, the ending of large-scale unified accounting, the unified distribution of the dividends of land shares in rural areas, and the creation of the public-private partnership policy in cities. Also important was the policy that eliminated the "dividing of the profits of enterprises into four parts,"[52] which granted fixed interest in accordance with the private shares approved by enterprises. This legislative package also reformed the laws governing urban private property and nationalized the private possessions of urban residents. This "policy packet" changed the structure of social stratification in China on a large scale. An independent class in pre-transformation Chinese society, the national bourgeoisie, disappeared as a coherent class after the transformation. However, the changes went beyond merely eliminating the bourgeoisie. Before the transformation, there was a property-owning class in cities, but after the transformation, this group was substantially disintegrated. According to the report, *The Current Situation of Urban Private Property and the Opinion on Socialist Transformation* (which was produced by the second office of the Secretariat of the CPC Central Committee), the ratio for the private possession of urban housing was very high until

52 The state policy of "dividing the profits of enterprises into four parts" was a practice applied to private businesses starting in 1953 in China, which meant that income taxes accounted for 34.5%, welfare funds accounted for 15%, public accumulation funds accounted for 30%, and bonuses obtained by capitalists accounted for 20.5%. This policy was abolished during the socialist transformation of industry and commerce in 1956 and was transformed into a fixed-rate policy. See Chen Mingxian, et al. *Studies on the People's Republic of China 1949-1989*. Beijing: Beijing Institute of Technology Press, 1989.

1956. If we take all privately possessed urban housing and the very small amount of foreign real estate as private property, then factor in areas under construction (except in Beijing, where private housing was calculated according to the number of existing rooms), then the ration for privately and publicly owned housing can be calculated for every city. In Beijing, 55.63% of housing was privately owned, 73.6% in Shanghai, 56.59% in Tianjin, 62.25% in Nanjing, 78% in Jinan, 86% in Suzhou, 80.25% in Wuxi, and 44.66% in Harbin. According to the policy at that time—apart from private houses occupied by their owners—all private housing had to be submitted to the state for management and, later, all of them were gradually turned into public housing. The government let these houses to the non-property-owning families in the city at very low prices. The result of the implementation of this policy was an equalization of urban housing to a great extent. Therefore, the orientation of the policy of private property transformation was very clear—housing equalization. A class that benefited financially from renting housing in cities gradually disappeared.

Classes and strata did not just disappear; they also emerged. After the socialist transformation of industry and commerce, urban state-owned and collective enterprises developed rapidly. People who worked in these enterprises, usually referred to as "employees," rapidly became very numerous. I once advanced the opinion that employees in state-owned and large collective enterprises were, in a relative sense, a typical social middle class in China prior to the market transformation. Compared with China's huge farming population—which amounted to 80% of the total population—or ordinary working people and the employees of small urban enterprises, the employees of state-owned and large collective enterprises had obvious advantages in economic status, welfare, and social status. But, the employees of state-owned enterprises and the middle stratum were obviously different in a general sense. Therefore, I referred to them as "middle class-like stratum." The "middle class-like stratum" was an important social stratum that emerged after the mid-1950s. This was also evidence of policy changing social stratification structure.

Another prominent example of policy changing social structure was the Cultural Revolution, which ran from 1966 to 1976. Policy changing social structure in the Cultural Revolution was embodied by "the sudden change of statuses." One could be a high-ranking official one day, but be deprived of all power and criticized the next. Such people's social position suddenly fell to the bottom. However, some raised their position by acting as political activists. Under the guidance of extreme leftist policies, the majority of Party and government leaders in power were deprived of power and criticized, which resulted in dramatic changes in the social structure. Not only did the political structure change greatly, but economic stratification also changed a lot. I once researched household economic history. This analysis showed that, from the perspective of income distribution and wealth allocation, wealth and income were noticeably shifted to the lower social strata during the Cultural Revolution. Both calculations of Gini coefficients and the analysis of property income distribution indicated that the gap between the rich and the poor was at its lowest point in sixty years during the Cultural Revolution. That is why I always think we cannot just talk about narrowing the gap between rich and the poor, since the equalization of wealth and income is not our goal. If the smaller the gap between rich and poor the better, then the minimal income distribution gap between rich and poor during the Cultural Revolution represented a very high degree of equality. Using the Equal Division Method (in which the entire population is divided into five equal divisions according to their average income), the available data indicates that, from 1949 to 1965, the lowest income group took generally between 6 and 7% of the total income. During the Cultural Revolution, however, the lowest income group obtained between 9 and 10%. It is clear that policy really changed the distribution of income and wealth during the Cultural Revolution. But, most reasonable people will agree that narrowing the income gap by such means was a very poor decision—even a disaster.

The policies of reform and opening, which began in 1979, once again proved that policy can change social structure. From November 10, 1978 to December 15, the CPC Central Committee held a working meeting. This was a very important meeting that dealt with a central adjustment of strategy and policy. Then, the Third Plenary Session of the Eleventh Central Committee was held from the 18th to 22nd of December. Deng Xiaoping delivered an important speech at the meeting that became the guideline for strategy adjustment. Both meetings decided to stop using the slogan "take class struggle as the guiding principle" ("以阶级斗争为纲") and shift the focus of work on to economic modernization. At the meeting, Deng Xiaoping proposed emancipating people's minds, correcting the serious leftist errors, and criticizing the "two whatevers" ("两个凡是"). In terms of economic policy, he also proposed "allowing some areas, some enterprises, some workers, and some peasants to earn more and live better first." As mentioned above, the policies implemented after the Third Plenary Session of the Eleven Central Committee were a "policy packet" rather than a single policy. This policy packet involved policies covering politics, economics, ideology, culture, and society in general. Needless to say, the implementation of this set of policies was intended, above all, to change the social structure.

Just after the meeting, on January 29, 1979, the CPC Central Committee made *A Decision on Casting off the Labels, 'Landlord' and "Rich Farmer" as well as the Issue Concerning the Family Origin of Their Children* ("关于地主、富农分子摘帽问题和地、富子女成分问题的决定"). This announcement stated that the majority of those living with the political label of "landlord," "rich farmer," "counterrevolutionary," or "bad man"—except for the few who adhered to reactionary ideals—should no longer be so labeled and be treated as people's commune members. This required the approval of "the masses" and the County Revolutionary Committee. The personal statuses of the children of landlords and rich farmers were made equal to all other people's commune members and these people were no longer discriminated against in school enrollment, job recruitment, joining the army, joining the league, joining the Party, or the allocation of work. As has been mentioned many times above, I believe that China was a society stratified by politics prior to the reforms; political status was the primary determinant of every individual's status. The redefinition of those at the bottom of the political stratification scale that occurred with the end of the Third Plenary Session carried a special meaning. The political standard for stratification was gradually weakened; the importance of political stratification decreased while that of economic stratification increased. In a certain sense, Chinese society was rearranged and reassembled, which definitely caused significant changes in its social structure.

A number of significant changes since then have been even more profound. Among them was an important adjustment of the policy governing the social status of intellectuals. Before the reforms, a person's status in society was mainly decided by his position in the political stratification structure. What, then, was the political position of intellectuals at that time? In "A Speech at the National Conference of the Communist Party of China on Propaganda Work" in 1957, Chairman Mao Zedong proposed, "Most intellectuals at present still belong to the bourgeois intelligentsia," which set the tone for the class attribution of intellectuals. In the political stratification environment of the 1950s, the label "bourgeois" was very dangerous. The bourgeoisie were the object of revolution and transformation. During the Cultural Revolution, the bourgeoisie were designated as being an enemy group. Intellectuals were regarded as an object of the transformation once they were positioned as members of the bourgeoisie. This was a group with a very poor political reputation. Additionally, they were referred to as "the stinking ninth category" ("臭老九") during the Cultural Revolution. After the Third Plenary Session of the Eleventh Central Committee at the end of 1978, the Central Government proposed "implementing the Party's policy on intellectuals." Under this new policy, the social status of intellectuals underwent a fundamental change.

The variable of state policy that changed the status of intellectuals was also part of a comprehensive policy packet, rather than a single policy. This made the packet's overall effect very strong. I believe that by observing the operation of "policy packets," the specific operation mechanisms by which policy changes the social stratification structure can be identified.

In this "policy packet," the most basic and critical element is the "basic tone theory." This sets the tone and direction for the policy packet, namely the judgment regarding the "class attributes" of intellectuals. In the past, the tone theory belittled intellectuals' political status as being "bourgeois." Later, to enhance the political status of intellectuals, the tone theory entitled intellectuals to be seen as "members of the working class." This tone theory originated from a number of discussions led by Deng Xiaoping, from which we can observe the operating mechanism of a specific policy influencing social stratification. In this instance, a leader delivered an important speech at an important ceremony or occasion. Though the speech itself was not official policy, its function was to set the tone for an impending change in policy. The future policy was bound to be formulated according to the direction of the tone. The most important speech in the redefinition of intellectuals as "members of the working class" was made in March 1978, at the opening ceremony of the National Conference of Science. Deng Xiaoping claimed that the vast majority of intellectuals had already become intellectuals of the working class and laboring people. Therefore, it could be said that intellectuals had already become a part of the working class. The difference between them and manual workers lay only in the social division of labor.

It was under the guidance of this tone theory that the Central Committee developed a series of specific policies from 1978 to the mid-1980s. These policies were later collectively referred to as the *Implementation of the Party's Policy on Intellectuals*. First, the Organization Department of the CPC Central Committee put forward *Several Opinions on the Implementation of the Party's Policy on Intellectuals* on November 3, 1978. This greatly expanded the foundation set by Deng Xiaoping and, through this document, the department also advanced a series of operative policies. These policies advocated effecting change in five key areas. They initially recommended further and overall evaluation of intellectuals and that arrangements be made for review, redress, and rehabilitation of all intellectuals who had been unjustly mistreated in the past. Additionally, the policies emphasized promoting intellectuals with good political awareness, work abilities, diligence, and the ability to keep good relations with the masses to appropriate leadership posts. They also advised appointing intellectuals and professionals to appropriate positions and introducing some specific measures to improve the working and living conditions of intellectuals. The document, which was issued by the Organization Department of the CPC Central Committee, turned the basic tone theory of Deng Xiaoping into operative policies; therefore, it was a very important step. However, the "Several Opinions" were still focused on the primary principles of the policies. The "policy packet" concept I proposed comprises a "basic tone theory," the "principles of broad policy," and "specific, enforceable policies." The following document outlined these more specific policies.

On September 15, 1986, the Organization Department of the CPC Central Committee issued the *Notice on the Inspection of the Implementation of the Party's Policy on Intellectuals*. It illuminated the specific contents of this inspection and described very specifically the details of the "policy packet" I mentioned. One of the more important policies was correcting the unjust mistreatment and falsified cases previously made against intellectuals. In former political movements, large numbers of intellectuals suffered such mistreatment. Another policy ordered an end to the unreasonable use of intellectual labor resulting from incorrectly assigned punishments. Steps were also taken to solve the problem of household registration for those intellectuals—as well as their spouses and children—who had formerly worked in cities but had been exiled to the countryside as a punishment. The provision of compensation to those whose salaries had been withheld or reduced in the course of the injustices perpetrated against them during the Cultural Revolution, and appropriate subsidies to those intellectuals who were wrongly classified as rightists, were also advocated. Additionally, this policy packet outlined the gradual rationalization of salaries in relation to general salary reform and the resumption of appointing experts to professional technical positions. Also, property confiscated during the Cultural Revolution was to be returned or compensation was to be provided for lost items. Any private houses confiscated or seized during the Cultural Revolution were also to be returned. The two policies relating to confiscated property were not designed especially for intellectuals, but because this type of problem affected many intellectuals, were an important aspect of the *Implementation of the Party's Policy on Intellectuals*. Also, a policy was implemented to correct those cases in which intellectuals who had performed part-time jobs and charged reasonable fees had been treated as economic criminals. These specific policies provided details about how to improve the status of intellectuals and how to improve their living and working conditions. Of course, the contents of this "policy packet" went far beyond those mentioned above. For example, a policy proposed the promotion and appointment of "revolutionary, young, well-educated, and professionally qualified" cadres. Among these qualifications, being well-educated and professionally qualified were prominent intellectual characteristics. Under such a policy, a large number of intellectuals who met with the above qualifications came into leadership positions at all levels after the 1980s. A "policy packet" is a collection of many policies. The above illustration only covers part of this particular policy packet.

In addition, in the tenth section of the preface to The Constitution of the People's Republic of China, published in 1982, the elementary social force of Chinese society was discussed: "The socialist construction cause must rely on workers, farmers, and intellectuals, and unite all the social forces that can be united." This expanded on the former two elementary social forces of workers and farmers by including intellectuals to make three social forces. Thus, the status of intellectuals was guaranteed by law.

After being promoted by the above "policy packet," the social prestige and status of intellectuals were comprehensively upgraded. In the mid-1980s, a popular term in the media was "respect knowledge, respect talent" ("尊重知识、尊重人才"). Later, by the 1990s, intellectuals had become a respected social group. It was very difficult for people to understand that such a group was once referred to as "the stinking ninth category." So, the change in intellectuals' status after the process of reform and opening up began could be regarded as a very good example of policy changing statuses in social stratification.

However, status declines have also been the result of the influence of policy variables after the 1970s. In 1983, for instance, according to the spirit of Deng Xiaoping's speech at the Second Plenary Session of Twelfth Session of the CPC Central Committee, the central government issued the "get rid of three kinds of people" policy. "Three kinds of people" referred to the adherents of anti-revolutionary groups under the leadership of Lin Biao and Jiang Qing during the Cultural Revolution, those who held deep-rooted factionalist ideas, and those who fought, smashed, and looted during the Cultural Revolution. Thus, prominent men who made their fortunes in the Cultural Revolution were removed. Their social status was significantly lowered. There are many other cases of this kind that I will not repeat here.

3. The Influence of Policy on Social Stratification Structure after Deng Xiaoping's Southern Tour in 1992

The policy adjustments that have taken place since the 1990s can be divided into two important periods. The first began after Deng Xiaoping's southern tour in 1992; a major adjustment and a significant influence on social stratification structure took place after this event. The second period began after 2003 when the new leaders of the CPC collectively advanced some new ideas for development. They promoted a slogan that advocated paying more attention to social fairness and implemented some new policies, which also impacted social stratification. An analysis of the first period, Deng Xiaoping's southern tour, and how it affected basic tone policy will be conducted here.

The most important variable of policy in the 1990s was triggered by Deng Xiaoping's southern tour. If we take into account the special social background of that time, we even can say the influence of the southern tour was equal to the effect of the start of the reform and opening-up process. According to the logic of the above analysis, the influence of major events and policies on social stratification can be divided into three stages: the "basic tone theory," the "principles of broad policy," and "specific, enforceable policies." Deng Xiaoping's southern tour speech can be categorized as a change in "basic tone theory." What kind of tone theory was it? It was a theory dictating that "development is of overriding importance," "economic construction is to be taken as central," and "a socialist market economy is to be established." After Deng Xiaoping's southern tour speech, a market economy was gradually established in Chinese society; the documents of the CPC Central Committee also formally proposed the concept of a socialist market economy. From then on, resources would be mainly allocated by the market, greatly reducing the proportion allocated by the government. However, market competition and the market allocation of resources are associated with a particular social stratification structure. A careful analysis of the variables of policies influencing the structure of social stratification since the mid-1990s to the early twenty-first century reveals two opposing influences. Suppose that the idea of "expanding the proportion of middle-income people in society" advanced by the central government was the goal. Many policy variables favorable to this aim can be found, but so can many unfavorable variables. Which variables in policy were more influential? From the perspective of sociology, this matter is one of theoretical hypothesis and verification. This requires empirical investigation and a lot of data analysis to verify which policy variables were more influential.

What were the specific influential factors? Since the 1990s, there have been two policy variables: state-owned enterprise reform after the mid-1990s and the introduction of a market mechanism.

First, an examination of the reform of state-owned enterprises will be conducted. From the perspective of relationships internal to state-owned enterprises, relatively speaking, the reform of state-owned enterprises since the beginning of the 1990s has mainly been beneficial for their managerial and technical staff because they have had some control of power in the enterprise and have been the managers and operators of reform. Even if some have chosen to leave the enterprise, they have had good prospects for finding other positions. For the workers or ordinary employees, however, conditions have been less favorable. Such people make up the majority of enterprise employees. After the mechanism of employment changed, a large number of such workers became unemployed.

At that time, policies regarding the reform of state-owned enterprises included: "break the three irons"[53] ("破三铁"), "downsize for efficiency," "lay off and reposition workers," "optimize and regroup," "invigorate large enterprises while relaxing control over small ones," etc. By the late 1990s, under the guidance of the new policies, great changes had taken place in the four systems of housing, medical security, pensions, and employment. It was unavoidable that such an important transformation would change resource allocation among people, thus influencing the structure of social stratification.

This was particularly the case in the fields of mining industry, heavy industry, and traditional industry; the proportion of workers or ordinary employees was rather large in these industries. A large number of these workers, therefore, became unemployed. According to the data of a national survey I performed in 1998, the urban unemployment rate exceeded 10% (if the recently unemployed were included). Due to the inadequacies of the social security system at that time, once unemployed, those who were "bought out" on severance payments lost their medical and pension security. One of my studies showed that there was a big difference between the 1990s and the 1980s. In the 1980s, physical labor was more profitable thanks to the reforms, and farmers' and workers' incomes increased significantly. After the 1990s, however, income distribution significantly favored managerial and technical personnel. The positive aspect of this was that the reversal of the wage differential between mental and physical labor disappeared, but the negative aspect was a visible decline in the statuses of workers and ordinary employees.

A discussion of the introduction of a market mechanism is also necessary. The direct result was that those who were closer to the

53 The "three irons" refers to the "iron rice bowl, iron income, and iron post," a description of the lifelong guaranteed jobs characterized by a neglect of work achievements and efficiency in state-owned enterprises that existed before state-owned enterprise reform in 1992.

market became the direct beneficiaries. Who, then, was close to the market? It was those who were engaged in industrial and commercial business. That was why huge numbers of people started businesses in the mid-1990s. However, because the market mechanism was newly established, many aspects of it were far from perfect. For example, the examination and approval procedures were rather complicated, and opportunities for market competition were not equally available. Therefore, many officials colluded with merchants, making money through approval procedures. Rent seeking was a common practice. Power and the market began to become intertwined, and corruption and the trading of power for money became very common. Additionally, Chinese society has prioritized the network of social relations since ancient times. The social relation system based on households and friends is extremely complex, creating a rather free environment for trading power for money. In fact, all East Asian societies bear this characteristic—family ethics being considered the primary criterion of one's worth. Although South Korea and Japan have become highly modernized, corrupt power-for-money transactions still frequently occur. Local state departments, the industry and commerce administration, the tax bureau, and the banks had the easiest access to market benefits. There was a saying about them. "Three wolves live in the district: industry and commerce administration, the tax bureau, and the banks. If you don't believe it, take a look at their houses."

It is indisputable that, through the introduction of a market mechanism, the industrial and commercial strata and the bosses who were the closest to the market benefitted a great deal. However, there were internal differences between them; they were stratified. After this period, they have been distinctively divided into two groups: big industry and business owners (or big bosses), and small industry and business owners (or small bosses). Up until then, there were only a few big industrial and commercial owners and bosses in China; the vast majority were small industry and business owners and small bosses. According to an investigation of the National Association of Industry and Commerce, by the end of 2006, China had 4.65 million private enterprises and 11.84 million private enterprise investors. Those in possession of capital in excess of one hundred million yuan accounted for only 2.2%, which equated to a very small number of individuals. Only two million of them enjoyed the median sum of capital for their stratum. This showed that, among

Chinese business owners, the vast majority were medium- and small-sized bosses. Actually, most private industry and commercial owners were self-employed or individual businessmen, not owners of private enterprises—the number of which has continued to rise in recent years. Currently in China, there are more than twenty-five million individual business owners and more than fifty-one million people engaged in individual business. If unregistered rural businessmen and those doing business in cities—including unemployed personnel—were added together, this number would be greatly increased. Therefore, an important question related to policy is whom should it favor? Should the policy be inclined to favor those rich in capital or those possessing small and medium sums of capital? This relates to the design of social stratification structure in the future. As mentioned above, the general idea of the central government was "to expand the proportion of medium income people in society," meaning that their intention was to make the middle income group the main body in the social structure of the future; the intention was that the top and bottom strata would be relatively small. An examination of the class structures of other countries reveals the existence of two types of class structure: the middle-class dominated structure of states such as those in Sweden, Denmark, Norway, Germany, and Austria, etc.; and structures within which the lower classes make up the majority, but a few upper-class capitalists possess most of the wealth. In this latter kind of society, the middle class tends to be very small and the majority of the population is of the lower class, as can be seen in many South American, African, Southeast Asian, and South Asian countries. Therefore, in order to gradually expand the middle class, policies favorable to small- and medium-sized businesses should be introduced. Enabling medium-sized and small businesses to develop is an important part of the process of enriching the general public.

As mentioned above, two major policy variables—the reform of state-owned enterprises and the introduction of a market mechanism—have also caused the multifaceted and multilevel reallocation of social and economic resources since the 1990s. Among them, the most important resources include housing, land, state-owned assets, mineral resources, etc. A detailed analysis of this reallocation process and the effects of reallocation of these resources on social stratification will be made below.

Though the process of urban housing reform began in 1979, the most substantive reallocation occurred after the mid-1990s. Within ten years, urban China had transformed from a public-housing-based society into a society in which the majority own their own housing. Housing is an important resource that is closely related to the interests of the public. As this was such a large-scale housing resource reallocation, the allocation of housing was certainly not completely equal and imbalances were common. In the past, houses were distributed by work units, but now houses are purchased through the market. In this respect, the principle of housing resource allocation changed greatly. The reallocation of housing resources has had two major influences on social stratification. The majority of people benefited from the urban housing system reform. At present, more than 80% of urban residents in China own houses; the rate of private housing ownership is higher than that in Europe and the U.S. A recent survey showed that, among Chinese urban house owners, about 70% obtained their houses through the work unit housing reforms and 30% obtained their houses through the market. Of course, as the market has progressed, the proportion of people purchasing housing has continued to increase. With work unit housing reforms, houses were sold to work unit employees at relatively low prices. The main reason for this was the consideration of employees' past contributions to their work units. Today, with the exponential rise in housing prices, the prices of houses purchased during the work unit housing reforms also rose significantly. Workers, especially older workers, have therefore benefitted from the housing reform policy. However, in the last ten years, urban property prices have risen sharply, resulting in a heavy burden being placed on the younger generation—most of whom do not yet own houses. For example, in the Beijing housing market in January 2007, a set of ordinary, previously-owned ninety-square-meter apartments cost 730 thousand yuan each, this calculation being based on the price of 8,113.87 yuan per square meter (new properties are more expensive). It would take twelve years for an ordinary family with an annual income of sixty thousand yuan to save this sum of money, assuming they do not eat or drink anything during this period. This is similar to the case of Seoul, in South Korea. As mentioned above, people there were divided into a rich class that owned real estate and a poor class that did not.

Since Decree No. 55, *The Provisional Regulations on the Urban State-Owned Land Transfer*, was issued by the State Council in 1990, land rental and batch leasing has become legal. The real estate market has developed rapidly since then. Real estate agents and the real estate business have grown rapidly and synchronously. It has proved much more efficient for real estate agents to develop real estate than for work units to take care of housing, as they had done under the planned economy. Urban residents no longer worried about an inadequate supply of housing. However, after the marketization of real estate, the process of allocating property and land resources underwent a number of fundamental changes. When batch land leasing became legal, people began to profit from land. The core of the urban housing system reform was the reform of the land system. After the land system was reformed, real estate business owners were able to own land and build houses. Housing prices are actually determined largely by land prices. The skyrocketing of urban housing prices is the result of skyrocketing land prices. Because the price of materials for building houses has changed little, the major factor causing changes in price is land prices. Who has profited the most from rising land prices? Han Jun, director of the Rural Section, State Council Development Research Center, produced a report in which he said that 40% of land revenue has been collected by real estate businesses. If that is the case, can it be said that real estate agents have profited too much? Thus, in the list of China's richest citizens, real estate agents make up a significant portion of the wealthiest people in the country. Taking this into consideration, China's land policy can be said to have a huge loophole because the allocation of land resources is advantageous for large capital and adverse for ordinary people. China's constitution says that urban land should be owned by the people; but, in actual operation, it is in the hands of real estate agents and mainly benefits the greater bourgeoisie. Land renting and leasing in China has always been a large field, excluding people who lack sufficient capital. Although land belongs to all people in theory, in the real world, land ownership is disadvantageous for those with medium and low incomes and advantageous for people with access to large quantities of capital. In short, taking the variable of land policy into consideration, this represents a further departure from the aforementioned principle of favoring the petty bourgeoisie.

The reform of state-owned enterprises reached a climax after the mid-1990s. Some state-owned enterprises became joint-stock companies, some went bankrupt, some went bankrupt but were able to utilize their remaining assets to set up new factories, some were sold, some were purchased, and some were contracted under the "invigorate large enterprises while relaxing control over small ones" ("抓大放小") policy. In short, various loopholes appeared one after another in the new forms that appeared after the transformation. Injustices were rife. When questioned by the Hong Kong professor Lang Xianping in his speech, "Green Kerr: Carnival at the feast of the state-owned economy stepping back and the private economy coming in," the president of Green Kerr, Gu Chujun, admitted to having frantically acquired a number of formerly state-owned enterprises. The acquisition technique was to "borrow hens to lay eggs" ("借鸡下蛋"), which was popularly referred to as "catching a white wolf with empty hands" ("空手套白狼"). The result was that Gu Chujun was arrested on the suspicion of the illegal acquisition of state-owned assets. Considered from the perspective of the fairness of public asset allocation, the loss of state-owned assets and a phenomenon of a small number of people seizing public property did occur in the reform of state-owned enterprises. Moreover, during the acquisition and merging process of state-owned enterprises, even the rules governing asset evaluation contained many loopholes. Therefore, the general tendency was that the interests of workers and the public were damaged and the influence of such policies on the structure of social stratification was negative.

The Chinese public has recently attacked policies concerning the allocation of mineral resources. According to the law, China's mineral resources are publicly owned and the state has never agreed to the privatization of minerals. However, if one were to take a tour around the country, one would frequently encounter the privatization of minerals—especially coal mines. Although mining is nominally contracted, in many places so-called reforms were made—such that many coal mines are contracted out to private owners. A lot of illegal behavior—such as power for money deals—have occurred in the process of contracting out mineral resources. Some have obtained state-owned mineral resources at very low prices (or even for free), and then sold them on the market. Such no-cost-and-huge-profit transactions have enriched a minority of people. Coal mine owners have now become a rich group and have attracted the attention of Chinese society. Some time ago, a number of coal mine owners from Shanxi province frantically purchased luxury cars and commodity houses in Beijing, shocking Beijing's citizenry. Two "contrasts" of these mine owners have made people rather uncomfortable. One strong contrast is the contrast between the coal mine owners' wealth and their lower-class backgrounds, which is very dazzling and has made people uncomfortable. Another is the contrast between the wealth of the mine owners and their exploitation and inhumane treatment of the miners they employ. Mineral resources are so unfairly allocated and the influence of this process on the social structure results in polarization, causing the middle stratum to narrow rather than widen.

In addition to the above four types of resources, another resource is that of organizational resources. Housing, land, state-owned enterprises, and minerals are all at the disposal of organizational resources. Organizational resources can allocate housing, land, state-owned enterprises, and minerals; this can be done both legally and illegally. The worst-case scenario is the corrupt use of organizational resources to control other resources, meaning the use of organizational resources for private benefit. Sometimes, it is done to benefit a small group and a few people. Regardless, the use of organizational resources is very complex. No matter how they are used, they influence the allocation of resources in society and ultimately impact social stratification structure.

The above analysis shows that, since the 1990s, the policy variables in the first period greatly accelerated the accumulation of resources by a small number of groups. In the process of key resources allocation such as housing, land, state-owned enterprises, and minerals, the public benefited in some respects. For example, work unit housing reform allowed many workers to buy their houses. However, the trend of a minority of the people gathering wealth became more prominent than that of wealth diffusion. The survey data indicate that during this period, the gap between the rich and poor widened at the fastest rate ever recorded in China.

There are many theories analyzing this period in the field of sociology. The most popular one explains it from the angle of market transformation, holding that China was a society dominated by a planned economy before the reforms. Economic resources and other resources were not allocated by the market, but were rather distributed by governmental organization and management. For example, housing was distributed and wages were adjusted by the state. The structure of income distribution was characterized by "equalitarianism" and "equalization." After the reforms, due to the gradual introduction of a market mechanism, economic resources and other key resources came to be mainly allocated by the market. The market was fair and advocated survival of the fittest, so great imbalances and differentiation came into being. Of course, as sociologists have also pointed out, the market system in China was still at an early stage of construction and was far from perfect; also, the market was only one of the steps in the process of market distribution. In countries with well-developed market systems, there are many links within the market economy—such as social security, social welfare mechanisms, and so on—that regulate distribution. These methods may eventually realize balanced development. Therefore, the problem in China was not the introduction of a market mechanism, but that the market system still needed to be gradually improved.

Chinese sociologists have presented four main theories to explain social stratification in this period. The first is the ten strata theory proposed by the research team led by Professor Lu Xueyi, which argues that under the influence of market reform, the structure of Chinese social strata changed, new strata emerged, and the positions and ranking of the original strata changed. According to the new hierarchy, these strata were: state and social administrators, managers, private enterprise owners, professional and technical personnel, office clerks, individual industrial and commercial households, commercial service staff, industrial workers, agricultural labors, and the rural and urban jobless (those who choose not to work, unemployed and semi-unemployed).

The second is "cleavage theory," which was advanced by a friend of mine, Professor Sun Liping of Tsinghua University. This theory emphasizes that social cleavage has been structural. The reason for social cleavage was the sharp market transition. Because the transformation to a market economy was too fast, a lot of social strata were eliminated or abandoned, such as disadvantaged groups in enterprises: the laid-off and unemployed. The cleavage in this theory is multifaceted, but includes cleavage in the reorganization of state-owned enterprises, in urban and rural structures, in production and consumption, and in culture, etc. The theory suggests that because "elements of different social periods existed in one society," social cleavage was multidimensional. In this theory, the stratification between the poor and the rich is also explained as a form of cleavage. If society was really so fractured and disconnected, would it not collapse? Scrutinizing this theory reveals that its prediction for the future of Chinese social development is not pessimistic. There is also an optimistic prediction of social transformation. So, this theory is more of a warning to society.[54]

The third is the "inverted T-shaped social structure" theory I developed. It will be discussed in detail in chapter 11, so this will not be rehearsed here.

The fourth is the "four interest groups" theory, or "clique theory," which was co-operatively proposed by Dr. Shen Yuan, Professor Sun Liping, and me. This theory, which is based on the losses and gains of different groups since the reforms, divides people into four interest groups, or cliques: the special beneficiary group, the ordinary beneficiary social groups, the relatively damaged group, and the bottom-most group. Since the reforms themselves were an adjustment of the social interest structure, they definitely enabled some groups to benefit and made other groups lose out. It was not impossible to achieve a universally beneficial outcome, but this would have been extremely difficult to realize. In recent years, reforms in the housing, medical, pension, and unemployment insurance systems have all inevitably damaged some people's interests. In order to make the reforms smooth, analysis is necessary to establish which groups benefitted from the reforms and which groups were negatively affected. In order for the reforms in China to progress smoothly, it is necessary to ensure that the reform measures do as much as possible to benefit more people. The "four interest groups" theory attempts to explain social differentiation at the present stage.

4. The Influence of New Policies on Social Stratification after the Slogan "Pay More Attention to Social Fairness" Was Proposed

After 2003, especially after the resolution of the SARS public health emergency, the administrative approach of the new generation of the collective leadership of the CPC Central Committee became gradually clear. For the principle policy and concept, two main opinions were put forward: scientific development and building a harmonious society. In terms of policy and ideology closely related to social stratification, the CPC Central Committee promoted the slogan "Pay More

54 Sun Liping. *Cleavage: China Society since 1990s*. Beijing: Social Sciences Academic Press, 2003. 14.

Attention to Social Fairness" ("更加注重社会公平"). On the basis of the above policy analysis, this can be viewed as a correction or development of the basic tone theory. According to the new basic tone theory, the central government emphasized allowing all people to share the benefits of reform and development. Such a new policy orientation was certain to influence the allocation of resources and social stratification structure.

First, the specific process by which policy and tone theory changed will be analyzed; this analysis will be followed by a discussion of how policy influenced social stratification.

The proposal to "pay attention to social fairness" was initiated at the Fourth Plenary Session of the Sixteenth Session of the CPC Central Committee in September, 2004. In October, 2005, when the *Suggestions for the Formulation of the Eleventh Five-Year Plan for National Economic and Social Development* was passed by the CPC Central Committee at the Fifth Plenary Session of the Sixteenth Session, this point of view was clarified. One year later, in the Resolution of the Sixth Plenary Session of the Sixteenth Session of the Central Committee of the CPC in October, 2006, it was stated that it was necessary to "pay more attention to social fairness, dedicate more effort to increasing the income of the low income group, gradually expand the proportion of middle income earners, regulate exorbitantly high income, resolutely forbid illegal income, and promote common prosperity." In contrast with the former slogan, "giving priority to efficiency and taking fairness into consideration," the adjustment is obvious. Worthy of note was that the CPC Central Committee never repudiated the slogan "giving priority to efficiency and taking fairness into consideration" in the document. "Pay more attention to social fairness" was a correction to the policy orientation in the former period, shifting the focus of policy towards fairness to a certain extent. But, it should be noted that it never passed the central line between fairness and efficiency. The central government never said "give priority to fairness and take efficiency into account" in the document. In other words, the policy's focus was largely balanced, but slightly inclined towards efficiency. This was because the central authorities had repeatedly stressed that development was of overriding importance; development was the priority. Therefore, the policy adjustment in the second period was more an act of fine-tuning than a denial of the past policy. The report of The Seventeenth National Congress of the CPC in 2007 also reflected such a policy orientation. Though the report repeatedly emphasized social fairness and justice, when it came to its income distribution policy, the principle of "correctly handling the relationship between efficiency and fairness both in initial distribution and redistribution, with more attention being paid to redistribution" was stated.

The policy orientation of focusing on fairness resulted in a series of specific policies: promoting the new socialist rural construction and strengthening rural reform; attaching importance to rural public utilities; increasing farmers' incomes; promoting the coordinated development of urban and rural areas; improving overall planning of urban and rural development; promoting coordinated regional development; creating active employment policies and harmonious labor relations; giving priority to the development of education and educational fairness, as well as the construction of democracy and the rule of law; improving the public finance system; gradually achieving the equalization of basic public services; putting more funds into the field of public services; deepening the reform of the income distribution system; accelerating the establishment of a social security system covering both urban and rural residents; establishing a basic medical and health system and harmonious culture, ideology, and ethics; constructing a service-oriented government; strengthening social management and public service functions; constructing urban and rural community; coordinating all the relations of interests; properly handling social contradictions, etc.

This trend in policy adjustment reveals obviously the increased emphasis of the policy variables on fairness, as an attempt was made to make the allocation of social resources more beneficial to a wider population. If all these policies can really work, then more resources and wealth will flow to the general public.

However, a new trend was found when we observed how the policies affected stratification during this period. The aforementioned three great historic events—the Socialist Transformation in 1956, the Cultural Revolution in 1966, and the Reform and Opening Up in 1978—were examples of policy variables that affected social stratification in a way that set the whole country on fire. Under the impact of a policy packet, China's social stratification structure suddenly changed. The policy adjustment after 2003 was a kind of fine adjustment, and the way it influenced stratification was different from past approaches; it was more moderate. How then, should we think about these two different methods? It is clear that important policy adjustments significantly impact social structure with rapid changes. However, this kind of sudden change may cause social conflicts and turbulence while fine adjustment tends to be more beneficial to the smooth progress of a society, as it avoids the negative effects brought about by turbulence and conflicts. The application of fine adjustment reflects the maturity of the ruling party, indicating that they rule the society by law and conventional practice. The influence on society of fine adjustments is small; the intensity and extent to which such policies influence social operations are also small. An example is given below to demonstrate the effect of fine adjustment on social stratification.

As mentioned above, after the reform of the urban housing system, real estate became the most important, even the main, property of urban residents. Housing differentiation became the main symbol of the economic interest differentiation of residents. Because of the sharp upward trend in housing prices, the differentiation in urban residents' housing interests revealed two extremes: one group obtained tremendous benefits from real estate, such as real estate agents; the other group gained nothing at all from real estate. The recent flare-up in housing prices has left the middle and low income groups, who did not benefit from the real estate market, in a very unfavorable position. In order to protect the interests of middle and low income earners, the government introduced two major policies, which are referred to as *State Eight Measures* ("国八条") and the *State Six Policies* ("国六条") in 2005 and 2006, with the intention of stabilizing housing prices.

The *State Eight Measures* was advanced by the Executive Meeting of the State Council hosted by the Chinese Premier Wen Jiabao on April 27, 2005. At the meeting, the real estate market was analyzed and a discussion was held on how to further strengthen macro-control of the real estate market. Eight measures were put forth to strengthen the guidance and control of the real estate market. The eight measures are as follows.

1. Strengthen the planning, regulation, and control of the real estate market and improve the structure of commercial housing. Local governments at all levels should confirm as soon as possible the construction scale, project layout, and the schedule for ordinary commodity housing and affordable housing in the next two years (2005 and 2006). This information also had to be announced to the public.
2. Strengthen the regulation and control of the land supply and improve stricken land management. On the basis of enforcing the overall planning of land use and land use programs, the governments of all regions should increase the land supply for the construction of low-cost, ordinary commodity housing and housing in the cities in which land and housing prices rose too fast. Stop land speculation behavior according to the law.
3. Strengthen the regulation of the prices of common commodity housing and affordable housing to ensure the effective supply of middle- and low-cost small apartment housing. The prices of affordable housing should be determined strictly according to the governmental guiding price.
4. Improve the urban system for renting affordable housing, securing the basic housing needs of the lowest income families. Expand the coverage of low-rent housing. The construction of an urban low-rent housing system should be listed in the assessment criteria for the provincial inspection of the target responsibility system of municipal (district) governments.
5. Regulate and control the real estate market through taxation and other economic means, especially strengthening the regulation of real estate transactions.
6. Strengthen financial supervision. Commercial banks should strengthen credit management of real estate loans and personal housing mortgage loans to prevent loan risk.
7. Rectify and regulate market order. Strengthen the rule of law and severely punish illegal sale behavior.
8. Strengthen market monitoring and improve the market information disclosure system. Strengthen the guidance of public opinion and enhance the transparency of policies.

The above eight measures were very specific and detailed. According to the above analysis, these policies belonged to the subset of specific policies guided by the "basic tone theory" and "principles of broad policy." If the previous procedures and experiences are considered, these specific policies should have strong restrictive power over social operation. Policy will play its role of influencing society and adjusting its structure. However, things did not happen as they were supposed to. The regulation and control by policy failed unexpectedly and housing prices in large and medium-sized cities kept rising in a frenzied fashion.

In these circumstances, the central government unveiled the *State Six Policies* for housing price control. The *State Six Policies* were put forward by the Executive Meeting of the State Council chaired by Premier Wen Jiabao on May 17, 2006. At the meeting, he stressed the need to further guide and control the real estate market. Six measures were proposed with the intention of solving existing problems.

1. Significantly adjust the structure of housing supply. Focus on the development of middle- and low-cost housing, middle- and small-sized ordinary commodity housing, affordable housing, and low-rent housing. Local governments at all levels should formulate and implement housing construction plans and specify the requirements for future housing structures, as well as their proportion.

2. Allow taxation, loans, and land policy to play a greater role in regulation. Strictly implement policies on housing development and sales, improve tax policy that applies during housing transfer, differentially and moderately adjust loan policies, and guide and regulate housing demand. Scientifically determine the availability of land for real estate development, strengthen the supervision of land use, and forbid hoarding land.

3. Reasonably control the scale and progress of urban housing demolition and slow down the excessive growth of passive housing demand.

4. Further rectify and standardize the order of the real estate market. Strengthen the supervision of the whole process of

real estate development and construction. Forbid the unauthorized changing of projects, illegal transactions, hoarding of housing, and bullying behavior.

5. Speed up the construction of urban low-rent housing, regulate the development of economically affordable housing, and actively develop the secondary housing and rental markets. Solve the housing difficulties of low-income families in a step-by-step fashion.

6. Obtain better statistics on real estate and the system of information disclosure. Strengthen the transparency of the real estate market. Release market supply and demand information comprehensively, accurately, and in a timely manner. Adhere to the correct guidance of public opinion.

No one could say that these policies were not specific. Compared with the *State Eight Measures*, the *State Six Policies* put more emphasis on the duties related to the government's supervision and regulation. The specific ratio requirements for housing structures were also outlined, thus greatly enhancing the strength of regulation. According to past experience, powerful policies like this are likely to immediately influence housing prices. However, what was the result this time? Housing prices continued to rise higher and higher. The *State Six Policies* did not work. Rising housing prices told their own story. According to data released by the National Development and Reform Commission and the National Statistics Bureau, housing sale prices in seventy large- and medium-sized cities rose compared with the same time one year previous. They were as follows: 5.5% in January 2006; 5.6% in April 2006; 6.6% in October 2006; 7.1% in June 2007; 9.5 % in October 2007; and 10.5% in November 2007. Neither the *State Eight Measures* nor the *State Six Policies* were able to control prices. Harsh policies not only failed to stabilize housing prices, but made prices rise higher. Why was this? Does it mean that the two propositions presented above—that of a "government-led society" and of "policies influencing the structure of social stratification"—have ceased to exist?

In 2009, the price of housing in large- and medium-sized Chinese cities soared once again. Faced with this situation, the General Office of the State Council issued a *Notice on Promoting the Stable and Healthy Development of the Real Estate Market* on January 10, 2010. The notice advanced eleven measures for controlling housing prices, known as the *State Eleven Measures.* These eleven specific measures were:

1. Speed up the construction of middle- and low-cost housing, as well as middle-sized and small ordinary commodity housing.
2. Strengthen the effective supply of land for housing construction and improve the supply, development, and use efficiency of land.
3. Strengthen the enforcement of differentiated credit policy. It was particularly made clear here that the ratio of the first payment for one's second house should be less than 40%.
4. Continue to implement the differential housing tax policy.
5. Strengthen real estate loan risk management.
6. Continue to rectify the real estate market order.
7. Further strengthen the management of land supply and commercial property sales.
8. Strengthen market monitoring.
9. Strive to solve the housing problems of 15.4 million low-income households by the end of 2012.
10. The central government will enhance its support of projects for low-income urban residents and properly increase the subsidy standard for low-rent housing in the central and western regions.
11. Further improve and implement the responsibility system that enables the provincial people's government to take overall charge while the municipal people's government is to take specific responsibility for stabilizing the real estate market and solving the housing difficulties of low-income families.

Here, the orientation of the price control policy of the *State Eleven Measures* was the same as the *State Eight Measures* and the *State Six Policies*. We should admit that skyrocketing housing prices in China's large and medium-sized cities in 2009 had much to do with the "four trillion yuan investment plan" for stimulating the economy, which was introduced by the state in order to alleviate the impact of the international financial crisis. So, I think my two propositions—"government-led society" and "policies influencing the structure of social stratification"—are still reasonable and can be verified. The reason why the *State Eight Measures, State Six Policies,* and *State Eleven Measures* could not control the soaring housing prices was that governmental policies often change. Because of the adoption of the plan to stimulate the economy, the government played a very important role in causing house prices to rise in 2009. Therefore, the I think policy continues to influence the structure of social stratification.

At present, two factors are not the same as they were in the past. The first factor is the market. After thirty years of market transition in China, especially since the late 1990s, the market now plays an independent role in society. The market has its own independent function and independent rules of operation. The market does not take orders from the government. The market is controlled by supply and demand. From this point of view, the concepts of a "government-led society" and "policies influencing the structure of social stratification" will change after the market transformation. But, this does not mean that after the establishment of market system, I will deny the characteristics of a "government-led society" and "policies influencing the structure of social stratification." We cannot draw a conclusion now; we need to observe further. The second factor is (as was mentioned above) that new policies have tended to have more of a fine-tuning style at this stage. What effect does fine-tuning have on social stratification; what is the difference between the large-scaled adjustments of past policies and now? Further observation is required.

The conditions are not right at present to make a comprehensive theoretical judgment because not enough time has passed since the adjustments in policy. Coupled with the fact that the policy adjustment has entered a fine-tuning stage, it takes time to assess the results of an overall observation of policy variables.

Overall, in the last thirty years of reform, China has perpetually been in a state of rapid social change. In order to summarize the rules of these changes, continuous theoretical exploration and innovation are necessary.

Chapter 7

Wealth Disparity and Social Equity

Wealth disparity and social equity have, for a long time, been a hot topic of social discourse. It has been the subject of heated controversy from the beginning of the economic reforms of the 1980s. There are a number of scientific approaches to conducting research on income distribution and wealth disparity. This study discusses them from the perspective of sociology. First it will deal with the relationship between sociology and income distribution and then it will make an analysis of wealth disparity and social equity.

1. The Relationship Between Sociology and Income Distribution

Formal academic sociology was established in the 1830s in France. During the past 170 years, many distinguished figures have become prominent in the field of sociology and have made great contributions of two types.

One type is that of theory. A great number of theories have been developed over the past 170 years. Sociologists such as Michel Foucault and Anthony Giddens, who have become very popular in China in recent years, are representative figures in the field of sociological theory. Well-known sociological theorists like them number in the hundreds and have created many theories. Every one of those theories, if examined closely, can be found to be similar to the formulas of mathematics and physics. However, those theories are expressed as a form of description. Economists have also created many theoretical models, but sociological models differ from those of economics; models of economics, such as demand and supply curves, focus upon quantitative analysis. Sociological ones also do this, but since social factors are often more resistant to analysis than economic factors, it can be more difficult for sociologists to build their models. Sociology and economics are contrastive disciplines. While economists examine purely economic factors, sociologists believe that there are other factors that have not been paid attention to, or things that cannot be explained by economics alone. Therefore, it is more difficult to construct sociological theoretical models.

The other great contribution to sociology is that of method. Among the various disciplines of social science, sociology has an advantage over others in the field—social research methods. Sociology specializes in basic research methods, such as social investigation methods and social statistics methods, including all kinds of quantitative and qualitative methods. Sociologists have devoted quite a lot of time to the study of research methods. For example, the study here pays great attention to method when analyzing income distribution and wealth disparity. Studying China's income distribution and wealth disparity requires appropriate methods and truthful data regarding people's incomes. However, obtaining accurate data can be difficult in China. For that reason, to conduct good research, one must first of all develop a research methodology. This is an important aspect that sociologists lay great stress upon in their research.

What are the prominent features of sociology in contrast to economics? Economics studies economic phenomena, which are very important in every society. That's why Deng Xiaoping advocated "centering on economic construction;" actually he was referring to the material support system of a nation. So economics necessarily plays a role in the creation of wealth. The core concept of economics lays stress on economic benefit and profit. If an entrepreneur cannot bring about profit, he is not good at marketing; if an economist talks quite a lot about economic theory but is useless in promoting the development of China's economy, then the economic theory he advocates is not a good theory. What then is sociology about? Actually, sociology exists in relation to economics. What sociology emphasizes is social benefit. Sociologists often criticize entirely economically focused points of view. They believe it is definitely right to create profit, but wonder if a society can function only by creating profit. Who will possess those profits? By the end of 2008, the savings of urban and rural residents in China totaled more than twenty trillion yuan; overall, this is a great accumulation of wealth. Unfortunately, the size of the low-income population is great, too. Farmers in rural areas are, generally speaking, still poor. A big gap between urban and rural residents, in terms of both income and consumption, still exists. That shows that it is not enough for a society to merely create wealth. After creating wealth, the question of whether the model of distribution is equitable or not must be raised. Basically, sociology explores and deals with the second part of the issue: equity and inequity. Though sociology deals with many social issues—such as social security, social welfare, and social reform—equity is at its core.

When sociologists discuss equity and inequity, their basic model is that of "social stratification." As is well known, social stratification is a universal feature of human society. China is a country with a recorded history of more than three thousand years, and five thousand years of civilization. Research has found that no single historical stage was without social stratification. Even before human history, there existed social stratification among the gorillas. The gorillas also have their leaders who, of course, are often the stronger ones. According to available research, when leaders become old, other gorillas will challenge their leadership, which usually involves a ruthless fight. After the cruel struggle, the stronger gorilla will become the leader and the others make obeisance to them. This method of studying society is called the observation and experimental method.

In social stratification, three social positions and social strata are most important. The first is called economic (or property) position. Sometimes it is not easy to research property, so we often study income instead, using it as a reflection of a person's property ownership. Altogether, they reflect people's economic positions. From this dimension, people are ranked into the wealthy, the penniless, the rich, the poor, a high-income group, a medium-income group and a low-income group, which are comparatively easy to measure. The second position is called the power position. Power refers to the ability to control others. Power stratification in China is clearly represented by executive ladders or administrative levels—such as ministerial level, vice-ministerial level, bureau level, department level, section level, or group level. The higher one is on the executive ladder, the more people one controls. So, this can be used to measure power. The third is social reputation. It is said that social reputation is not as substantial as economic position—but actually, that is not so. Famous people—such as singers, performers, and professors—are supported by their reputations. They would suffer a drastic decline if they became unpopular. Property, power, and reputation are different. One could have power and property, but not have a good reputation or high prestige. John Davison Rockefeller (a famous American businessman in petroleum circles) was very rich; his assets were ranked top in the world at that time. He tried to run for president many times, but did so in vain. He possessed property and power; why didn't people vote for him? This had to do with reputation. If a person has high prestige, people will vote for him; if he does not, people will not vote in his favor. Ayatollah Khomeini (a religious leader in Iran) had no property or political power in his early life. Later, because of his increasingly impressive reputation, he became a greatly influential political and religious leader. He overthrew the Shah just because he enjoyed considerable popularity among Islamists. The importance of reputation cannot be denied. Reputation is somewhat queer as it refers to other people's evaluation of a person; it is subjective. The three social positions mentioned above are not equal to one another. Since social stratification can be observed from different dimensions, why do we focus mainly on income distribution and wealth disparity? Because, according to Marxist Economics, economy is the foundation. If you cannot explain the economic dimension clearly, it will prove very difficult to interpret the other dimensions. In developing countries like China, one's economic position is of great importance; property and income are indispensable to survival. Wealth disparity and economic inequality are therefore the primary concern of people.

2. Income Distribution and the Wealth Disparity in China

When referring to income distribution and wealth disparity, people tend to ask how big the gap between rich and poor is in China. To answer this question accurately, we must first explain how to measure it. One internationally recognized measurement method is the Gini coefficient. Its maximum value is 1 and its minimum value is 0. It is the same as the Lorenz curve in the field of economics. It realistically reflects inequality by measuring the distribution of property and income within a population. Suppose there are one hundred people in a country; if we divide the property in this country into one hundred even portions, then the absolute mean distribution is, of course, that every person has one portion. In such case, the Gini coefficient is 0. Now, suppose that among the same one hundred people, one person is king and the other 99 people are his subjects. If the king gets 100% of the property in the country while the other 99 people have nothing, the Gini coefficient is 1. In fact, it is impossible for these two absolute extremes to occur. So, the value that reflects the income distribution will be more than 0 and less than 1, with a value closer to 0 meaning greater equality, and a value closer to 1 signifying a wider gap between rich and poor. According to the principle of the Gini coefficient, the degree of inequality in any country can be reflected by a number between 0 and 1.

Before the economic reforms, how big was the gap between rich and poor in China? According to the report of the World Bank Economical Study Group, the Gini coefficient of Chinese urban residents was 0.16, the lowest index in the world at the time. The Gini coefficient usually exceeds 0.2, even in areas with a narrow gap between rich and poor. How, then, did China narrow the gap between rich and poor urban residents to such a low margin? At that time, a low income and coupon systems were applied in cities. Food, goods, and daily essentials were evenly distributed through the issuing of coupons. Therefore, daily necessities were evenly distributed to everyone through a planned economy and a well-developed system of organization and management in the cities. Thusly, an almost even distribution was realized.

The Gini coefficient used above refers to the disparity between urban residents. If we measure the disparity among farmers, the Gini coefficient was higher. According to the statistics of the World Bank, in 1979 the Gini coefficient of rural residents was 0.31. If we combine the urban and the rural residents, the Gini coefficient would be 0.33—a much higher index than that of urban residents. That showed that the gap between cities and rural areas was already sizeable before the economic reform. A Gini coefficient of 0.33 is still low compared with the Gini coefficient of the whole world, but after the 1980s, the Gini coefficient of China climbed quickly. Research conducted in 1988 by Zhao Renwei of the Chinese Academy of Social Sciences proved that the Gini coefficient of both urban and rural residents in China had risen to 0.382.

In 1994, through an analysis of data gathered in a national investigation, it was found that the Gini coefficient of urban and rural residents in China had reached 0.43-0.44, which attracted the attention of the world. It is necessary for us to contrast this with other countries. We know that in most of the developed world, the gap between rich and poor is not very wide: the average Gini coefficient is below 0.4. The Gini coefficient of some northern European countries such as Sweden and Denmark was even lower—about 0.25. The highest index Gini coefficient was found in South American, African, and South Asian countries such as Brazil, Chile, Peru, and Argentina, etc. The Gini coefficient of many of them was over 0.5, and a few even exceeded 0.6. According to international experience, it is advisable to keep the Gini coefficient within 0.3–0.4, which is termed "medium inequality." In the research of 1994, therefore, I proposed that Chinese income inequality had passed international medium inequality. If there is a line at which we should take warning, it is 0.4. An index above 0.4 means that the gap between rich and poor has become risky. After 1994, according to the responses of researchers, more voices could be heard discussing wealth disparity. Additionally, although the gap between rich and poor in China was not the widest in comparison with other countries, such an accelerated widening from a point of almost absolute equalitarianism to a large gap between rich and poor was a real concern.

What is wealth disparity like in China in the twenty-first century? According to the State Bureau of Statistics, the Gini coefficient of the per capita income of rural and urban households in China in 2006 is 0.47. The statistics from other researchers and scholars show that in 2008, the Gini coefficient of per capita income was no less than 0.5. Such an accelerated widening weighs heavily on our minds. I think that this upward trend in particular resource allocation is manifested in three characteristics. First, wealth has become more and more centralized. That is to say, the economy in cities—especially in centralized metropolises—has become more prosperous. However, the development in the marginal regions—especially in rural areas—as well as the living conditions of farmers, have become worrisome. The rich in a few centralized metropolises and the poor in many small towns and remote areas form a sharp contrast. This research also indicates that, in terms of both income and personal savings deposits in the bank of rural and urban residents, the higher the position of the small group at the top of the social ladder, the higher the relative proportions of their ownership. Secondly, the polarization between town and country and regional differentiation are obvious. At the same time, the gap between developed coastal provinces and interior developing provinces, and the difference between the East and West of China, are also large in terms of income and consumption. Third, some low-income groups and disadvantaged groups are excluded from the social security system. In rural areas, people follow the old practice of "bringing up sons to support them when they are old" ("养儿防老"). So, in fact, in rural areas people seek social security through their own families. The government has already predicted that by 2020, the social security system will serve all urban and rural residents.

However, is it absolutely an evil when the gap between the rich and the poor widens? If it is, why do people not want to return to the days before the economic reforms when there was little wealth disparity? From the perspective of sociology, it seems we should seek a reasonable disparity as wealth can never be distributed evenly in any society. What, then, is reasonable disparity? If the income gap is narrow and everyone has the same income "no matter how much or how little, how well or how poorly they have done in their work," then society will lose its motivation to work. Just like water, it will not flow if it does not at some point fall. For that reason, it is not necessarily a good thing for the Gini coefficient to be too low. That was the problem of Chinese society before the economic reform.

What is interesting is that when the gap between rich and poor widens, economic growth usually accelerates—as can be seen in the last thirty years in China, as well as in international experience. This basic principle states that every country has to choose between two kinds of strategy. One strategy is to enforce reform to stimulate production by laying off workers, widening the gap, and rewarding actors with high income to boost economic growth and raise labor efficiency. The other strategy focuses on social justice, believing that there should not be too many disadvantaged groups in society, emphasizing retention of workers at their posts, reduction in unemployment, encouragement of enterprise in offering opportunities to vulnerable people, and provision of income to the unemployed. In that case, social welfare and social security are enforced. But overly extensive social welfare and social security will damage the general enthusiasm for work, economic stimuli, and social dynamics, since people will get paid even when they do not work. Therefore, every country should properly regulate its dynamic mechanisms for developmental impetus and its welfare mechanisms for securing justice. Otherwise, it may encounter either a shortage of economic dynamism or too wide a gap between rich and poor.

3. Wealth Disparity and Social Stability

Wealth disparity is noticeable because it is a matter of social stability. What concerns socialists is whether wealth disparity will result in social problems. People frequently question whether the present gap between rich and poor will lead to social instability. Actually, the complexity of sociology lies in the objective facts of a society and the way they are interwoven with and influence the subjective states of mind and social receptivity of people. This cannot be accurately measured as in physics, where we can trust that water boils at one hundred degrees centigrade. Social phenomena are not like that, as we can see with the objective index of wealth disparity. As mentioned above, setting a Gini coefficient of 0.4 as a warning line just indicates a kind of possibility. Some people believe that social problems will emerge as soon as this value is reached since this is the warning line value. Social phenomena, however, are not as simple as natural phenomena; they are particularly sophisticated and there exists no absolutely objective index.

Whether a society is stable or not, whether social conflicts occur or not, is greatly influenced by three elements, or indices. The first is the unemployment rate. If a large proportion of the population is jobless, social conflicts will occur. Some people are prone to predict that if unemployment exceeds 10%, social conflicts will occur. In fact, in some European countries, the unemployment rate has reached this level, but no major social conflicts have occurred. In the analysis of the national statistical data of China, it was found that in 1998, the unemployment rate in China had reached that level also. At that time, state-owned enterprises were under reformation and a great number of workers were being laid off. If the laid-off workers, the unemployed, and those waiting for a job were added together, their number exceeded 10% of the population. But, nothing significant occurred. I advanced a few theories, such as "potential employment," "family complementation" and "life chances" to explain why nothing serious happened in China when the unemployment rate was high.

The second element is the inflation rate. The social turbulence occurring in China by the end of the 1980s was closely related to high inflation. Inflation devalues currency, and so is injurious to the interest of almost everyone in the society. In addition, Chinese people prefer saving money, which makes them vulnerable to the effects of inflation. The Chinese society is therefore greatly sensitive to inflation.

The third is wealth disparity. The influence of wealth disparity on social stability is even more complicated. It is closely related to people's capacity for psychological endurance. People of different nations and cultures tolerate wealth disparity to different degrees. For example, in South America and Africa, the gap between the rich and the poor has been wide for a long time and sometimes the unemployment rate and inflation rate are also high. But, both culturally and psychologically, the general capacity to tolerate wealth disparity is rather strong. What about the capacity for psychological endurance of the Chinese people? Both cultural tradition and the structure of the Chinese social mind are rather complicated. It would not be correct to make a judgment about the limits of their tolerance by simply saying, "they are tolerant" or, "they are not tolerant." China is a nation of great flexibility. Sometimes it seems unable to tolerate inequalities. In Chinese history, peasant uprisings have been common. According to statistics, China has had the most peasant revolts of any country in the world. The leading cause of peasant revolts has been economics. In Chinese history, the cultural attitude of "inequality rather than scarcity is the cause of trouble" ("不患寡而患不均") is deeply rooted in people's minds. It is therefore really hard to say that Chinese people are very tolerant of wealth disparity.

In Chinese philosophy, one can find intolerance for conspicuous figures. A Chinese saying goes, "destruction pursues the great" ("木秀于林,风必摧之,堆出于水,流必湍之"). This means that if a person is conspicuously excellent or dares to stand out, they will be attacked by others. This is a widely shared belief in Chinese culture. The sayings, "it is the most beautiful bird that will be shot" ("枪打出头鸟") and "exposed rafters are the first to rot" ("出头椽子先烂") are true reflections of Chinese philosophy. In China, keeping a low profile is culturally ingrained. For example, Chinese calligraphy pays particular attention to concealing the peaks in each stroke. Public prominence is not acceptable, while holding back much and revealing little is seldom to blame. If Chinese people could tolerate disparity, then they would compete to be conspicuously excellent. However, to be conspicuously excellent is considered unacceptable and risky.

Sometimes, however, the Chinese are quite tolerant of disparity. In the later Qing Dynasty, for over three hundred years, the Han nation was ruled by other nations. But, this was accepted and the Chinese people were even satisfied during some years. So, in terms of disparity, people's opinions can be quite flexible. Sometimes people have a high tolerance for it, but sometimes they have a low tolerance. Thus, it is very difficult to make judgments. To analyze Chinese society, we should therefore be prudent about specifics. We should analyze the social background and adopt a flexible approach for different social phenomena.

The key to maintaining social stability is to ensure the basic living conditions of all Chinese people and gradually make the majority of the Chinese people well off. Accordingly, the Chinese society is stable at the moment. Over the past thirty-plus years, the living standards of Chinese people have improved drastically. In a national survey I did, I compared my data with that produced by a professor from the University of California in a survey of Russia and Eastern Europe, and I found that during the investigations we both asked the same questions of interviewees. We both asked whether their living conditions were better or worse compared with ten and twenty years ago. It turned out that most of the Chinese people thought that their living standards had improved, while in Russia and Eastern Europe, most people thought their living standards had worsened. By contrasting these statistics, it can be easily concluded that Chinese society is still stable.

4. Market Transformation and Wealth Disparity

The analysis in the previous passages indicates how wide the gap between rich and poor is by presenting social facts. In the following passages, we intend to find out the causes of wealth disparity. What are the leading causes of a big gap between the rich and poor? Research has found that the widening of the gap between the rich and poor during the past thirty-plus years is related to the introduction of a market mechanism to China. A market has three parts: a commodity market, a labor market, and a financial and credit market. China started experimenting in the early 1980s to introduce a commodity market. In the middle and the late 1980s, Deng Xiaoping repeated several times that we should make breakthroughs in commodity price reform. He was talking about the commodity market. A commodity market means that commodity exchange occurs through the medium of cash, not coupons. In the past, a high level of equalitarianism was achieved in China through the use of exchange coupons. Though we had a commodity market at that time, the prices of most commodities were controlled by the government, not by the market. The most fundamental feature of the commodity market when the reforms took place was that of shortages. A Hungarian economist developed the concept of a "shortage economy," which certainly characterized the Chinese economy before 1985. So, when the commodity market was introduced in 1985, it instantly caused serious inflation. Because the prices of commodities were decided by supply and demand, prices would rise when demand exceeded supply. So, as soon as the commodity market was introduced, prices skyrocketed and inflation occurred. For quite a while, many Chinese people had a strong desire to rush out and buy—to snap up—once rationed consumer goods. The obstacles for commodity price reform were very difficult to break through in the 1980s. Chinese people, however, were quite creative and overcame these difficulties within a short period of time. By the middle of the 1990s, the commodity market in China suddenly had surplus commodities. This proved that, given a transition period, the diligent Chinese people could overcome any difficulty.

We have not yet broken through the obstacles of labor market reform, though several attempts have been made since the mid-1990s. For example, university graduates used to have guaranteed job assignments. Beginning in the 1990s, recruiting meetings between employers and university students were tried. University graduates were encouraged to find jobs by themselves through these meetings; now, universities were no longer responsible for assigning jobs to graduates. The free mobilization of labor is good, but China is a country with a labor surplus. The opening of the labor market resulted immediately in the problem of unemployment and lay-offs. This was rather risky. Recently, measures have been adopted in labor market reform, including enhancing the communication of information about employment opportunities, constructing an information system for employment, promoting pre-job or career training, and improving the employment market.

The reformation of the financial and credit market is underway. It is risky to implement such reforms because the financial market allows the free flow of capital. The inflow and outflow of international capital is not completely within the control of the government. Therefore, this process should be conducted with prudence.

Why should we introduce a market economy? The reason is not complicated. From 1949 to 1979, we experimented with a planned economy, only to find that without a market economy, the economy loses its vitality. In 1949, the per capita GDP of China was two thousand dollars less than that of developed countries. By 1979, this gap exceeded five thousand dollars. So the planned economy as attempted by China proved ineffective, which is why we introduced a market economy. The introduction of a market economy has caused some problems, but one should keep in mind that the market is more capable of resource allocation than a planned economy. Before the economic reforms, Chinese people always thought that the government was the most capable authority in resource allocation. When university students graduated, they followed the job assignment they were given by the government, believing that government arrangements were the most suitable. Today, free labor mobility has shown its great advantages; just as the old saying goes, "no matter how much one plans things out, life always intervenes" ("人算不如天算"). But, the market is sometimes cruel. Before the economic reforms, state-owned enterprises could get subsidies from the government, with "equalitarianism and discriminate transfer of resources" ("一平二调") rather than the regulation of the market. For example, a wealthy enterprise in Shanghai would be required to hand in extra money to the government so that the government could send that money to subsidize a poor enterprise. At the time, this was considered acceptable. But in a market economy, if an enterprise is competitive, it can develop into a giant corporation. If it is not competitive, it will go bankrupt. Therefore, one should not be surprised to see wealth disparity following the ruthless wealth allocation of the market. Those enterprises that can meet the needs of consumers will do well and develop rapidly, but those who cannot will declare bankruptcy and lay off large numbers of workers. The market is merciless. The implementation of a market economy inevitably results in the widening of the gap between the rich and poor.

A theory called "Kuznets's Inverted U-curve" is internationally used to explain wealth disparity. As all economists know, Simon Kuznets—a Nobel Prize winning economist—collected statistics from many countries to analyze changes in wealth disparity and found that they formed a curve. Assuming the vertical axis represents inequality, and the horizontal axis represents per capita GDP, the relationship between inequality and per capita GDP appears as an inverted U-curve. This means that in some undeveloped poor countries, the gap between the rich and the poor is not very big. When the per capita GDP rises to USD500–1,000, however, the gap widens sharply. After it peaks, further growth of per capita GDP causes the gap between the rich and poor to gradually narrow. This theory explains the historical process of the relationship between economic growth and wealth disparity in many modern countries. To interpret from this theory why the gap between rich and poor in China grew so rapidly, we can say it was because China was located at the rising part of the curve. According to Kuznets's Theory, when per capita GDP increases to a relatively high level, the gap between the rich and the poor will narrow to some extent. Does this mean we should just wait, since Kuznets's Inverted U-curve theory says that the gap will shrink naturally with the further growth of GDP? Actually, the theory is not as simple as it appears. A close study of the underlying mechanism of the "inverted U-curve" of Kuznets's Theory is necessary.

It is impossible for a nation to distribute its wealth simply through the market; many procedures are required. Market distribution is the first step, but there are many stages of distribution after that. When a nation starts to establish a market mechanism, the mechanism is imperfect and full of loopholes. Its welfare and security systems will also be underdeveloped. At this stage, therefore, the gap widens faster. When a nation becomes rich, it will tend to invest more in social welfare and security. Then, wealth is redistributed and transferred many times. It should be realized that the market is fairly capable of distribution, but the market does not accomplish a final and equal distribution. Wealth distribution that proceeds only through the market is unfair. It should be regulated with more procedures.

Before the reforms, the dominant theory held that a market-based distribution of wealth was not conducive to equity. So, we did not go through the market, but rather followed government directives to achieve initial equitable distribution. After the reforms, however, people realized that it was wrong to deny the function of market distribution. Is initial distribution by the market reasonable? It definitely is because a market distinguishes the strong from the weak, the winner from the loser. The problem before was that no one admitted that a difference between strong and weak existed, believing that people should not be so divided and that some money of the strong should be given to the weak. But, this was not right. The strong should be allowed to overcome the weak in market competition so that successors can reap the benefits of achievement, which will push economic development forward and enhance social efficiency.

Although survival of the fittest is a must, if the income of the winner is too high and the gap between the rich and the poor is too wide, the situation becomes unfair. A well-established market should include redistribution with multiple procedures. Taxation is the most important redistribution mechanism. Progressive taxation in great amounts can transfer some wealth of the rich to the poor. Donation is also a very important mechanism. There is a theory named "marginal benefit" in economics. For wealthy magnates, the marginal benefit of their wealth is diminishing. But if they donate it to the poor, it will bring the poor high marginal benefit. Many countries with market economies, therefore, have established systems to encourage donation. A well-established market utilizes multiple procedures of wealth redistribution including tax adjustment, social welfare, social security, charity, and donation. Governments should make social policies to guide the redistribution of wealth.

5. Wealth Disparity and Social Equity

As mentioned previously, economics emphasizes efficiency while sociology lays stress on justice. In fact, these are two analytical dimensions of the same issue. When talking about efficiency and equity, three schools of thought must be mentioned. The first school is liberalism, represented by Milton Friedman, author of *Capitalism and Freedom,* who places priority on efficiency and advocates sufficient free competition. Friedman believes that high efficiency can only be present in a society with sufficient competition and that only high efficiency can create more wealth, so that people will have more to distribute. So, he advocates sufficient competition and freedom as well as the abolition of social welfare under centralized control. He gives priority to efficiency. The second school is represented by John Rawls of Harvard University, author of *A Theory of Justice.* By justice, he means equity, believing that all social resources should be evenly distributed, provided that "unequal distribution" cannot be made favorable to everyone. The two theories above are completely opposing. The third school is represented by Arthur Okun, who advocates that an equilibrium point should be found between efficiency and equity. Though efficiency and equity are contradictory, a balance point can be found under certain circumstances. Arthur Okun questions whether inequity promotes efficiency. He argues that excessive inequity leads to worker revolt ending both inequity and inefficiency, or vice versa. For instance, if a heavy duty is levied on entrepreneurs for the sake of equity, it will result in low enthusiasm for investment and operations, as well as low social production. As a result, people's income will decrease. With this loss of efficiency, the social welfare fund will decrease or be damaged and social equity will not be achieved in the end.

When it comes to China, two basic groups confront us: the rich and the poor. The questions to be answered are: how to deal with the rich group? How to treat the poor one? Is the relationship between the two fair? From the standpoint of social stratification, what kind of wealth distribution is fair? Deng Xiaoping once said that we ought to allow diligent people to become rich first. But it turns out that diligence alone cannot make people rich. Diligence is very important, but many farmers who are diligent about farming cannot make enough money by their labors alone. Therefore, in the study of inequity and stratification, the questions to be answered are: what kind of stratification is acceptable, assuming inequity and stratification are inevitable? The most common answer is that we should allocate relevant social positions to those with corresponding abilities. For instance, if one operates an enterprise successfully through one's own abilities, then being stratified into a high social position is acceptable. But, if that success is not achieved through one's ability, people will become dissatisfied with stratification. It is a common expectation that high quality personnel should be placed in high positions. Being placed in a low position because of low ability is to be expected. The crucial question is, however: does one become rich through one's own abilities? For example, Bill Gates left university before graduating, but he was shrewd in assessing the market and saw his chance in the field of software. So, he succeeded and, currently, he is one of the richest men in the world. The example of Bill Gates illustrates that if one is of high abilities, wise, and operates effectively, one's wealth will generally not be envied by others. If one emulates him, one could also try to found a "Microsoft." However, if one's wealth is obtained by corruption, smuggling, drug trafficking, and delinquency, a sense of inequity will be aroused.

Additionally, if one's wealth is acquired through inheritance only, that wealth will not be regarded as being fairly obtained. Recent research shows that during the last thirty-plus years of reform, the influence of inheritance on people's success has decreased to some extent and the importance of factors of achievement have somewhat increased. That is a positive transformation. Achievement refers to one's own efforts; for instance, enrolling at university, obtaining certificates, and operating enterprises are all the fruits of one's own efforts. The wealth gained through one's own efforts is regarded as having been fairly obtained. On the basis of this criterion, the quality of the rich has been improved to some extent during their accumulation of wealth. According to data gathered by the Entrepreneur Investigation System of the State Council, the educational level of private entrepreneurs in China has improved in that the proportion of those with MBA certificates has increased. Recently, a surprising number of Chinese private entrepreneurs were found to have undergraduate, postgraduate, and MBA degrees. This was not the case in the 1980s. As data show, the average educational level of private entrepreneurs was rather low then, even lower than that of common people. By the 1990s and in the twenty-first century, the credentials of many private entrepreneurs had greatly improved. If we look at trade instead of diplomas, we find that many private enterprises belong to newly emerging industries. In the 1980s, the so-called "ten-thousand yuan households" —the class that became rich first—made their money through retail since the level of economic development was still low at that time. But in the 1990s and the twenty-first century, a number of highly trained individuals have entered into business, which reveals positive changes in the society.

"Elite theory" is often adopted in sociology to analyze the wealthy classes. "Elite" refers to competitive individuals within the society. Over the past thirty-plus years, China has cultivated groups of economic elites. On this basis, we should not be hostile to wealthy groups. Of course, the wealthy group includes high quality individuals—if they were not, they would be eliminated from the group. This elimination system is named elite circulation. The theory holds that society has a system for eliminating those without high qualities and for raising up those with high qualities into the upper class. The wealthy group is not something bad, but a corresponding elimination system is necessary to ensure those with high quality enter the upper class and those with low quality move to the lower class.

The need for such a system has been understood in China since ancient times. The imperial examination system, for example, was intended to achieve elite circulation. It was established in 587 BCE during the Sui Dynasty and lasted for more than 1,300 years. That system was the best in the world at that time. In Europe, a hierarchical system was in place, with title and wealth being inherited. China was a leading country for many years just because of this imperial examination system. When we study the profile of imperial examination and enrollment in the Ming and Qing Dynasties, we find that a rather considerable proportion of those who passed the examinations came from poor families. This indicates that the imperial examination system really achieved elite circulation. Later, Western countries transformed their systems and adapted the imperial examination system from China to suit their own needs. The United Kingdom established its civil service examinations, which were equivalent to the imperial examination system. Moreover, Westerners overcame the shortcomings of the imperial examination system by selecting elites within a limited scope and popularized the "market mechanism." The "market mechanism" is much better than the imperial examination system. It is equivalent to an all-encompassing economic examination for the whole of society. Those capable of business and management take part in the examination. If one passes, one becomes elite. The market is also a place for elite circulation, so opening the market up to enterprises will lead to elite circulation, allowing capable entrepreneurs and real elites to improve their social status. It should be realized that it is not a bad thing to become rich while China is still a poor country as a whole, and that the so-called millionaires of China can never be compared with entrepreneurs in multinational corporations. To realize elite circulation, competitive opportunities should not be given only to those close to oneself, but also to unfamiliar people so that true elites are allowed to stand out. Among senior managers, those who are not elites should be eliminated to allow those competitive in management to take their places: that is the core of the whole matter. Unselectively attacking the rich, the wealthy, and entrepreneurs should therefore be avoided.

In addition, support and assistance policies should be implemented to aid the poor. There are two kinds of poverty: absolute poverty and relative poverty. Absolute poverty tends to be concentrated in adjoining poverty-stricken areas—such as regions populated mainly by ethnic minorities, remote areas, and mountainous areas such as Yunnan, Guizhou, and the Northwest regions. China's absolutely poverty-stricken population in rural areas in 1993 was 80 million, but this had decreased to 14.87 million by 2007. The relatively poverty-stricken population is largely made up of unemployed urban workers. In recent years, a system for providing minimum subsistence allowances for urban residents has been implemented to relieve the poor in cities. But in rural areas, no solution has yet been found. Recently, local government agencies have attempted to solve the problem of poverty in rural areas but have found the large population of rural areas to be a highly prohibitive factor. The solution to poverty may be higher taxation for those with high incomes and compensation for those with low incomes through a social security system. But, according to an analysis of taxation, the majority of individual income tax is still paid by ordinary employees. So, the high progressive tax and individual income tax do not actually "rob the rich and give to the poor" ("劫富济贫"). Another potential approach to solving the problem of poverty is to establish various social security systems. For instance, the recent medical system reform, housing reform, and unemployment compensation in China should take the interests of the low-income population into special consideration.

Anti-corruption is also very important in narrowing the gap between rich and poor. The gap between rich and poor in China is partially caused by violation of the law and of discipline. It should be admitted that such problems are hard to control because, due to institutional deficiencies, violation of the laws and regulations during the process of market transformation can hardly be prevented. In addition, education can play a role in solving the problem of wealth disparity. The dominant function of education is to transmit knowledge, or as the old saying goes, "to transmit wisdom, impart knowledge, and resolve doubts" ("传道、授业、解惑"). Parallel to this, education has some other latent functions. The first is that it can decrease the birth rate: national statistics show that more highly educated demographic groups have lower birth rates. The second is that it can reduce poverty: in a modern society, poverty is, to a great extent, caused by a lack of knowledge, technical skills, and education. Namely, it is "a poverty of education" ("文化贫困"). Once, in the mountainous areas of Guizhou in northwestern China and the regions inhabited mainly by ethnic minorities, many people were encouraged to move out of those areas to escape poverty. However, they later went back and lived the same life as before. The key to ending poverty is education—particularly basic education, technical education, and skills training. Only by gradually raising the general educational and technological level, enriching people's knowledge, and changing their minds, can poverty be eradicated.

Chapter 8

The High-Income Group in China

During the last thirty-plus years of economic reform, one of the most notable changes that has taken place in China's social stratification has been the rise of a high-income group. In studies of the polarization between rich and poor, the "rich"—or to put it in a technical way, the high-income group—have been a constant focus for researchers, as well as the whole of society. More than thirty years ago, when Deng Xiaoping proposed "allowing some people to become rich first," this became the subject of much controversy all over the country. People debated who should be the ones to become rich first. At that time, nobody knew who would. Now, after more than thirty years of reform, who has become the high-income group? To answer this question, an attempt is made in this chapter to offer some analysis.

1. How to Conduct a Survey on the High-Income Group

As is commonly acknowledged in the field of sociology, the study and survey of the high-income group in China has now become an extremely difficult task. Back in the early 1980s, when sociologists began to make door-to-door questionnaire surveys on such issues as people's income, surveys were comparatively easy to conduct in both urban and rural areas. Since the telephone was not yet a popular household item, even in urban areas, surveyors were not required to make appointments by telephone beforehand. They could just go and knock at the respondents' doors. The interviewed residents were generally at ease with such "uninvited visitors" and were much less defensive—for, at that time, they seldom possessed any valuable property. Furthermore, it was a time when a sense of privacy was still weak among Chinese people. It was a common practice, even among strangers, for people to ask each other about their income. For instance, a conversation with the person sitting next to you on a train would most likely start with these two questions: "What do you do for a living?" and "How much do you earn each month?" Therefore, to ask about one's income in a door-to-door survey was quite acceptable. Since people tended to have very simple sources of income during that time, calculating income was less complicated as well. At that time, the average income level of Chinese people was not high. The highest level, especially in rural areas, was that occupied by families that earned ten thousand yuan per year—the so-called "ten-thousand yuan households." Since most respondents, not earning much, did not mind enquiries about their income, the figures collected were generally reliable.

It was not until the mid-1990s that most Chinese families in urban areas had telephones installed. Because of that, making a phone appointment before visiting people in their homes became necessary, even among old friends. The "uninvited visiting" of just-go-and-knock-at-the-door was no longer acceptable. Since people's *living standard* had shown a perceptible improvement, many residents were much more defensive and aware of their privacy than they used to be. Door-to-door surveys were not as easy to conduct as before, but were still possible with proper arrangements. I organized several nationwide sample surveys, for example, in both urban and rural areas in the 1990s. In the urban areas, we contacted the neighborhood committees, which managed to inform the residents and make some arrangements before the survey. When we went to the respondents' houses, we would show them special cards identifying us as surveyors, as well as our IDs or student cards, and then offer them some gifts. Our surveys thus went smoothly and met no resistance. But, as income levels rose, some respondents who had become very rich were not without reservations in answering questions concerning income. In order to make sure that the information we obtained was accurate, we had to include a few questions intended to reveal lies in the survey, such as questions about expenditure items. They were usually inserted unobtrusively in the questionnaires to detect possible lies about income. Other dishonesty checks included a recording of the major durable goods owned by the family, which can indicate the income level of the family, and a recording of the general impressions of the family's financial status in writing by the interviewers at the end of the questionnaires. During that period of time, surveys both on urban and rural household income were mostly conducted successfully, even in major cities like Beijing and Shanghai. The results showed that there were some high-income families in China in the mid-1990s; we even visited families with a yearly income in excess of ten million yuan.

In the late 1990s and even more so in the twenty-first century—given the constant improvements in neighborhood environment, housing conditions, and people's living standards in China—people developed a much stronger sense of privacy and a sharper awareness of safety. Door-to-door surveys have almost become a "mission impossible." As the result of the significant change in the administration mode of urban areas, the neighborhood committees no longer hold the authority they used to in neighborhood affairs. Instead, property management companies, which are hired by and provide services to residents, have begun to play an increasingly important role in the new communities. These companies, out of regard for the residents' rights and privacy, will normally reject any visiting surveyors. Additionally, most urban residents have become owners of their residences since the housing system reforms of the late 1990s. They have become much more aware of their rights and interests, believing that no stranger should enter their residences without permission. In upscale communities where high-income residents gather, it has proved impossible to visit respondents in their homes. Although a number of door-to-door surveys have been conducted in residents' homes in urban areas during this period, the data collected indicate that upscale neighborhoods are usually not included. The data are therefore incomplete and merely reflect middle and low-income groups, seriously biasing the results of the survey. In major cities like Beijing and Shanghai, upscale neighborhoods are so well protected by the security staff hired by the property management companies that it is impossible even to enter the neighborhood, let alone visit individual residences. I once tried to enter one such neighborhood by cooperating with the local government, but even the requests from the governmental high administrative department were rejected for a simple reason: the residents did not want to be interviewed. There was no way at all to conduct our survey since the property management companies refused to offer any information about the residents. I once made an investigation into how many surveyors were rejected in such neighborhoods and the results indicated that the rate of rejection was rather high—over 85%—even if the survey was conducted with the help of the government. That is why I think door-to-door surveys are no longer appropriate or helpful in the study of the high-income group in urban areas.

Despite these issues, a comparatively effective survey was conducted by the Department of Urban Society and Economy of the State Bureau of Statistics of China. Tens of thousands of samples were drawn from urban families marked as consistent survey respondents. These families were required to keep household accounts as part of the legal obligations of a citizen. We must say, the data gathered through this survey was representative of the whole nation, so the statistical data about the high-income group subjected to analysis later in the chapter will not be without value. Information obtained through surveys facilitated by governmental legal intervention has the advantage that respondents are legally obliged to answer all questions and cannot refuse. However, since it is a kind of compulsory survey conducted by the government, it remains unknown whether the data the respondents keep regarding the accounts of their household income are reliable. Researchers have to take into consideration the respondents' possible reservations and the validity of the statistics.

If the door-to-door survey is no longer an appropriate research method for the high-income group, are there any other alternatives? The data to be analyzed later in this chapter were extrapolated from a non-door-to-door survey of the high-income group by me, which was conducted from October to December of 2007, in which I was assisted by Li Yanjiang, Shi Changhui, and Li Peilin. The respondents were required to answer questions about their expenses on items such as vacations and the like. Therefore, there was not much in the results of the survey that could be helpful for us in analyzing the social status of the high-income group. However, if we take into consideration the unusual challenge of the survey itself, the data are still significant in the research of the high-income group, even though they were quite limited. Instead of a conventional door-to-door survey, we adopted two other methods. The first was to conduct the survey in places where the high-income group usually gathered for collective activities. The respondents of the survey were mainly entrepreneurs and company managers who were taking educational or business training courses, or EMBA courses outside of the cities in which they worked. Since they were often in distant training centers that had nothing to do with their companies and they were required to answer the questions anonymously, their answers tended to be more truthful because the respondents were under little pressure. The second method adopted is called "snow-balling." The first step is to assemble a list of people whose income is high and then distribute questionnaires to them online through intermediates who are normally acquainted with high-income respondents. These respondents were also requested to provide answers to the questions anonymously, reducing the stress the respondents were put under. Consequently, 312 questionnaires were retrieved and 202 of them were found to be valid. Although the sample size was not large, the data proved to be invaluable as the respondents' membership in the high-income group was certain. Additionally, most areas in China were covered by the survey, including thirty-eight cities and regions, among which Beijing had the most samples (taking up 40.6% of the total), while 18.4% of them were from northeastern China, 8% from northern China, 18% from eastern China, 7% from Guangdong, and 8% from central and western China. I feel obliged to point out here that the samples collected in this non-door-to-door survey of the high-income group might not be representative of the whole nation—both because the sampling was not a random sample and sample size was small. But, as I mentioned earlier, surveying the high-income group itself is both difficult and rarely conducted, so our survey can, without doubt, provide some reliable information in the research of the high-income group.

2. Who Belongs to the High-Income Group

When we speak of the high, middle, and low-income groups, we speak in a relative sense. The high-income group is not a group with a clear boundary from other groups. When people mention "individuals with high incomes," "the rich," or "wealthy people," they refer to a special group with much higher incomes when compared with others. Of course, there are people even richer than this group, and the terms usually used for them are "magnates," "tycoons," and "billionaires." They are few in number in every country in the world. They are not the subject of our study here.

Now, a comparison of the descriptions of the high-income group in different datasets will be appropriate.

One of the first things we are going to look at are the statistics released by the State Bureau of Statistics of China. The survey system of the State Bureau of Statistics of China is the best in China and has the largest research team of any Chinese research organization. Since there is a significant income gap between city and countryside in China, big cities are indubitably where the majority of the high-income group live. Therefore, we will focus on the statistics from the urban household sample survey.[55] In 2006, the State Bureau of Statistics of China accumulated 56,094 large samples of urban households, among which were 5,571 households that were identified as the highest-income households in the urban areas. The survey found in 2006 that the urban high-income group had an annual per capita income of 91,266 yuan. This is not a small figure for a household. However, the annual per capita income for each person in the family was 34,834 yuan, and the average monthly income per person was 2,902 yuan. Taking the rising prices in cities in recent years into consideration, less than three thousand yuan per person per month can hardly be categorized as high-income. So we turn to even richer groups, again, from among the 56,094 samples. We divide the sample households into twenty groups, among which the lowest-income group earned less than 10,000 yuan annually, and the richest group had a yearly income of over 100,000 yuan. The lowest-income group accounted for 4.85% of the total sample and the highest-income group 2.07%. From the respective percentages of the two groups in the total sample, they represented the two ends of the twenty groups. The highest-income households had an annual per capita income of 134,528 yuan, with an annual income of 41,140 yuan per person, or 3,428 yuan per person per month. Although such high-income groups do not seem to have especially high incomes, they were the highest-income groups that we could find in the open data of the State Bureau of Statistics of China. However, compared with the lowest-income group, the highest-income group had a much higher living standard. The average income of this highest-income group is 13.9 times as much as that of the lowest-income group. With regards to food consumption, the Engel Coefficient for the lowest-income group was 24.8%, and 49.6% for the highest-income group, which clearly indicates different standards of living. Although they are relatively the highest-income group, this group can hardly be absolutely defined so or labeled as "rich folk" or "wealthy men."

Secondly, according to *China's Individual Income Tax Law* revised in 2005, starting from January 1, 2006, individuals whose yearly income was larger than 120,000 yuan would be required to proportionally pay income tax before April 1st of the next year. That brought about the

55 The Editorial Department of City in China Yearbook. *City in China Yearbook 2007*. Beijing: Chinese Academy of Social Sciences, China Research Institute of Urban Development, 2007. 137–41.

criterion of more than 120,000 yuan annual income or 10,000 yuan monthly income for determining high-income status.

Now, we will make a comparison of the income of the richest households who had a yearly income of over 100,000 yuan in the survey data of the National Bureau of Statistics of China with those that meet the above criterion. Obviously, the criterion income refers to the income of those people with the ability to engage in economic activity, not the average household income. Following this criterion, from the highest-income group of the aforementioned twenty groups, we should take only those who were working into account, and exclude those who were not working. By doing this, the average income of those who were working in the highest-income households was 66,270 yuan per person per year, or 5,523 yuan per person per month. Although the average income is much higher after this applying this criterion, it is still far below 120,000 yuan per year (or 10,000 yuan per month), which was the benchmark for individual income taxation. Therefore, people who met the standard were indeed of a high-income group. According to the statistics promulgated by the State Tax Administration in 2007, up to 1,628,706 individuals paid individual income tax and their annual per capita income was 310,000 yuan, which was much higher than that of other groups. These high-income individuals came mainly from the highly paid trades or businesses such as banking, insurance, securities, tobacco companies, electric companies, telecommunications companies, petroleum companies, petrochemical companies, aviation, real estate, manufacturing, hospitals, and colleges and universities. Most of them (81% of the taxpayers) came from cities and provinces like Beijing, Shanghai, Guangdong, Jiangsu, Zhejiang, Shandong, Fujian, Tianjin, Shenzhen, Ningbo, Qingdao, and Xiamen. However, the statistics of the State Tax Administration also indicates that most of the tax declarers were salaried staff. So, where were the bosses? Furthermore, how many bosses were there in China whose yearly incomes exceeded 120,000 yuan?

This has much to do with the private-entrepreneur class. In the 1980s, business owners began conducting private economic activity in China. Small in number, they were mostly self-employed. Few of them could be regarded as officially defined "private entrepreneurs." It was not until Deng Xiaoping's Southern Tour in 1992, when the fear—both theoretically and with reference to policy—of the development of a private economy gradually disappeared, that the private-entrepreneur class developed rapidly. On informal occasions, people still tended to use the term "individually run" instead of "privately run" in their description of private economic activities and private enterprises. It is worth mentioning that the Sixteenth National Congress of Communist Party of China in 2002 made an amendment to The Constitution of the Chinese Communist Party (CPC) in the first article concerning the qualifications for party membership applicants such that those qualified were modified from "Chinese workers, peasants, soldiers, intellectuals, or other revolutionaries over eighteen years old" to "Chinese workers, peasants, soldiers, intellectuals, or other outstanding individuals in other social classes over eighteen years old." This important modification helped the ruling party solve a problem that had long troubled it regarding whether the class of private business owners could be admitted into the CPC or not. The modification also indicated that the CPC, as the ruling party from then on, would absorb economic elites who emerged from this new type of economy. By 2007, China had a recorded 13,272,750 private entrepreneurs. However as mentioned above, only 1,628,706 people declared a yearly income of over 120,000 yuan in 2007, and most of them were salaried workers, not private entrepreneurs. This means that most of the private entrepreneurs did not pay income taxes as individuals whose yearly incomes were larger than 120,000 yuan.

Then, what are the facts about the income of private entrepreneurs? According to the data extrapolated from the sample survey on private enterprises in China, the annual per capita income of private entrepreneurs is about 186,000 yuan, with an average household income of about 259,900 yuan.[56] Since this questionnaire was completed by private entrepreneurs, their annual per capita income is above 120,000 yuan. Therefore, the high-income group in China is estimated to contain more than 10 million people, far exceeding the 1.63 million who had paid income taxes based on this standard of individual yearly income in excess of 120,000 yuan.

Thirdly, it is appropriate to take a look at the data extrapolated from the small sample survey on the high-income group I conducted.

---

56 *A General Report and Analysis on the Data Extrapolated from Sample Survey on Private Enterprises in China 2006 (II)*. By Research Team of the Sample Survey on Private Enterprises in China. Published by *Internal References of Private Economy, 2007-030-23*. Ed. Research Association of China's Private Economy.

The samples in this survey can be roughly sorted into five vocational groups: private entrepreneurs, top-level management staff, middle-level management staff, professional technicians, and ordinary office workers. It was found that the high-income group mainly includes the first two vocational groups: private entrepreneurs and top-level management staff. Although some middle-level management staff and professional technicians have high incomes, their proportional representation in the group is decreasing compared with the former two vocational classes (table 8-1 and table 8-2).

Statistics from the small sample survey indicate that 88.6% of private entrepreneurs and 75.6% of top-level management staff have a yearly income of more than 150,000 yuan. If those whose yearly income exceeds 100,000 yuan are included, the proportion of the two groups would reach 95.4% and 92.3% respectively. For that reason, the private entrepreneurs and top-level management staff studied are clearly of the high-income group, which conforms to the general societal opinion of them. This ratio is similar to the CFIC's data. Therefore, it seems quite apt for China's Individual Income Tax Law to classify those who earn 120,000 yuan per year into the high-income group. Additionally, the data of the small sample survey also indicates that nearly half of all middle management staff earn more than 150,000 yuan per year. It is noteworthy that, in most cases, there is not a clear line differentiating top-level and middle-level management staff. So, it is acceptable to combine the two classes into the high-income group. In comparison, professional technicians earn much less than top-level and middle-level management staff. Although their income has increased since the mid-1990s, it is much lower than that of the top-level and middle-level management staff. This is why they do not normally belong in the high-income group.

Table 8-1 Percentage of Individual Annual Per Capita Income of Five Vocational Groups/Classes (%)

Income (Thousands of yuan)	Private Entrepreneurs	Top-Level Management Staff	Middle-Level Management Staff	Professional Technicians	Ordinary Office Workers
<50		1.3	6.1	33.3	38.9
50~80	2.3	1.3	6.1	16.7	11.1
80~100	2.3	5.1	15.2		11.1
100~150	6.8	16.7	24.2	41.7	22.2
>150	88.6	75.6	48.5	8.3	16.7
Sample Size (Number of people)	44	80	33	12	18

Table 8-2 Percentage of Annual Per Capita Household Income of Five Vocational Groups/Classes (%)

Income (Thousands of yuan)	Private Entrepreneurs	Top-Level Management Staff	Middle-Level Management Staff	Professional Technicians	Ordinary Office Workers
<200					
200~300	11.4	25.0	60.6	91.7	55.6
300~400	9.1	11.2	18.2	8.3	33.3
400~500	9.1	8.8	3.0		
>500	70.5	55.0	18.2		11.1
Sample Size (Number of people)	44	80	33	12	18

3. Age, Education, and Social Class of the High-Income Group

First, we should examine the features of the age of the high-income group—that is to say, we must analyze the influence of age on income. Do young people have higher incomes than older people, or vice versa? Before the economic reforms, the allocation of resources and the distribution of income in our country had nothing to do with the principle of market competition. At that time, a unified post-rank salary system was applied to every walk of life all over the country, no matter if it involved state-owned work units (entities), collective-owned work units, public institutions, or government offices. An increase in wages was subject to a unified order given by the government. The system was, to some extent, seniority based—meaning that young people were always paid less than their elders. The older and the more qualified one was, the better one was paid. The young could not get promoted until the old retired. If demonstrated in a figure, the relationship between age and income would be a rising oblique line, starting from the bottom on the left and ending at the top on the right. Since the older generation tended to occupy high and authoritative positions, the young generation had to stay in low positions and were long oppressed. Most of the young generation just dawdled away, killing time, lacking the motivation for work, which led to low labor productivity.

However, in the mid-1990s, I found through data analysis that the correlation between age and income underwent a peculiar change. In some central cities, young people were better paid than the middle-aged and the old.[57] The data under analysis was representative as it was obtained from a strict PPS door-to-door sample survey. The data also showed that an age group of twenty- to thirty-year-olds had higher incomes than any other age group. This was a situation totally opposite to the one that existed before the reforms. What caused this surprising change? I believe that it had much to do with the transformation of the economic system in the mid-1990s, which caused large-scale unemployment and the laying-off of many middle-aged and older people. As the market mechanism was introduced into China, competitive young people with college degrees, diplomas, and an ability to speak foreign languages became employed by well-paid, foreign-founded, and overseas-funded enterprises or new industries. Meanwhile, older and middle-aged people, with their outdated techniques, were crowded into lower-income positions. That is to say, middle-aged and older people who failed to adapt to the changes in society experienced a relatively drastic drop in income in the 1990s. This might help to explain the change in income. What, then, is the situation today?

As shown in the data extrapolated from our small sample survey, young people no longer tend to occupy the best paid positions. Instead, middle-aged people have become the overwhelming majority in the

57 Li Qang. "Market Transformation and Intergenerational Alternation of China's Middle-class." *Tragedy and Management* 3 (1999): 35–44.

high-income group: thirty-one- to forty-year-olds make up 35% of the high-income group, and forty-one- to fifty-year-olds make up 39.6%. Together, these two groups make up 74.6% of the total. People aged thirty years old or younger and older people aged from fifty-one to sixty years old, however, make up only 13.2% and 11.7% of the high-income group, respectively. In the high-income group, the proportion of young people and old people are very low. Based on that, we could determine that the special period of system transformation in China is over. It is widely acknowledged that in a market economy, the correlation between age and income forms an inverted U-curve, with newly employed young people having lower income, those around forty-five years old having the highest income, and the elderly having lower income. This indicates that income is closely related to market competitiveness: middle-aged people, with their rich experience, are the most competitive and often have the highest incomes. Considering this correlation between age and income, China can therefore be said to have already reached a stage where resources are allocated according to the law of the market. The relationship between age and income for the five vocational groups or classes in the small sample survey is shown in table 8-3.

Table 8-3 Percentage of Five Vocational Groups/Classes by Age (%)

Age	Private Entrepreneurs	Top-Level Management Staff	Middle-Level Management Staff	Professional Technicians	Ordinary Office Workers
<=30	2.3	6.3	15.6	18.2	47.0
31~40	31.8	34.3	43.8	54.5	29.4
41~50	61.4	40.5	31.2	18.2	11.8
51~60	4.5	17.7	9.4	9.1	11.8
>=61	0	1.2	0	0	0
Total	100.0	100.0	100.0	100.0	100.0
Sample Size (Number of people)	44	80	33	12	18

Secondly, let us study the relationship between education and income. Who has the higher income, the well-educated or the poorly educated? In the early years of economic reform in the 1980s, when "small private-business owners," "ten-thousand-yuan-income households" and private entrepreneurs began to come into their economic strength, their low educational level and high income formed a striking contrast. At that time, those who ran private businesses and enterprises were usually poorly educated people, whereas the well-educated cadres or intellectuals still worked within the planned economic system and earned a relatively low income. This prominent phenomenon, known as "the reversal in wages between mental laborers and physical laborers," caused strong discontent among the well-educated. The opinion that "only the riff-raff make good money" ("不三不四的挣大钱") spread far and wide in the society. It was not until the mid-1990s, when fundamental changes took place in the relationship between education and income, that the educational level of the highest income group was significantly raised.

To begin with, the educational background of private entrepreneurs—the group with the highest-income in the late 1990s and twenty-first century—has changed greatly, forming a distinct contrast with the past. According to a nationwide sample survey conducted in 1993 by the Research Group on the Private-Entrepreneur Class in China,[58] only 5.5% of private entrepreneurs had any undergraduate or graduate education. Among them, only 0.6% had a master's degree; 11.7% of them had graduated from junior college; 35.9% had graduated from senior high school, vocational high school, or technical school; 36.1% (the largest proportion in the samples) had graduated from junior high school; 9.9% had graduated from primary school; and 1% were illiterate. However, this changed completely in the twenty-first century. A nationwide sample survey of private entrepreneurs by CFIC in 2004 showed that they tended to have much better schooling than they did in 1993.[59] From the 2,998 valid samples, 20.7% had received undergraduate education or above, which was nearly four times as many as in 1993. Among them, 5.7% had a master's degree, which was nearly ten times as many as in 1993; 31.1% of them had graduated from junior college; 33.6% had graduated from senior high school, vocational high school, or technical schools; 12.9% had graduated from junior high school; and only 1.7% had just primary schooling. Their educational level was undoubtedly much higher than the average educational level of the working population nationwide. However, within

58 *Private Economy in China Yearbook 1978–1993*. Hong Kong: Hong Kong Economic Guide Press, 1994. 117.
59 *Private Economy in China Yearbook 2002, 2004–06*. Composed by the Chamber of Commerce and Industry of China, et al. Beijing: China Zhigong Press, 2005. 30.

that group, a comparison by vocation reveals that better education did not necessarily equate to a higher income. On the contrary, the private entrepreneurs with the highest incomes tended to have less education when compared with other professional groups (table 8-4).

As shown in table 8-4, education was neither directly proportional to income, nor did related tests show a positive correlation. Compared with the middle-level management staff, top-level management staff had less schooling. I have deduced that this was simply because of the characteristics of the period during which they received their education. It was not until recent years that the higher educational system in China underwent rapid growth, while most of the top-level management staff received their education decades ago. So, it is natural that they had a lower educational level than the new generation did. Moreover, though professional technicians were paid much less than private entrepreneurs and top- and middle-level management staff, they were of a much higher education level. Briefly put, within the high-income group the influence of education on income was not significant.

Table 8-4 Comparison on Education Level of Five Vocational Groups/ Classes (%)

Schooling	Private Entrepreneurs	Top-Level Management Staff	Middle-Level Management Staff	Professional Technicians	Ordinary Office Workers
J	2.3	0	0	0	0
S/V/T	4.5	1.2	0	0	0
TS	4.5	5.0	0	0	5.5
JC/U	61.4	55.0	39.4	41.7	66.7
Master/PhD	27.3	38.8	60.6	58.3	27.8
Total	100.0	100.0	100.0	100.0	100.0
Sample Size (Number of people)	44	80	33	12	18

J: Junior High School;
S: Senior High School;
V: Vocational High School;
T: Technical School;
ST: Secondary Technical School;
JC: Junior College;
U: Undergraduate

In studying private entrepreneurs, one needs to look at their social origins. Which social classes did the highest-income class originally come from? If we compare the social origins of the private entrepreneurs before and after the economic reforms, we find that a prominent change took place. Similar to the above analysis of education, in the early years of the reforms, those traditionally high in social positions were rarely found self-employed in their own businesses, private businesses, or private enterprises. The pioneers who dared to do business in the market economic system were mostly people on the margins of society. The investigation in 1993 showed that many private entrepreneurs in the late 1980s to early 1990s were former farmers and workers. Of the rural private entrepreneurs, 53.5% used to be farmers; of the urban private entrepreneurs, 25.2% used to be workers and 17.2% farmers; the two made up 42.4% of total urban private entrepreneurs. What professions and which social classes, then, did private entrepreneurs come from in the twenty-first century? According to surveys conducted in 2004 and 2006, the proportion of private entrepreneurs who formerly worked as administrators, cadres, or professional technicians had increased significantly. A survey in 2004 indicated that 15.6% of private entrepreneurs were formerly the heads of state enterprises or public institutions,[60] 10.4% of them used to be ordinary cadres, 14.2% had been contractors or tenants of state-owned or collective-owned enterprises, 2.8% had been village cadres, 1.4% cadres at county level, 13.0% professional technicians, and 10.6% supply and sales staff: in total, 68% of private entrepreneurs used to be administrators, cadres, or professional technicians. A survey conducted in 2006 showed that 67.4% of private entrepreneurs were formerly cadres, heads of state-owned or collective-owned enterprises, professional technicians, supply and sales staff in companies, or individual businessmen.[61] The vocational evolution in the social origins of private entrepreneurs reflects another side of the great change in group composition that occurred during the reforms over a period of twenty years.

60 *Private Economy in China Yearbook 2002, 2004–06*, Composed by Chamber of Commerce and Industry of China, et al. Beijing: China Zhigong Press, 2005. 30.

61 *A General Report and Analysis on the Data Extrapolated from Sample Survey on Private Enterprises in China 2006 (II)*. By Research Team of the Sample Survey on Private Enterprises in China. Published at *Internal References of Private Economy, 2007-03-23*. Ed. Research Association of China's Private Economy.

4. The High-Income Group and Housing

After over thirty years of reform, Chinese society has substantially changed such that ordinary people have started to possess their own property. It was proposed by the CPC's Seventeenth National Congress in 2007 that wider ownership of property should be facilitated. *The Property Law of the People's Republic of China* passed in the same year also provided a specific legal protection of private property.

What, then, is private property? It is largely represented by housing. Housing reform in the urban areas of China has been a hard process, involving a change from the public ownership of housing before the economic reforms to the private ownership of today. Various measures have been taken by the central government to advance housing reform in urban areas, such as the policies of "pilot house sales" ("试点售房") and "subsidies for raising rent" ("提租补贴") that came in the early stages of the reform, and the "buying instead of renting" ("以售带租") policy that came later on. It was not until 1998 that substantial housing reform took place. That year witnessed the issue of the State Council's 23rd document, which explicitly marked the end and abolition of nationwide welfare-based distribution of public housing and its replacement with monetary-based housing distribution. Because of the inertia of the old system, the real abolition of the welfare housing distribution system varied from area to area and it took two more years to see its actual end. Regardless, the traditional practice of welfare housing distribution had completely disappeared by the time that China had entered the twenty-first century. Most public welfare housing was sold to employees based on their length of service and work-related contributions, or to whomever was dwelling there, at a discount. Since then, China's urban citizens have bid farewell to public housing ownership and walked into an epoch of private housing ownership. According to the *Statistical Report on the General Situation of China's Urban Housing in 2005* by the Ministry of Construction, the proportion of houses that were privately owned was about 72% in 2002 and as high as 81.62% in 2005. Who possessed these privatized houses? Statistical data indicate that most of them were sold to ordinary employees through housing reform, which was referred to as "(privatized) housing from public housing." According to statistics, by the end of 2003, the total floor space of "housing from public housing" was about eight billion square meters, or 67.54% of the total floor space of all housing in China. Although, the ratio of "housing from public housing" to other forms of housing varied in different cities, there was consistently more than 50% of "housing from public housing" in every city. Therefore, it can be said that most of the private housing owned by ordinary people today came from the housing reform. Employees became house owners of formerly publicly owned houses at cheap prices, so the housing reform was indeed an extensively beneficial reform.

At the same time, per capita floor space, of course, has been increasing. Before the economic reform in 1978, the *per capita floorage of city and township residents* was only 6.7 square meters. It also grew before the housing reforms took place, in the 1980s, but rather slowly. It was not until the marketization of the housing system in the 1990s that the per capita floor space grew quickly. In 2000, it reached 20.3 square meters. In 2007, the *per capita floorage of city and township residents* had reached 28.2 square meters. Additionally, the quality of housing and facilities has greatly improved.

The No. 18 Document issued by The State Development and Reform Commission in 2003 proposed changing the basic structure of housing supply so that "most families will gradually purchase or rent ordinary commercial housing" ("逐步实现多数家庭购买或承租普通商品房"). It meant that most families' demands for housing would be met by the market. Thereupon commercial housing became the major means by which citizens came to own a house. Of course, this did not exclude half-commercial housing such as "economically affordable housing" ("经济适用房") or "price-capped housing" ("限价房"). In this way, the housing of *city and township residents* went from being provided by the government or a work unit to being purchased by the individual. Table 8-5 below shows the fast transition of housing into an individually purchased commodity from the 1990s to early 2000s.

Table 8-5 Statistics of Individually Purchased Commercial Houses in China (1991–2003)[62]

Year	Total floor space of commercial housing purchased nationwide (10,000 m2)	Floor space of the individually purchased housing (10,000 m2)	Percentage of the individually purchased housing (%)
1991	2,745	926	33.73
1992	3,812	1,456	38.20
1993	6,035	2,943	48.77
1994	6,118	3,344	54.66
1995	6,787	3,344	49.27
1996	6,898	3,666	53.15
1997	7,864	5,233	66.56
1998	10,827	7,792	71.97
1999	12,997	10,408	80.08
2000	16,570	14,464	87.29
2001	19,938	18,250	91.53
2002	23,702	22,793	96.16
2003	29,778	28,714	96.43

As demonstrated in table 8-5, since the 1990s, more and more commercial housing has been privately bought by citizens instead of being purchased by work units or the government, as in the past. As early as 1991, only 33.73% of commercial houses were sold to private buyers. But, the ratio of individually purchased housing grew rapidly in the middle and late 1990s until it reached 96.43% in 2003. Work units and public institutions were, therefore, generally no longer involved in purchasing houses. After 2004, housing prices rose rapidly. Nowadays, in big cities like Beijing, Shanghai, Guangzhou, and Shenzhen, an apartment costs between one million yuan and ten million yuan, as do commercial properties purchased by residents. However, for ordinary citizens whose houses are just used for living in, rather than for speculative trading, there is no benefit in this price surge. For the high-income group, since they may own multiple apartments, there is the opportunity to make huge commercial profits by treating housing as an investment. *The Report on Life in China, 2006* published by the Research Association for Contemporary China and the Master Card

62 Source: State Bureau of statistics of China. *China Statistic Yearbook 2004.* Beijing: China Statistics Press, 2004. 254.

International Organization claims that more than 40% of those whose yearly income exceeds 110,000 yuan own two or more apartments and that 22% of this group is planning to buy more houses in the coming year.[63] What, then, is revealed about the high-income group's housing in the small sample survey I conducted?

As is shown in table 8-6, the two most prominent vocational groups within the high-income group—private entrepreneurs and top-level management staff—prevail in possessing commercial housing. Over 90% of private entrepreneurs own two or more properties; 67.4% own three or more houses, 30.2% own four or more, and a few even have ten. Among top-level management staff, 83.2% have two or more apartments, nearly 50% have three or more, and 18.2% have four or more. However, most professional technicians have only one residential apartment.

Table 8-6 The House Owning Rate Among the Five Vocational Groups (%)

How Many Houses / Apartments Do You Have?	Private Entrepreneurs	Top-Level Management Staff	Middle-Level Management Staff	Professional Technicians	Ordinary Office Workers
1	9.3	16.8	37.5	66.7	72.2
2	23.3	33.8	62.5	33.3	16.7
3	37.2	31.2	0	0	0
4	9.3	9.1	0	0	11.1
5	11.6	7.8	0	0	0
6	2.3	1.3	0	0	0
8	2.3	0	0	0	0
10	4.7	0	0	0	0
Total	100.0	100.0	100.0	100.0	100.0
Sample Size (Number of People)	43	77	32	12	18

Very few individuals did not answer this question.

So, the data obtained from this small sample survey reveal that private entrepreneurs and top-level management staff, representatives of

63 The Embarrassment behind the High Private Housing Rate: Public Housing Is the Short Slab in Construction. Web: Window of Houses. July 10, 2006. Source: *Shenzhen Economic Daily*. http://www.ehomeday.com/NEWS/2006-7/20067102733.htm

the high-income class, obtained a great many houses within a short time. This means that the aforementioned data regarding per capita floorage no longer make sense. Superficially, the *per capita floorage of city and township residents* had reached 28.2 square meters, but it would not make any sense to discuss *per capita floorage* if the richest possess as many as ten houses while the poorest cannot afford even one. The price surge in urban housing in recent years seems to benefit all house owners, but, actually, most ordinary employees who have bought a "house from public housing" will not choose to sell their houses, and so would not benefit from this. However, many private entrepreneurs and top-level management staff possess a lot of houses that are not used for residence but as investments or for commercial transaction. The price inflation of housing virtually matches the inflation in the number of house-related business transactions. The gap between the rich and poor, reflected in housing conditions in China, is significantly large.

Nevertheless, as has been repeatedly emphasized above, the housing reforms of the 1990s were both generally beneficial and caused wealth disparity. The housing reform in the 1990s was beneficial to middle-aged and elderly ordinary employees, yet it also accelerated wealth disparity as the rich classes came to possess so much property that housing prices soared.

Chapter 9

Theories of the Middle Class and the Status Quo of the Middle Class in Western Countries

Recently, the middle class has become a prominent issue in China, a trend linked to the rapid growth of this important stratum in the country. It was after World War II in the United States and Europe, and after the creation of economic prosperity among the so-called "Four Little Dragons" in Asia in the 1970s, that research on the middle class became widespread. Therefore, merely by reflecting on the degree to which research has been focused on this group, one can infer that the middle class or middle stratum in China has finally entered a phase of rising to ever greater prominence. This trend has been caused by China's rapid economic growth over the past thirty-plus years.

Before probing into China's middle class, it is necessary to define what is meant by "middle class," explain the relevant theories, and summarize the development of the middle classes in the West. After the general state of the middle class is understood, we shall investigate to develop a clearer picture of the middle class in China in this chapter.

In Chinese "*zhong chan jie ji*" ("中产阶级") is the most common translation of the term, "middle class." In Chinese, the technique of Chun Qiu ("春秋") writing (a writing style of the Spring and Autumn period characterized by a detailed description or narration of political events with praise and blame implied) has produced many other terms such as middle stratum, middle level, middle grade, intermediate stratum, intermediate class, and social middle layer, which can all be used in different circumstances. All these terms have different precise meanings in Chinese but they are covered by one single noun phrase in English. They are therefore treated equally in this book. Basically, this book will utilize two terms—the middle class and the middle stratum—to address international and domestic issues respectively.

Past theoretical research into the enormous middle stratum has been insufficient. Some researchers have treated it with simple negation, holding that the term "middle class," or "middle stratum," blots out class boundaries. However, from the perspective of sociology, the existence of the middle class is a social fact; regardless of how it is termed, it definitely has functions in relieving social contradiction, which makes the study of its functions necessary. Since Western countries can use this intermediate structure to maintain societal stability, why cannot China use it in the same way?

Since the middle class is a concept foreign to China, it is necessary to introduce the various theories developed in the West relating to the concept of middle class.

1. The Origin of Theories of the Middle Class

Theories of the middle class deal with a trend common in modern society—the effects of a constantly growing middle class. When the majority of the population are members of the middle class or the middle class is able to dominate a society, the lower classes, or the low-income earners, will become a minority group. At this point, the society becomes characterized by stable structures because the middle classes tend to identify strongly with dominant societal values, serving to alleviate social opposition and contradiction. However, if the middle class accounts for only a small proportion of the population, the lower classes are certain to be large in size, which may put the society at risk of instability because the low-income groups tend not to identify strongly with society and are more likely to develop anti-social sentiments. In addition, without the middle class acting as a buffer, contradiction and conflict can easily erupt between the other two classes.

Karl Marx, a renowned theoretician, realized the aforementioned social function of the middle class relatively early. Of course, he elaborated on that in a negative view of middle class theory. He pointed out in *The Communist Manifesto* that the collapse of the intermediate stratum, leading to the majority of the intermediate stratum joining the proletarian group, will cause social polarization—a trend he saw as a necessary precondition for major social conflict, instability, and revolution in capitalist society.[64] The "intermediate stratum" is the term used in the Chinese version of this masterpiece, while the term "middle class" is the term used in the English version. As a result of Marx's

64 Marx and Engels. *Selected Works of Marx and Engels*. Vol.1. Beijing: People's Publishing House, 1972. 259, 261–263.

theories, Marx and other scholars came to agree that the loss of the middle class would intensify social contradiction.

What especially captures our attention is that Marx twice put forward the point in *Theories of Surplus Value* that the middle class will continuously expand and asserted that the development process of bourgeois society is to that end alone.[65] Marx criticized scholars who held opposing views. He also mentioned this trend in *Das Kapital (III),* in which he indicated that the expansion of industry and business requires a tremendous increase in the demand for management, leading to the emergence of a class of enterprise and business managers,[66] but Marx did not elaborate on the social position of this stratum.

Of course, the one who truly expounded on the notion of the middle class was not Marx; from the perspective of the literature, Eduard Bernstein is the writer who adapted Marx's theories.

Eduard Bernstein set forth the earliest set of comprehensive views on the middle class in a speech in 1909. On the basis of contemporary salary statistics in Germany, he indicated that statistical analyses of salary trends showed an increase, rather than a decrease, in the size of the high-income and moderate-income groups and that the growth of the two groups noticeably outpaced that of the population.[67] He put forward the notion that the polarization in social relations predicted by Marx was not taking place, and that "there is no remarkable narrowing of any stratum between the working class and the very wealthy people;"[68] he also vividly illustrated middle class expansion by using graphs of the social structure. Judged according to his words, Bernstein believed that class conflicts tend to ease off rather than escalate.

The German sociologist Emil Lederer first expanded upon views of the middle class from an academic perspective. In 1912, he published his first work on the middle class, *Private Sector Clerical Workers in Modern Economic Development,* in which he presented research on a new social stratum that included staff members, technicians, office workers, and so forth. Lederer named them "the new middle class," predating by forty years the first references to "the new middle class"

65 Marx. *Theories of Surplus Value (III).* Beijing: People's Publishing House, 1976. 63.
66 Marx. *Das Kapital (III).* Beijing: People's Publishing House, 1975. 437.
67 Research Office of the International Communist Movement of the Compilation and Translation Bureau of the CPC Central Committee. *Debate on Eduard Bernstein by German Social Democratic Party.* Beijing: Sanlian Bookstore, 1981. 22.
68 Research Office of International Communist Movement of the Compilation and Translation Bureau of the CPC Central Committee. *Debate on Eduard Bernstein by German Social Democratic Party.* Beijing: Sanlian Bookstore, 1981. 25.

by American sociologist C. W. Mills. Lederer held that the new middle class, comprised of staff members and technicians, was a special social stratum with special class interests distinct from those of industrial workers, leading labor unions to distance themselves from this new class. He thought that the traditional working class was made up of manual workers who differed greatly from the new middle class. Afterwards, Lederer collaborated with others to publish *New Middle Class* (1937), a work that fully explored the origin, development, social status, and social function of the middle class.

After World War II, due to the war's tremendous effect on social structure, the number of white-collar workers continued to grow, making the middle class a popular subject of research. During this period, many fine academic works were published on the topic. The most important works are Mills' *White Collar: The American Middle Classes*, David Lockwood's *The Black Coated Worker*, Theodor Geiger's *Class Society in the Crucible*, and Fritz Croner's *The Salaried in Modern Society*, which present in-depth research on this newly arisen social stratum. Four characteristics are common to all these works. Firstly, based on a large number of empirical studies, the European and American middle class can be considered to have been researched relatively comprehensively; secondly, all groups composing the middle class were classified for this research, including professional technicians, managers, and office workers; thirdly, the characteristics, lifestyles, and social status of the middle class were analyzed; and, fourthly, schools of thought concerning the middle class were developed.

A core issue of discussions dealing with the middle class is, from the perspective of social stratification, determining how to incorporate the middle class into the existing social structure; that is to say, where do we locate the middle class in the social hierarchy and what is its relationship with the other classes? Two schools hold opposing opinions on this issue. One thinks that the middle class is closer to the bourgeoisie in status—forming an extension of the bourgeoisie (or the ruling class) because the middle class, by and large, undertakes four functions in the division of labor: supervision, providing expertise, administration, and sales. These functions have one thing in common: they were undertaken by employers as part of their work in the past (and today in some cases). As a result, the intermediate class is categorized as being part of the ruling classes. Fritz Croner is a representative of this school. The other school maintains that the main role of the middle class is to serve as employees, so they are closer in status to the working class. Theodor Geiger and C. W. Mills are two representatives of this school. Mills holds that the new middle class is made up of wage earners in an economic position belonging to an employee stratum without independent property and that, "from their objective status, we can see that the mass of white-collar workers has become increasingly similar to average workers with respect to their position in structure, being without property and with increasingly similar incomes. All the status factors that distinguish white collars from average workers have been irreversibly weakened until now."[69] Of course, Mills is also aware that, despite having the same employment status, members of the middle class do not identify with the working class—which, he says, leads to their inner feelings of being at odds with their economic status.

These various views of the middle class are largely a result of the vast interior diversity of the middle class, making the problem more complicated. Therefore, whether the middle class is one unified group or made up of a number of subsets is a topic very worthy of discussion.

2. A United Middle Class or a Divided Middle Class

The significance of studying whether the middle class is a united group or a divided one is closely associated with the social function of the middle class. If it is a united group, its function of stabilizing society will have a sturdy foundation. If it is divided, the group itself will experience conflicts and they will struggle to stabilize society.

Since the 1960s and '70s, a good many scholars have taken up the investigation of this issue, leading to substantial differences of opinion to which this section provides a brief introduction. The influential British scholar, Anthony Giddens, has earned worldwide fame thanks to his research in recent years on "the constitution of society," and "the third way," but in the 1970s his major research subject was social stratification. Giddens's thoughts about the middle class are mainly reflected in his book, *The Class Structure of the Advanced Societies*, in which he develops a standard for defining the middle class—market capacity. He asserts that the three social classes should be differentiated based on

69 C. Wright Mills. *White Collar: The American Middle Classes.* Hangzhou: Zhejiang People's Publishing House, 1987. 333.

market capacity and suggests that the possession of the means of production, the possession of education or technology, and the possession of manual labor are the three most important forms of market capacity. This gives birth to the three basic classes of developed capitalist societies: the upper class, the middle class, and the lower class. He holds that there are clear-cut boundaries between the three sorts of market capacity and that the boundaries defining the middle class are equally clear. What distinguishes the upper class from the middle class is the fact that the upper class possesses the means of production while the middle class does not. What distinguishes the middle class from the lower class is the fact that the middle class possesses education and technical certifications while the lower working class does not.[70]

But if Giddens sees such immense diversity within the middle class, why, then, does he persist in considering the middle class to be united? He believes that the diversity of social groups stems from two aspects: one being differences in market capacity and the other being differences in divisions of labor. Market capacity is defined as technical knowledge required by the market and technical abilities acknowledged by society. Does this mean that anyone possessing these abilities belongs to the middle class? Giddens certainly does not think so. The middle class prevents members of other classes from joining by means of social closure. The market capacity of technical work is limited by social closure in form of professional gate-keeping of admittance and all specific technical associations exclude members of other classes. He said that division of labor does give rise to differences, but that these cannot be considered class differences when compared with those that market capacity brings forth. Therefore, despite the tremendous differences within the middle class, the far more important differences are the results of the division of labor. Giddens points out, for example, that from the perspective of labor division there are huge differences between a draftsman and a social worker while, from the angle of market capacity, they are similar in many respects.[71] He also criticizes Dahrendorf's views on the middle class, as he believes that what the middle class shares is its role as a class to serve and establish a bridge between the rulers and the ruled. As far as Giddens is concerned, this definition can never be used to distinguish the middle class from other classes. All in all, by drawing from the theory of market capacity, Giddens has explained why the middle class is a united class.

70 Anthony Giddens. *The Class Structure of the Advanced Societies*. London: Hutchinson, 1973. 107–108.
71 Anthony Giddens. *The Class Structure of the Advanced Societies*. London: Hutchinson, 1973. 186–188.

Nevertheless, there are few proponents of Giddens's viewpoint of a "united middle class" and a good many opponents. For instance, K. Roberts, F. G. Cook, S. C. Clark, and E. Simonov often refer to "the fragmentary middle class," believing the so-called middle class has gradually divided into different strata located at distinct positions in the social stratification system. They conducted an empirical study of stratum awareness among British white-collar workers. Their results show that firstly, only 27% accept the concept of "the big middle class" and consider themselves to be members of it. "The big middle class" refers to the idea that the middle class forms the main body of society, with a small number of people in both the upper and the lower classes. The real income of this group of white-collar workers actually stands in the middle. Secondly, 19% believe that the middle class is quite a small social stratum, with an enormous working class underneath and an upper class with a small number of people on top. They think they belong to the marginalized middle class and most of them are small businesspeople. Thirdly, 15% hold that there are a number of rungs on the social ladder—more than four, for example—and that no class in the middle exists or that there is no class stratification whatsoever. This group of people consists of well-educated and well-paid white-collar workers possessing marketable expertise. Fourthly, 14% believe that the main body of those constituting the social classes is the working class, which lies at the bottom of the stratification system. This group thinks that it is part of the working class and that its position differs a lot from that of upper management and professional technicians. This group of people is made up of low-status white-collar workers with low pay and few opportunities for promotion.[72] Accordingly, their conclusion is that the middle class will become increasingly fragmented and gradually divide into different classes. Nicholas Abercrombie is even more radical, thinking that the middle class is moving towards a state of polarization with one part rising to the top and the other falling to the bottom. This opinion coincides exactly with Marx's view of the middle class' proletarianization. Generally speaking, those who deny that the middle class is a united group tend to treat society with varying degrees of radicalism.

John H. Goldthorpe, a very important British scholar of social stratification, also disagrees with theories of a united middle class. He names the class at the top: "a service class," which includes advanced professional technicians, management staff, officials,

72 K. Roberts, et al. *The Fragmentary Class Structure*. London: Heinemann Press, 1977.

managers of large-sized corporations, large land owners, and so forth. In other words, the middle class is conjoined with the upper class. Goldthorpe disapproves of the use of the term "middle class" but does employ the phrase "intermediate classes." His so-called intermediate classes include office workers, private service providers, petty proprietors, petty businessmen, and general technicians, among others, whose position is subordinate to that of "the service class." In his subsequent research, Goldthorpe then came to believe that, due to the differences in employment states and employment relations—such as in pensions, opportunities for promotion, and pay raises—there are vast differences between the different strata. He therefore tends to further divide the concept of class and regard class structure as a transitional mode with multiple classes instead of a united middle class. Of course, his opinion is fairly moderate, reflecting a new manner of approach to an easing of social contradictions.

3. Research on East Asia's Middle Classes

3.1 Principal Aspects of Hsin-Huang Michael Hsiao's Research on East Asia's Middle Classes[73]

Hsin-Huang Michael Hsiao was the first to analyze the historical background of the new East Asian middle classes. From the 1960s to the 1990s, some East Asian countries and regions—including Singapore, Hong Kong, Taiwan, and South Korea—experienced rapid economic expansion, giving rise to newly industrialized economies in which a "new middle class" of staff members, management, and technicians could grow and the "old middle class" of petty proprietors and petty bourgeoisie were impelled to strengthen their positions. East Asian middle classes have played significant roles in politics in society. Hsin-Huang Michael Hsiao's research on East Asian middle classes is based on a large-scale empirical investigation of Taiwan, Hong Kong, South Korea, and Singapore, that mainly focuses on four aspects.

The issue of class boundaries is examined, dealing with the question of how to distinguish the middle class from other classes. He also investigates the social characteristics of East Asia's middle

73 Hsin-Huang Michael Hsiao, ed. *East Asian Middle Classes in a Comparative Perspective*. Taipei: Institute of Ethnology, Academia Sinica, 1999. 3-4.

classes in terms of family background, marriage, and homogeneity. Another important feature of his study is his attempt to determine how East Asia's middle classes self-identified in terms of class. He also develops a general picture of their lifestyles. Finally, Hsiao examines these groups' political attitudes towards the concepts of nation, social justice, and democratic transition, as well as how active they are in political movements and voting.

3.2 Hsin-Huang Michael Hsiao's Analyses of East Asian Middle Class Characteristics[74]

This research introduces five key characteristics of East Asia's middle classes.

To begin with, the middle classes in East Asia are intrinsically intergenerational. The first generation of East Asia's middle classes was born after World War II, mostly coming from worker and farmer families. For example, in South Korea, 33.3% of those whose parents were farmers became part of the new middle class and 51.4% joined the old middle class. In Hong Kong, 36.1% of the children of workers became new middle class members and 44.1% became part of the old middle class. In Singapore, 32.7% of the progeny of worker families entered the new middle class and 46.2%, the old middle class. This generation led a hard life when young and later lived in multi-class families; as a result, the first generation of the middle class had vague ideas about class identity and how to treat other classes, and tended to side with workers in labor disputes. For instance, those who identified themselves as members of the working class accounted for 15.9% of the new middle class and 24.9% of the old middle class in South Korea, 10.3% of the new middle class and 28.8% of the old middle class in Hong Kong, and 31.7% of the new middle class and 31.3% of the old middle class in Singapore. Therefore, a large proportion of the middle class did not agree that their status was middle class, seeing themselves as working class. This research also proves that this mindset has a great deal to do with their family origin.

74 Hsin-Huang Michael Hsiao, ed. *East Asian Middle Classes in a Comparative Perspective.* Taipei: Institute of Ethnology, Academia Sinica, 1999. 4–48.

It is important to remember that the middle class is the main beneficiary of East Asian economic growth and the benefits of market development have allowed them to enjoy higher living standards. The class has developed a system of symbols to indicate its social position. These individuals tend to search for spouses within their own class and are inclined to be conservative in order to maintain their social position. The aforementioned economic characteristics of East Asia's middle class are shown in table 9-1.

Table 9-1 Some Characteristics of East Asia's Middle Classes

Percentage of Those with the Following Characteristics	South Korea		Hong Kong		Singapore	
	New Middle Class	Old Middle Class	New Middle Class	Old Middle Class	New Middle Class	Old Middle Class
Master's Degree or Higher	70.5	23.6	31.7	3.0	15.9	6.1
Monthly Income Over USD1,250	84.5	74.1	91.2	66.7	69.8	48.5
Annual Household Income over USD36,000	63.6.	54.1	74.7	26.2	44.6	27.3
Own Car(s)	61.4	41.3	20.4	34.8	53.7	63.6
Have Credit Card(s)	87.3	53.8	86.4	59.1	64.9	39.4
Own a House	63.2	48.0	59.2	54.5	20.4	87.9
Own a Second House	5.9	1.8	23.3	19.7	3.0	18.2

Members of this class tend to think that their status is on the rise and are becoming aware of their own independent interests and taking up a strategy of favoring neither labor nor capitalists. Self-proclaimed defenders of democracy, they carry forward the environmental movement, the women's movement, the consumer protection movement, and actively participate in the democratic process. Governments, capitalists, and labor therefore all seek to engage with this class.

Questionnaire surveys have inquired about what middle class people themselves think about their role in society and it turns out that that they offer positive replies to questions emphasizing their role while they tend to be negative when assessing how much they care about politics. See table 9-2.

Table 9-2 Self-Evaluation of Social Roles by East Asia's Middle Classes

Percentage of Those Agreeing with the Following Statements	South Korea		Hong Kong		Singapore	
	New Middle Class	Old Middle Class	New Middle Class	Old Middle Class	New Middle Class	Old Middle Class
The middle class contributes a lot to society.	53.6	52.2	85.0	87.7	65.1	60.6
The middle class is the pioneer of democracy.	54.5	46.2	50.0	53.6	41.3	30.3
The middle class pays more attention to political freedom than to economic growth.	25.0	33.6	12.4	16.1	23.5	15.2
The middle class is not concerned with politics.	42.3	54.5	28.3	24.6	24.2	28.1
The middle class faces excessive taxation.	59.5	38.8	73.5	65.5	43.7	48.5

This research also raises the issue of whether the middle class is a united class or a divided one. Research proves that vast differences exist within East Asia's middle classes, which are made up of a good many strata, including the new middle class (employed professionals, technical workers, and management), the old middle class (petty employers and petty bourgeoisie), and the marginal middle class (operational non-manual workers). The new middle class is characterized by their sympathy for the working class, the old middle class by conservatism, and the marginal middle class by a tendency to take collective action.

When class origin is analyzed, it is found that (compared with the new middle class) more members of the old middle class and the marginal middle class come from workers' and farmers' families. When the middle classes evaluate their own class status, a higher proportion of the old middle class and the marginal middle class regard themselves as being of the working class and, therefore, an inferior class.

In terms of market capacity, the new middle class is obviously equipped with more education than the marginal middle class and has a much higher monthly and yearly income than both the marginal middle class and the old middle class. In terms of cultural activities and lifestyle, the new middle class has a greater inclination towards modern lifestyle trends such as going to concerts, playing golf, joining social clubs, traveling abroad, and so forth.

The middle classes of various countries and regions in East Asia differ from each other because of cultural diversity. The middle classes of Singapore and Hong Kong are politically conservative and show little interest in political movements while the middle classes of Taiwan and South Korea have a high level of political consciousness and often attempt to transform the political system in areas connected to them. For instance, Singapore's middle class acts moderately in terms of supporting labor and opposing major capitalists, showing a firm stand in their support for the government and a minimal interest in getting involved in politics.

3.3 Cheng-Shu Kao's Research on the Middle Class in Taiwan[75]

Cheng-Shu Kao has researched small and medium-sized enterprises in Taiwan in order to determine the four basic characteristics of the middle class in Taiwan.

The first characteristic is a high degree of mobility. Professor Kao discovered that the middle class in Taiwan is able to move relatively freely between government organizations and privately owned enterprises—and vice versa—with the former being the main direction for this flow. This is somewhat similar to the trend of government officials going into business in mainland China. This fluidity can be more obviously perceived when compared with Japan and South Korea. Japan emphasizes lifetime employment, allowing the middle class little fluidity, while South Korea is characterized by large enterprises, which also have low mobility rates. In Taiwan, a striking example of mobility is that many people begin their careers as employees before going on to start their own businesses. In the course of their employment by a company, they acquaint themselves with business, methodologies,

75 Cheng-Shu Kao. "Observations on the Middle Class in Contemporary Taiwan." *Middle Class in Changing Taiwanese Society.* Ed. Hsin-Huang Michael Hsiao. Taipei: Juliu Publishing House, 1989. 21–31.

and market networks before they become the chairmen of their own companies. A large number of chairmen in Taiwan have demonstrated a high level of mobility. Ninety percent of Taiwanese enterprises are small and medium-sized. What Professor Kao depicts above is also common in contemporary mainland China among many chairmen and bosses, which may have some connection with the Chinese pattern of familial relations playing a key role in economic development.

Another type of mobility that Professor Kao mentions is longitudinal mobility, indicating that small and medium-sized business owners—most of whom come from farmers' or workers' families—worked for others in the past before becoming self-employed. By means of running businesses, they become petty employers and accumulate wealth, with an annual turnover of three to four hundred million Taiwanese dollars.

The second characteristic is vagueness, which is related to mobility. Because of its high degree of mobility, the Taiwanese middle class shows a comparatively low degree of structuration. This means its members have indistinct ideas about class identity: members of the first generation in particular tend to identify with the class they originated from.

The third characteristic is diversity. This again involves the issue of whether the middle class is a united class or a divided one. Professor Kao holds that there is a great deal of diversity within the middle class in Taiwan. For instance, small and medium-sized enterprise owners differ a great deal from civil servants; the heterogeneity between them outweighs their homogeneity. Even within the group of small and medium-sized company owners, there are tremendous differences between different industries.

The fourth is realism. This means that their identification and action orientation tend to be based mainly on realistic interests rather than ideology. They tend to focus primarily on obtaining more business opportunities and other advantages. They also tend to be more concerned with economic factors—such as exchange rates, stocks, and the export market—than with political ideology.

4. The Status Quo of the Middle Class in Western Countries

The formation of the middle class in Western countries was part of a prolonged period of historical development. The early Western middle class was academically called the "old middle class" and was mainly composed of petty business owners, petty businessmen, petty employers, and petty landowners. In the late nineteenth century and early twentieth century, capitalism shifted from free competition to a more monopolistic state; small enterprises were swallowed and replaced on a wide scale by large organizations able to establish monopolies. Huge monopolistic organizations in industry, transportation, finance, and business, among others, started to employ a great many white-collar workers engaged in administration, management, sales, accounting, finance, and so forth. Moreover, the rise of joint-stock companies also gradually separated management from the owners, giving birth to the management group. In this period, "the old middle class" gradually went bankrupt one after another and downsized day by day. Conversely, the "new middle class" rapidly became an important group. According to statistics, in 1931, the "white collars" of the "new middle class" in Britain accounted for 23% of the economically active population. The most important difference between the "old middle class" and the "new middle class" was that the former class could not form an enormous group because it was made up of independent proprietors—the employers that hired others—while the latter was composed of the employed and so was able to bloom into a huge social group.

The status of the "new middle class" did, however, decline to some degree, compared with that of the "old middle class." In the rapidly growing group of white-collar workers, only a small proportion of the top ranking figures are able to maintain close contact with business owners and have power over major issues. Most staff members are highly distant from factory owners and the board of directors, making them average employees. Their employment security and their chances for upward mobility decrease, as do their incomes. Take Britain from 1913 to 1978 as an example. In 1913, the average income of a British professional technician was three to four times as high as that of a blue-collar worker, but had dropped to only 1.7 times as high by 1978. In the same period, the ratio of American white-collar workers' incomes to that of blue-collar workers underwent a similar change. Before World War II, the number of professional technical intellectuals in the white-collar stratum—including engineers, teachers, and scientific researchers—was small, but their status was stable.

There was a great degree of employment security and a good many opportunities for upward mobility.

After World War II, the technological revolution and the development of national monopoly capitalism caused the social structure and labor structure to undergo significant changes. As the number of nonmanual workers increased, the white-collar group rapidly expanded and became the dominant group in society.

In America, the total number of white-collar workers rose from 16.08 million in the 1940s to 51.05 million by the end of the 1970s—twice as many as before. Among these, the professional technical group grew the fastest, from 3.879 million to 15.423 million, becoming appropriately three times as large as before. The proportion of white-collar workers rose from 31% of the total labor force in the 1940s to 50.9% in 1979. The same is true of Britain: the total number of British white-collar workers rose from six million in the 1940s to 11.60 million in 1978, at which time white-collar workers accounted for 44% of the total labor force.

By the 1970s, the number of white-collar workers in all major developed countries amounted to 40% or more of the population (see table 9-3).

Table 9-3 Percentage of White-Collar Workers and Blue-Collar Workers in the Economically Active Population in Developed Countries in the1970s[76]

	The U.S.	Canada	West Germany	France	Holland	Sweden	Norway	Japan
Time of Survey (year)	1976	1979	1978	1975	1977	1979	1979	1975
Percentage of White-Collar workers (%)	50.4	46.9	45	40	45.9	46.5	43.1	41.9
Percentage of Blue=Collar workers (%)	43	44.8	53.7	44.8	46.7	50.6	54.2	55.7

White-collar workers stayed the socially dominant group until the end of the twentieth century throughout the beginning of the

76 Source: The Labor Organization of the United Nations. *Labor Statistics Yearbook* 1979–1980. Labor Statistics Handout. Department of Labor of the United States of America, 1977.

twenty-first century (see table 9-4). In 2001, the number of white-collar workers in developed countries generally exceeded 50%, as was the case in the United States, Britain, Germany, and Japan. In the U.S., the proportion of white-collar workers even reached 60%. This clearly demonstrates that in developed countries, white-collar workers have become the dominant group in these societies and, normally, the more prosperous the economy is, the higher the proportion of white-collar workers becomes. This also indicates that, with economic development, an increase in the number of white-collar workers is, to a certain extent, inevitable.

Table 9-4 Percentage of White-Collar Workers and Blue-Collar Workers in the Economically Active Population in Some Developed Countries in 2001[77]

	The U.S.	Britain	Germany	Japan	Italy	Australia
Percentage of White-Collar Workers (%)	59.9	52.2	52.4	51.3	45.3	44.3
Percentage of Blue-Collar Workers (%)	40.1	47.6	46.0	48.0	54.4	56.9

However, with the expansion of the white-collar group, the social status of white-collar workers in the original sense shows a tendency towards further differentiation. On one hand, the development of joint-stock companies and the socialization of capital has led to higher positions for the top levels of management and this group is often actually in charge of large companies and enterprises; on the other hand, the work and income of low-level white-collar workers has grown increasingly similar to those of workers. Moreover, because of the growth of tertiary industries, quaternary industries, and all kinds of "information" industries, those with professional technical knowledge increasingly become those who undertake production. This middle-class group that originally remained at the upper-middle level has also experienced a downturn. As McColloch pointed out, until the 1970s, American engineering technicians had very slim chances for promotion, so their positions tended to remain unchanged over a long period of time, leading them to lose confidence in the process

77 Source: Extrapolated from *Labor Statistics Yearbook*, 2001. United Nations Labor Organization.

of promotion.[78] But, this difference within the white-collar group really indicates continuity and the class's multilayered nature; even white-collar workers at the lowest level usually have some promotion opportunities, meaning that the group retains a certain level of unity and integrity. There is, however, a degree of controversy among Western sociologists regarding this issue.

Throughout the 1960s and '70s, some moderately developed countries, such as South Korea and Japan, formed sizeable middle classes and the trend of their social structure evolution closely resembled that of Europe and America.

Then, under what historical conditions does the middle class develop? What are the fundamental conditions for the formation of a large middle class? It is generally accepted that a number of conditions are necessary. To begin with, the industrial structure must change. When the focus of industrial structure shifts from industrial manufacture to tertiary industries, then information-based industries, banking, financial insurance, communications, education, health care, and public services tend to undergo rapid growth. These newly arisen industries require large numbers of professionals and managers, resulting in the emergence of a group of professional technicians and management. The second condition is the change of occupational structure. Changes in occupational structure are influenced by three important factors besides the aforementioned industrial changes. Firstly, the separation of capital ownership and management must continuously widen, giving birth to a management class with the power to manage property rather than a right to property. Also, as science and technology become increasingly important for social production, professional and technical elites equipped with higher educational degrees and expertise become necessary. Additionally, the widespread use of machinery leads to a massive reduction in the need for physical labor, so professional machine operators increasingly occupy traditional blue-collar jobs. Occurring in concert, these factors bring about a rapid growth in the size of the white-collar stratum throughout society and a relative decline in the number of blue-collar workers. Another necessary factor is the adjustment of income distribution. With the socialization of capital and the decentralization of stocks, big

78 McColloch. *White Collar Workers in Transition: The Boom Years, 1940–1970*. London: Greenwood Publishing House, 1983. 16, 117.

asset owners have a lower proportion of capital income as it is shared between strata. Moreover, guided by the concept of a "welfare state," social security has made tremendous progress; it objectively serves to adjust income distribution, and progressive taxation systems close the income gap further. These systemic improvements all help to generate an enormous medium-income stratum in capitalist societies. The development of education is also an important factor. The appearance of the white-collar stratum requires a corresponding major development in education, a necessity that can be perceived from the levels of education generally possessed by different types of staff. Take the U.S. as an example. The schooling of professional technicians and management normally spans fifteen to sixteen years, that of office workers twelve to thirteen years, and that of blue-collar workers only eleven to twelve years. Contemporary developed capitalist countries all underwent a transition from elite education to mass education. At present, in countries such as the U.S., the percentage of the population receiving higher education is more than 30%. Educational development lays a crucial foundation for the emergence of the white-collar stratum.

The formation of the middle class exerts a far-reaching influence on society, it being an essential force for maintaining social stability in contemporary Western countries. There are a number of reasons for this. Firstly, from the perspective of social structure and social action, the middle class acts as a buffer between the upper class and the lower class, greatly alleviating social contradiction and reducing the likelihood of social conflict. The professionalism of the middle class leads its members to form all kinds of "occupational groups" conducive to communication between the middle class and the lower class. Secondly, the middle class tends to have extensive social contact, providing the basis for different groups to coordinate conflicting interests in an organized manner. Thirdly, ideologically speaking, the central position of the middle class allows the group to represent a moderate and conservative ideology, leaving less room available for radical ideas. This provides the ideological basis for social stability. Fourthly, in terms of economic stability, the middle class is also the main determinant of social consumption. The income stability of the middle class allows for smooth and steady consumption: when the middle class is in the majority, a steady and huge consumer market is guaranteed. This provides the foundation for economic stability.

Looking over the course of history, the middle class holding a dominant position can be seen to have been the most important structural reason for the capability of modern capitalist societies to move towards social stability. This has vital significance for China's current social development.

Chapter 10

The Middle Stratum in China

1. The Middle Stratum Classified from an Economic Perspective

As mentioned at the beginning of the previous chapter, there are various expressions used for the middle class in China; the term "middle stratum" is considered more acceptable nowadays. This term is therefore frequently adopted when analyzing the situation in China.

It can be easily intuited that the middle stratum is the social group whose living standards, property, and status are of an intermediate level. The notion of a middle stratum therefore intrinsically implies the stratification of social groups from high to low. Accordingly, with the middle class, there are the upper and lower classes. Because conceptions of "upper," "middle," and "lower" are themselves relative, definitions of the middle class are comparatively indistinct and there is no clear dividing line between the upper, middle, and lower classes.

Another entirely different method of classifying social groups is to examine social stratification from the perspective of social relationships. For instance, when economic relationships are examined, social groups are divided into property owners and the proletariat, bosses and employees, private entrepreneurs and their employees, foreign enterprise bosses and their employees, state-owned enterprises managers and their workers. Such classification emphasizes the relationships between different groups. According to this method, there can be no transition between different groups and the dividing line is obvious: people are either in one group or another. However, overlapping and transition between the lower, middle, and upper groups obviously exist. The vagueness of this kind of classification of social classes is beneficial to the relaxation of conflicts and the resolving of contradictions, which is necessary for a functioning society. It is therefore understandable that discussions of the middle class have become common in newspapers and that this subject has been widely adopted by people.

Moreover, since the middle class, or middle stratum, comes from the classifications of upper, middle, and lower, it follows that all countries possess these classes. But, the different economic levels that exist in different countries mean that what comprises the upper, middle, and lower classes will differ greatly from country to country.

Since the concept of the middle class itself is comparatively indistinct, various practical definitions of the middle class are available. The simplest kind is a definition based on income. The following table tells us the situations of the middle class in some countries and districts according to people's incomes.

Table 10-1 Indexes of Income Standards and Percentages of the Middle Strata in Some Countries and Regions*79

	Annual Household Income (in Ten Thousands of USD)	Percentage of Middle Class (%)	Percentage of Subjective Identity	Per Capita GDP of 2002 (Calculated Using Current Prices)	Total Population of a Country or Region (in One Hundred Millions)
USA	4~20	80		37305	2.8
Britain		65		24819 (in 2001)	0.59
Germany	3~8	50	75	25427	0.842
Sweden	4~10	55	80	25979	0.0887
Japan	4.4~6.8			37000	1.27
Singapore		90		26500	0.04
Hong Kong	2~4.1			23800	0.0682
Taiwan	2~3.8			12900	0.224
Korea	2~3.6			10260	0.48
India	0.07~0.3	29.2		468	10.27
Mainland China	5.37~16 (in Ten Thousands of RMB)	10	35.9	1000	12.845

Table 10-1 shows that, since conceptualizations of the middle class are relative, the standards for what constitutes a middle class are relative as well. The standard for the annual income of a middle-class family in America ranges from 40,000 to 200,000 USD, while the standard for the annual income of an Indian middle class family is only 700 to 3,000 USD. The standard for the annual income of a middle-class family in China was put forward by the Research Group of

79 Source: Research Team, Academy of Macroeconomic Research, National Development and Reform Committee: *Strategic Research on Enlarging the Proportion of Middle-Income Earners*. Beijing: Academy of Macroeconomic Research Institute, National Development and Reform Committee, 2003. 6–7.

Macroeconomic Research Institute of the National Development and Reform Committee. This organization holds that the annual income of a member of the middle class in China is based on a per capita annual income ranging from 34,000 to 100,000 RMB, such that a per capita disposable income of a middle-class household ranges from 18,000 to 54,000 RMB, and an annual household income ranges from 53,700 to 160,000 RMB.[80] Considered from the perspective of income, these standards can be said to be in accord with China's national situations and seem to be appropriate.

The advantage of classifying and calculating the middle stratum according to income is that such a classification is simple and easy to implement. But, in fact, this means of classification is no more than a statistical game because statistically divided groups are totally different from groups found in real societies. For example, in one department of a university, a teacher's annual income is 30,000 RMB, which is below the annual income standard for the middle class, so this teacher cannot be considered a member of the middle class in the statistical sense. In reality, however, university teachers employ professional techniques and have educational diplomas, so they are typical members of the middle class. For another example, a migrant worker collecting waste in a city may obtain an income of more than 3,000 RMB per month because of his diligence, capability, and popularity; thus, his annual income is above the standard of 34,000 RMB. This worker would therefore be a member of middle class in the statistical sense. But, according to social reality, he is by no means a member of the middle class. So, the middle class in a statistical sense is one thing, but it is something else in a socially interactive sense. The real concern here is the real, socially interactive social group.

This shows why we discuss and study the middle class in China. A real social stratum has its own significance with relation to social interaction, social integration, social attitudes, and social actions. According to the experiences of different countries, middle classes are social groups with specific interests in politics, economics, culture, and ideology, with their own social identity. In China, if this group develops into the main body of society, it will enable Chinese society to undergo stable development, relieving some of the structural strains on Chinese society. Classifications based on standards derived from

80 Research Team, Academy of Macroeconomic Research, National Development and Reform Committee. *A Strategic Research on Enlarging the Proportion of Middle-Income Earners*. 6-7.

statistics and income, which have no close relation with actual social status and interaction, are not of great significance, as members of the middle class are not all socially associated with each other. This kind of classification is therefore of little use. So, more attention should be paid to the indexes and elements reflecting social interactions when studying the middle class in China. Besides income, other indexes such as education and occupational status should also be taken into account. In this sense, the standards for what constitutes the middle class should include the following three elements: the income and living condition of this group should be fairly high; the professions of the group should be mainly of the white-collar type; and they should be comparatively well-educated. An analysis of the latter two standards of education and profession will be made as below.

2. The Middle Stratum by Education and Occupation

According to the standards put forward above, members of the middle class should be comparatively well-educated. Generally speaking, they will have received college educations. So, among the enormous Chinese population, how many people possess the necessary qualifications for membership in the middle class? On the whole, the proportion of qualified people possessing the required educational background to be in the middle class is rather small in China. Since the next population census (the sixth population census) was to be carried out in July 2010, we will use data from the fifth population census, which was carried out in 2000. According to this census, the average level of schooling (by year) of the Chinese population over fifteen years old was 7.85 years, and 7.42 years for over twenty-five-year-olds, so the average level of schooling was between primary school and junior high school level. The fact that the rural population, which makes up the majority of the Chinese population, is generally poorly educated leads to the low average educational level of the Chinese population. Since there are huge differences between rural and urban China, failing to conduct research separately on rural and urban populations can easily result in confusion. In the following study, we will therefore separate the data of the countryside from that of urban areas (see graph 10-1). As is shown in graph 10-1, most of the population in both the countryside and the cities are educated to a level between primary school and junior high school. In cities, the percentage of those who received postgraduate education, both male and female, (among fifteen- to sixty-four-year-olds)[81] are 0.5% and 0.2% of the total population, respectively, which is really infinitesimal. The percentages of those who received a university education are 6.8% and 3.9%, and the percentages of those who received junior college educations are 9.8% and 8.1%. This means that in cities, the percentage of those who received higher education—taking the highest level of male education, including junior college education, into consideration—amounted to 17.1% of the male population from fifteen to sixty-four years old. The situation worsens in rural areas, where not a single person received a postgraduate education; the percentage of those who received university education was 0.5% and 0.2%, and 1.9% and 1.0% for junior college education. So, even by analyzing the highest percentage of male education, including junior college education, those who received higher education accounted for only 2.4% of fifteen- to sixty-four-year-old males in rural areas, a far cry from what can be considered middle class.

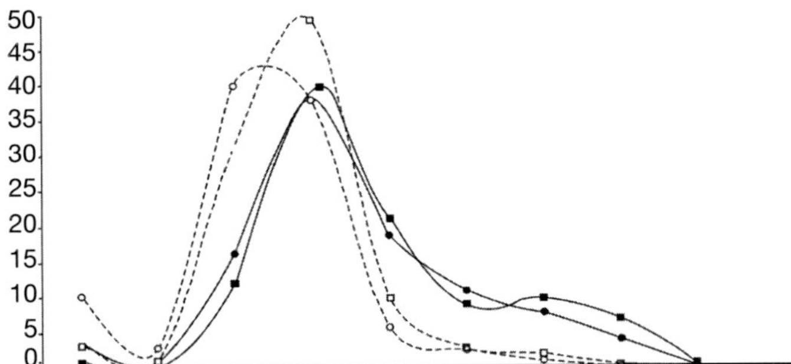

	Illiterate	Literacy Class	Elementary School	Junior High School	High School	Technical Secondary School	Junior College	University	Postgraduate
Urban Male	0.9	0.2	12.0	39.9	20.9	8.9	9.8	6.8	0.5
Urban Female	3.3	0.8	16.2	37.9	18.7	10.7	8.1	3.9	0.2
Rural Male	3.3	1.1	31.3	49.2	9.9	2.9	1.9	0.5	0.0
Rural Female	10.1	3.0	39.7	37.8	5.7	2.5	1.0	0.2	0.0

Graph 10-1 Education by Gender in City and Countryside

81 Statistics for the analysis in this chapter is based on the population of fifteen- to sixty-four-year-olds.

If we take the developing trend of educational levels for each group into consideration, the middle class has some potential for development. Graph 10-2 attempts to show the potential for the development of different age groups in urban and rural areas.

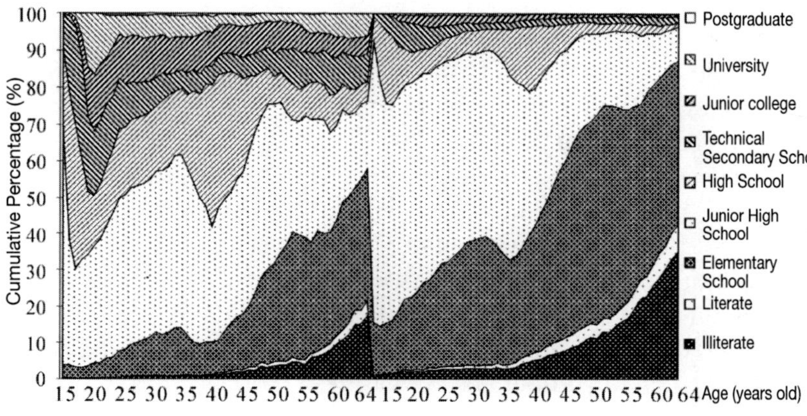

Graph 10-2 Education by Age in City and Countryside

Graph 10-2 is based on the statistics of China's fifth population census, which was carried out in 2000; the age range runs from fifteen to sixty-four years old. The horizontal axis displays the age groups and the vertical axis conveys the cumulative percentage for different educational levels. The graph on the left side contains data from cities, while the right side presents data from villages and small towns in rural areas. The graphs demonstrate that the general educational trend of those receiving higher education is still rather small, with striking differences between urban and rural areas. Another feature shown in graph 10-2 is that the educational levels of each age group have undergone dramatic rises and declines. What could have caused this? Close scrutiny reveals that the educational levels of the Chinese people have been significantly impacted by three special historical eras.

The first era was the Cultural Revolution, when education in both urban and rural areas underwent a dramatic reduction in scale. As is shown in graph 10-2, the peak value of the proportion of the population receiving middle-level education was reached shortly after the foundation of the People's Republic of China and then dropped sharply. The percentage of those around forty years old who had received high school education was fairly high, while the number of people between thirty-five and forty years old with high school educational backgrounds obviously declined and the lowest figures are for those aged thirty-four, meaning those born in 1966. From 1969 to 1973, high school education was almost entirely suspended. After that, because of the recovery of the education system towards the end of the Cultural Revolution (In 1971 and 1972, China reinstituted the College Entrance Examination and there were high school graduates after 1974), the educational levels of Chinese people started to rise again.

The second era was the period following the Cultural Revolution, during which the rate of educational recovery remained high. First, the number of people with high school educations increased with each passing year in both urban and rural areas. In 2000, the population of eighteen-year-olds with high school educations peaked. So, it can be predicted that, in view of this developmental trend, a fundamental change of the educational level of Chinese people will be observable within several decades. Second, the number of various educational institutions above the high school level has been expanding year by year, and this developmental trend has proved consistent both in urban and rural areas. In cities, technical secondary school education, junior college, university, and postgraduate education have all been becoming increasingly prevalent. By 2000, the percentage of eighteen-year-olds who had received higher education (including tertiary education) was over 30% in cities. If this speed of development is maintained, there is the potential that the educated urban middle class will develop into the largest group in Chinese society. Educational expansion in rural areas mainly comes from technical secondary school and junior college education; other forms of higher education are still rare. Graph 10-2 also indicates that the number of those in their twenties who had received high school education has surpassed all previous records. In other words, based on data from those currently in their twenties, the educational crisis and the shortage of talent caused by the Cultural Revolution have been overcome.

The third era is the period following the policy of expanding higher education, during which the percentage of those in their twenties who had received higher education increased dramatically—especially the percentage of those under twenty-five receiving higher education.

During the peak period, this figure exceeded 30%. So, it can be predicted that, with the constant expansion of higher education in China and the natural elimination of population, the educational qualifications of the majority of the urban population will meet the basic requirement for being considered middle class.

In rural areas, the policy of higher education expansion has not significantly upgraded the educational level of much of the rural population. As the graph clearly demonstrates, even after the expansion of higher education, none of the rural population between the ages of twenty and twenty-five received higher education. The reason for this is that the policy of higher education expansion was implemented without any policy for encouraging well-educated people to remain where they were born—meaning that the most intellectually capable people in rural areas rapidly moved away to cities. As is shown in the graph, people of the same age in urban and rural areas contrasted sharply in their educational levels after the higher education expansion. The number of people receiving higher education increased significantly in cities, but no improvement was evident in rural areas. It is easily foreseeable that if the situation continuously follows this developing trend, a massive middle class will quickly take shape in the cities, while in rural areas, the middle class will remain tiny. So, in order to accomplish the strategy of expanding the Chinese middle class, the structures of city and countryside must be significantly altered. This necessitates a discussion of occupational structure.

In a modern society, occupational status is the primary determiner of social status, so the middle class is primarily made up of groups in the middle occupational stratum of the social ladder. Generally speaking, if the occupational system of a society is fairly stable—especially when the occupational structure of a society is solid—it is hard for fundamental changes to occur in a short time. Occupational structure is derived from industrial structure. The occupational structure of the middle class is therefore reliant on the formation of modern industrial structures. Occupation shapes individual behavior and lifestyle; only after a middle-class professional structure is formed can the political attitudes, behavior, and lifestyles of a middle class gradually take shape. Therefore, the actual formation of a middle class depends largely on the formation of modern occupational structures. In addition, in view of the relationship between occupation and educational status, occupation can be said to manifest social status while education is social status in potential, being one of the most important qualifications for entering into some occupations. As has been discussed, the educated urban population is rapidly growing, forecasting the trend of constant growth in the number of people stepping into the occupational structure of the middle class.

What is the occupational composition of China's middle class? As stated above, the experiences of a number of countries indicate that middle classes are mainly made up of the following four occupational groups: managerial staff, professional and technical personnel, office clerks, and business service personnel. These people are also referred to as white-collar workers. White-collar workers, or white-collar groups, are the most predominant component of the middle class in a modern society.

What then is the status quo of Chinese white-collar groups? Table 10-2 shows the changes in the numbers of managerial staff, professional and technical personnel, office clerks, and business service personnel in China during a period of nearly twenty years.

Table 10-2 Changes of the Percentages of the Main Occupational Groups in China from 1982 to 2000[82]

Occupations	1982	1990	2000
Directors of State Organizations, Party Organizations, Enterprises, and Public Institutions	1.56	1.75	1.67
Professional and Technical Personnel	5.07	5.31	5.70
Clerical Personnel and Similar	1.30	1.74	3.10
Business and Service Personnel	4.01	5.41	9.18
Operators of Productive and Transportation Equipment and Similar	15.99	15.16	15.83
Productive Personnel in Agriculture, Forestry, Animal Husbandry, Fisheries, and Water Conservancy	71.98	70.58	64.46
Individuals Defying Classification	0.09	0.04	0.07
Total	100	100	100

Table 10-2 shows the changing sizes of the four white-collar groups within the employed population of China from 1982 to 2000.

82 Source: Data extrapolated from: 2000 National Population Census in China (Volume 2), 1990 National Population Census in China (Volume 2) and 1982 National Population Census in China by the State Census Office under the State Council.

From the data of 2000 alone, the directors of state organizations, party organizations, enterprises, and public institutions or managerial staff account for 1.67% of the employed population of China. At this point, professionals and technical personnel made up 5.70%, clerical personnel and similar constituted 3.10%, and business and service personnel accounted for 9.18%. We cannot, of course, define these four groups as being the Chinese middle classes, as the situation is more complex than this. Some directors of state organizations, party organizations, enterprises, and public institutions belong to the top leadership and high-level group; only the middle-level executives belong to the middle class. Business and service personnel in China cannot be simply termed white-collar either because, during the period after the founding of the People's Republic of China, this group was treated with scorn or prevented from developing. Meanwhile, because commercial services—mainly grain stores, groceries, and department stores—did not at that time require much technology, almost all of these businesses' employees were engaged in manual labor. Therefore, the proportional size of the middle class among the employed population in China is not a simple sum of the proportions of the four groups. After a detailed calculation, it is believed that in 2000, the middle class accounted for around 13% of the employed population, while in 2009, this percentage had risen to around 16%.

Internationally speaking, this percentage is rather low. Based on occupational classification, the percentages of white-collar groups in some advanced countries are 73.5% in America, 66.9% in Britain, and 63.2% in Germany. This clearly demonstrates that the social structures of these countries are dominated by their middle classes. Compared with these advanced countries, China lags far behind. Long-term structural adjustment is required to make up this gap, not short-term solutions.

Of course, although the percentage of the middle class in China is rather small considering China's enormous population, the overall size and influence of the Chinese middle class are remarkable. As mentioned above, the percentage the middle class composes of the population capable of economic activities is estimated to have been around 16% in 2009, a total of 0.128 billion people—despite the apparently small percentage, this is a lot of people. No one can afford to neglect the influence of the middle class in China, whether in economics, politics, ideology, or culture.

How, then, should we develop a strategy for expanding the middle class? Concerned that increasing the number of managerial personnel and leaders might increase administrative cost, the countermeasure of streamlining institutions has been adopted instead of expanding the group of managerial personnel and leaders. This has been occurring in spite of the fact that expanding the middle class—especially the number of technical professionals, as well as business and service personnel—is beneficial to society and conforms to the trend of modernization. At present, there is a shortage of operational technicians on the market; policies should be implemented as countermeasures to increase investment and provide guidance and training in this field. It is also necessary to facilitate access to certification for operational technicians and to not erect barriers such as the household registration system. As for the business and service personnel, efforts should be made to improve their educational levels as well as their professional and technological skills.

3. The Five Cliques of the Middle Class in China

Taking the integrated factors into consideration, five important cliques could be considered as the constituents of the middle class in contemporary China.

The first clique is the most typical of the middle class: professionals and technical workers, including the traditional intellectual class, among whom a small number of top-level experts belong to the upper class. In the sixty-year history of the People's Republic of China, other types of middle class have all been through ups and downs. For example, since the mid-1950s national industrialists and businessmen have on one occasion almost disappeared as a group. Conversely, the professional, technical, and intellectual class has consistently remained middle class in terms of wages, welfare, and working conditions, and its members obviously hold higher positions than ordinary peasants and manual workers. Intellectuals were, however, held in lower esteem as "the stinking ninth category" during the Cultural Revolution. After the reforms, great changes also took place within this group. Traditionally, intellectuals stay within state-dominant institutions (usually representing state power and supported by state finance), but recent rapid market developments have attracted some intellectuals to step out of state-dominant institutions. Many of them now work in private enterprises, foreign enterprises, and Hong Kong, Macao, and Taiwan-invested enterprises, and a small but significant number of them are engaged in freelance work. In terms of the nature of their jobs, they still belong to the professional class, possessing professional knowledge and skills which are typical of the white-collar class.

The second clique is made up of managerial personnel of all kinds, including the great mass of governmental cadres. Among these, managerial personnel and top-level leaders belong to the upper class. The proportions of such top-level managerial personnel and CEOs are quite small, however, and the majority of managerial personnel still fall into the middle class category. Managerial personnel—no matter what kind of organizations they belong to—are undoubtedly superior to non-managerial staff in terms of economics and power, as well as professional and market status. The cadre stratum in China—nowadays referred to as the "public servant" stratum—might have lower wages and other incomes than enterprise managers, but positions as public servants remain the hottest opportunity for many university graduates. This is because such positions provide steady incomes, more chances to ascend to higher positions, and better housing conditions, medical treatment, and pensions. Additionally, they play an important role in political, economic, and spiritual life in Chinese society. In brief, most managerial personnel in China are either public servants or managerial and administrative staff in enterprises and institutions. Although there are some differences in income and political status between the two groups, they share the significant commonality of being the most important parts of the middle class in China.

The third clique is the so-called "new middle stratum," the contemporary characteristics of which will be stressed. This stratum originally came into being in the mid- to late 1990s, when I was studying the problem of lay-offs in traditional industries. At that time, a strange phenomenon was observed: the average incomes of young people were surpassing those of the middle-aged in some big cities, which I referred to as "the intergenerational replacement of the middle class."[83] In 2004, a survey conducted by the Urban Survey Corps of the State Bureau of Statistics confirmed this finding after carrying out investigations in ten provinces

83 Li Qiang. "Market Transformation and the Intergenerational Replacement of the Middle Class." *Strategy and Management* 3 (1999): 35–44.

and cities across China. The survey indicated that the incomes of young people between twenty-five and thirty-five years old were higher than those of any other age group. Most of these individuals had received higher education and generally engaged themselves in newly emergent professions or were technologically focused. Such young people make up the majority of the so-called "new middle stratum."

An article in a newspaper stated that:

> Insiders had made a domestic consumption list for the middle stratum, which was calculated on the basis of the consumption of middle class couples. It found that expenses for food and drink amounted to between fifteen hundred and five thousand yuan per month; investment for medical insurance ranged from six hundred to fifteen hundred yuan per month; "topping up" ("充电") and further education cost from three hundred to five thousand yuan per month; transportation and communication ranged from five hundred to one thousand yuan per month; clothing and beauty treatment expenses ran from one thousand to five thousand yuan per month; and entertainment and leisure cost between six hundred and thirty five hundred yuan per month. In summary, the report found that to live a "middle class" life in modern China, a couple's income needed to be at least ten thousand yuan per person per month, with each family earning a combined household income of twenty thousand yuan per month. The highest household incomes for this class reach forty thousand yuan per month. According to the analyses and judgments of economists and sociologists, five types of people are representative of the "middle stratum" in China. These groups are entrepreneurs engaged in scientific and technological development, Chinese managerial personnel in foreign companies based in China, middle and upper level managerial personnel in state-owned financial industries, and technical professionals of all kinds—especially those working as intermediaries and those who are individual and private entrepreneurs.[84]

84 See "Five Kinds of People Will Enter into the Middle Stratum, Minimum Monthly Income of a Couple Should Reach Ten Thousand Each" originally published by *Beijing Business Today*, and republished anonymously at http://www.sina.com.cn, Feb. 9, 2003.

The preceding description is much more characteristic of the "new middle stratum" rather than the whole Chinese middle class.

The fourth clique is the staff of state-owned enterprises, joint-stock enterprises, and other enterprises, companies, and institutions with good economic returns. "Staff," a concept predating China's market reforms, I once defined as a "quasi-intermediate stratum" believing that the staff of state-owned enterprises belonged to a typical middle stratum before the reforms. Although quite a few state-owned enterprises with poor economic returns transformed their systems or went bankrupt after the reforms, in enterprises with good economic returns the economic and social status of the staff remained constant. So far, the model of this clique has remained comparatively similar to that before the reforms, when daily necessities, housing, medical treatment, and retirement pension were all guaranteed (commonly known as the "iron rice bowl," or secured employment). At present, this employment model is becoming far less common; but, as a result of social inertia, there are still some people living under such a pattern of employment. Although more enterprises and companies have made market-oriented reforms, the assimilative capacity of the traditional model remains powerful. Some joint-stock enterprises; companies with limited liability; foreign enterprises; enterprises operating in Hong Kong, Macao, and Taiwan; joint-venture enterprises following the traditional pattern of, for instance, purchasing houses and then renting them to their employees at low prices under the prerequisite that the enterprises have good economic returns, thus enable their employees to enjoy comparatively stable economic positions.

The fifth clique is made up of a great many entrepreneurs and owners of small and medium-sized industries as well as small and medium-sized industrial and unusual commercial enterprises that cannot be compared with other professional types. Some individuals even fall into a sixth type of professional classification—such as those who prosper from successful management in rural areas, owners of small and medium-sized industries and businesses, independent businessmen, and managers of small and medium-sized companies in cities. The constituent parts of this clique are the most complicated of the middle stratum, as they feature the recombination of Chinese social strata in a social transformation. As stated above, this clique is the fastest growing of any in the middle stratum, which is significant for China because the former four cliques are basically made up of urban inhabitants. Only in the fifth clique are there a considerable number of rural residents, which means that if this group develops rapidly, the social structure of rural areas will undergo a corresponding change. So, promoting the development of the clique comprising businessmen and managers of small and medium-sized companies and enterprises is of strategic importance for China. From a cultural prospective, the Chinese nation also has such a traditional familial economic model that fits well with the management model of small and medium-sized industries and commercial enterprises.

4. The Social Functions and Responsibilities of the Chinese Middle Stratum

Through the preceding analysis of the aforementioned five groups, it can be clearly seen that the middle stratum plays a decisive role in any country. Members of the middle stratum are not only important undertakers of social production, social management, and economic activities, but also the creators and disseminators of ideas and transmitters of traditions.

From the perspective of social structure, the middle stratum assumes important social functions. Societies all over the world dominated by middle classes show that large middle strata are a powerful social force for maintaining social stability. As is stated above, the middle stratum is a buffer layer between the upper class and lower class in a society. When it becomes the main constituent of a society, conflicts between the upper class and lower class will be hindered; social contradictions, and the degree of social strain will be greatly relieved, which is the main political reason for achieving social stability. The middle stratum also tends to stand for mild and conservative ideologies, so their occupation of a leading position provides little room for extremely wild thinking and conflicting ideas, which is an important ideological reason for pursuing social stability. Additionally, the middle class is the most important group for guiding social consumption: when the middle class becomes the largest group in a society, their lifestyles guarantee a vast and stable consumer market, which is an economic motivation for focusing on social stability. Though the national conditions of China differ from other countries, the function of a middle class in maintaining social stability remains similar. So "gradually expand the proportion of middle income people" ("逐步扩大中等收入者比重") has become an important Communist Party strategy, as is indicated in the documents of the Central Committee of the Chinese Communist Party.

At present, since China is at a critical moment of development, the social responsibility of the middle stratum has become particularly important. Over thirty years of reform have brought astonishing growth in wealth along with many social problems. Reforms are profound transformations in social systems, social norms, and social concepts. Like a pendulum, changes in the location of its fixed point can significantly change the swing of the pendulum bob. It is not, therefore, unusual that a great number of irregular and illegal social behaviors have become more common. What matters is that the middle stratum in facing such a key moment should assume its weighty responsibilities with regards to ideals, faith, and morality. China's middle stratum, bearing a sense of social responsibility, should take on its correspondent social responsibilities.

Up until now, the proportional size of China's middle stratum has remained comparatively small. If all overlapping parts are eliminated from the analysis, the aforementioned five cliques of the middle stratum amount for roughly 16% of the whole employed population in China. As previously discussed, further development of the middle stratum demands a number of basic conditions. Industrial transformation is necessary because middle strata generally emerge during processes of transition from secondary industries to tertiary ones; only during transformations of industry will a middle class reach a certain size. A transformation of professions is also required: in a modern society, modern professional structures need to replace traditional professional structures, allowing a sizeable white-collar stratum to develop and expand. It is also necessary for more people to receive higher education, allowing a populous middle stratum to gradually develop. Universities are institutions that produce middle strata, since one of the most important functions of a university is to enable people to accept the dominant norms of the middle class. The development of a Chinese middle class also depends on other comprehensive social conditions. These include the accomplishment of urbanization with Chinese characteristics, the final transformation of the agricultural population into a modern urban population, alleviating the polarization between rich and poor through the strict regulations of taxation, and establishing a social security system that covers the majority of the Chinese society.

Chapter 11

T-Shape Social Structure and the Low-Income Group

The high-income group was discussed in chapter 8, the middle-class group in chapters 9 and 10, and the low-income group, which occupies a comparatively inferior position, will be analyzed in this chapter.

When analyzing the relationship between social stratification and the 2008 economic downturn in chapter 1, it was pointed out that the current situation of overproduction and under-consumption exists because of China's disproportionally large low-income group, which lacks money and consumptive capacity. This means that expanding China's domestic demand will require adjustments in its social structure.

Sociologists commonly employ analogy-based methods to analyze and interpret social structure. For example, the oft-mentioned "pyramid-shaped" structure attempts to describe a society with a small upper class, a middle class[85] in transition, and a very large lower class. Another example, usually called the "olive-shaped" or "spindle-shaped" structure, refers to a society in which the middle class is the principal body, with both upper and lower classes being relatively small. As discovered through analysis of China's fifth national census, Chinese social structure is neither olive nor pyramid-shaped; instead, it can be described as an inverted "T-shape." The specific approach adopted in this analysis is the measurement of International Socio-Economic Index of Occupational Status (ISEI). In the employed Chinese population, a huge group of relatively low socio-economic status was found to have consistent ISEI scores. Visualized, this group is the horizontal bar of an inverted Chinese character "丁," while the vertical stroke of the character represents the groups made up of a variety of classes occupying different socioeconomic positions.

1. ISEI and the Inverted "T-shaped" Social Structure

Sociology, as was mentioned above, employs various methods for analyzing social status, including both qualitative and quantitative

85 As is in chapter 10, conceptions of middle class, intermediate class, and middle strata will be consistently expressed using the term middle class in this chapter.

analyses. Class analysis, for example, is generally qualitative in focus. Marx employed relationships based on ownership to analyze class, then classified society into two main groups based on who possessed the means of production and who did not: the capitalist class and the employed working class. By using this analytical model, modern developed countries can indeed be dissected into two huge groups: the employer group, which possesses the means of production, and the employee group, which accounts for the vast majority of the population (including all levels of managers). However, an analysis that places high status employees like top-level managers, lawyers, doctors, and professors into the same class as waiters, garbage collectors, and street cleaners can hardly be said to be a good explanation for social phenomena and social problems. To avoid this pitfall, Erik Olin Wright, a Neo-Marxist, attempts to develop Marx's views by adding two more variables: organizational capital and technical capital. By classifying people based on these three criteria, he grouped Western society into twelve classes. Wright claims his classification is Marxist, but he is actually inspired by Weberian ideas of multivariate classes. The advantage of qualitative analyses lies in the interpretation of macrostructures, while the main disadvantage is that only the macrostructure is analyzed and many small social classes are neglected.

Another method of quantitative analysis is description, which attempts to interpret the subtle nuances distinguishing different social statuses. Prestige measurement, for example, can identify the superiority or inferiority of different occupational statuses by scoring each occupation. But, this approach is based on subjective assessment rather than objective standards. The method of analyzing groups by income is an example of analysis based on objective standards. For example, according to the statistics of a sample survey released by China's State Bureau of Statistics, rural Chinese households in 2007 can be grouped based on their annual net income as follows: those with a per capita annual income of less than 100 yuan made up 0.53% of rural households, 100–200 yuan 0.09%, 200–300 yuan 0.31%, 300–400 yuan 0.19%, 400–500 yuan 0.25%, 500–600 yuan 0.34%, 600–800-yuan 1.18%, 800–1,000 yuan 1.65%, 1,000–1,200 yuan 1.97%, 1,200–1,300 yuan 1.18%, 1,300–1,500 yuan 2.64%, 1,500–1,700 yuan 3.16%, 1,700–2,000 yuan 5.21%, 2,000–2,500 yuan 9.73%, 2,500–3,000 yuan 9.89%,

3,000–3,500 yuan 9.17%, 3,500–4,000 yuan 8.34%, 4,000–4,500 yuan 7.40%, 4,500–5,000 yuan 5.98%, 5,000 yuan and over 30.94%.[86] According to the above classifications, low-income households with a per capita annual income lower than 800 yuan make up a very small proportion of the total rural population—a mere 2.71%. Such a method of grouping makes it possible to analyze income levels and the size of each group, but purely statistical grouping cannot express real social interactions or social relations.

To compensate for the deficiencies of grouping only by income, researchers developed the "socioeconomic index," a system originally proposed by the Canadian scholar Bernard R. Blishen[87] and the American scholar Otis D. Duncan.[88] It scores the weighted average educational and income levels of each professional group and sometimes also takes into account factors such as employees' ages, or even their fathers' wealth. Although the "socioeconomic index" is somewhat similar to the prestige measurement index (which is subjective), it is objective because its rankings and scores are the result of the aggregation of a huge number of socioeconomic indicators. In the process of internationalizing these indicators, the efforts of D. Treiman were particularly significant. He integrated eighty-five sets of data on occupational prestige from sixty countries including developed and traditionally underdeveloped societies from every continent and developed the "Standard International Occupational Prestige Scale."[89] Thereafter, Treiman, Ganzeboom, and de Graaf jointly brought forward an "International Socio-Economic Index of Occupational Status," from which the ISEI scores used here are derived.[90] When formulating the index, Treiman, et al, utilized thirty-one sets of statistics from sixteen countries, including both developed and undeveloped countries in order that the index would be internationally representative. They adopted the International Standard Classification of Occupations and

86 National Bureau of Statistics: *China Statistical Yearbook 2008*. Beijing: China Statistical Publishing House, 2008. 339.
87 B. R. Blishen. "The Construction and Use of an Occupational Class Scale." *Canadian Journal of Economics and Political Science* 24 (1958): 519–31.
88 O. D. Duncan. "A Socioeconomic Index for All Occupations and Properties and Characteristics of the Socioeconomic Index." *Occupations and Social Status. Ed.* A. J. Reiss. Glencoe, IL: Free Press, 1961. 109–61.
89 D. J. Treiman. *Occupational Prestige in Comparative Perspective*. New York: Academic Press, 1977.
90 Harry B. Ganzeboom, et al. "A Standard International Socio-Economic Index of Occupational Status." *Social Science Research* 21 (1992): 1–50.

internationally standardized the indicators of education and income when developing the index, overcoming the problem of cross-country differentiation.

When converting the occupation index into ISEI scores, this study conducted occupation clarification to ensure that the conversion was as accurate as possible. Since China's National Bureau of Statistics and National Bureau of Standards released the standards for occupation classification in 1986, China's classification of occupations has been revised several times, such that it now tends to be close to accepted international standards. Nevertheless, each country's occupational categorization is unique because no two countries have the same national conditions. The occupational classification used here was derived from China's 2000 census. Compared with international standards for classifying occupations, occupational categories in China are not as numerous as in many other countries and tend to be simply defined. This is probably due to inadequate definition of occupations. But, for many occupations, it is easy to find appropriate counterparts so that the transfer of scores is not affected since China's system of occupational classification originated with the international standards. For a few occupations for which it was difficult to find counterparts, my student Zhang Haihui and I discussed and analyzed the specific nature of these occupations and finally matched them all to the international standard classification. It should be clarified here that early studies of occupational status in China found that it was modulated by the work unit system; so, by extension, individual inequality was also modulated by inequalities between work units. But, work unit status is not considered when the transfer to the ISEI is conducted here because the development of China's market economy has significantly decreased the extent to which work units influence the status of individuals. Additionally, the high rate of new work unit formation has made it difficult to assess their social status. Therefore, the application of work unit status will be avoided here to prevent further errors.

After transferring the data into the ISEI, the status scores of different occupational groups are as follows (table 11-1):

Table 11-1 "Standard International Socio-Economic Index of Occupational Status" (ISEI) of Chinese Population Aged from 16 to 64

ISEI Score	Number of People	Percentage
16.00	1744	3
20.00	6655	1.0
22.00	1151	2
23.00	405659	63.2
24.00	5322	8
25.00	1428	2
26.00	951	1
27.00	3114	5
28.00	4579	7
29.00	18739	2.9
30.00	30004	4.7
31.00	9520	1.5
32.00	4443	7
33.00	7645	1.2
34.00	10948	1.7
35.00	2530	4
36.00	6955	1.1
37.00	26089	4.1
38.00	11578	1.8
39.00	3456	5
40.00	2795	4
41.00	2589	4
42.00	131	0
43.00	9959	1.6
44.00	754	1
45.00	7817	1.2
46.00	445	1
47.00	3245	5
50.00	3046	5
51.00	4551	7
52.00	1036	2
53.00	776	1
54.00	1414	2
55.00	212	0
56.00	74	0
57.00	91	0
58.00	320	0
59.00	34	0
60.00	19	0
61.00	2563	4
64.00	261	0
65.00	631	1
66.00	7178	1.1

ISEI Score	Number of People	Percentage
67.00	105	0
68.00	7387	1.2
69.00	13786	2.1
70.00	95	0
71.00	72	0
73.00	963	2
74.00	478	1
77.00	2343	4
78.00	17	0
79.00	45	0
82.00	143	0
83.00	177	0
85.00	481	7.5
88.00	2804	4.4
90.00	218	3
Total	641547	100.0

We were astonished by these results, which showed that China's fundamental social structure was even worse than the so-called "pyramid-shaped" structure and could be described as an inverted "T-shape." Of the entire population, 64.7% had very low scores, marking a sharp division from the other groups, which formed a column. This radical distinction can be seen in the figure below.

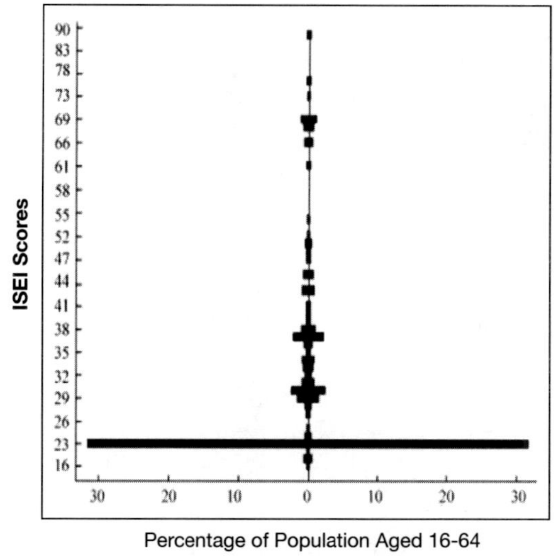

Figure 1: Structure Diagram of Chinese Socio-Economic Status by ISEI Score

An introduction will be made to some major groups that constitute the above ISEI scores as follows.

The group that scored twenty-three made up 63.2% of the employed population; most of these people were agricultural workers. This group included those engaged in field or trellis crop farming, agricultural and sideline product processing, animal husbandry, and livestock and poultry breeding. It also included nonagricultural workers such as garbage collectors and cleaners. The majority, making up 91.2% of the group, were however those engaged in agricultural labor—farmers in the traditional sense. Of the above 63.2%, 48% were farmers and only 5.2% were engaged in other types of work. These people constituted the large group at the bottom of China's inverted "T-shaped" society, which is a reflection of a harsh reality in China—the huge size of its lower class.

Another group scored 29-31, and was one of the two comparatively large groups among those constituting the column in the structural diagram. This accounted for 9.1% of the employed population, most of whom were construction workers, earthwork construction workers, concrete preparation workers, scaffold builders, geological exploration workers, coal and metallurgical minerals miners, building material makers, metal processors, porters, rickshaw pullers, takeaway delivery staff, haulers, lumberjacks, butchers and meat processers, fur processing workers, shoemakers, and handicraftsmen, among others. This group of people mostly comprised migrant workers, farmers who became township enterprise workers, and those whose actual social status and living conditions were similar to those of farmers.

The group that scored 33-38 was another large part of the column, accounting for 10.3% of the employed population. This group mainly comprised highway, railway, and water transport workers; employees of other transport services; clerks working in the commerce and service industry; service staff in the catering industry; factory personnel for electromechanical products, electronic products, and mechanical power equipment; machinery equipment repairmen; production and assembly personnel for electronic components and equipment; instrument repairmen; arts and crafts makers; music instrument makers; assembling, operation, and repair personnel for living and production power facilities; as well as semi-skilled workers in the chemical and processing industries. This group, positioned between the middle class and blue-collar class, could be called the upper blue-collar, or the lower white-collar.

Other groups made up even smaller proportions of the population. Among them, particular mention should be made of the following groups.

The group scoring 43-45 made up 2.3% of the employed population. This was a typical white-collar class mainly made up of primary school teachers, kindergarten teachers, nurses, ordinary administration and operation personnel, clerks in offices, clerks in enterprises, service staff in businesses, salesmen, sales exhibition personnel, purchasers, and sale personnel.

The 68-69 group, which constituted 2.6% of the employed population, was composed of managers or executives in all kinds of productive, sales, and service enterprises; science and technology professionals; planning and design personnel; engineering and technical personnel in electronics, power, broadcasting, film, television and traffic; high school teachers; educational personnel in various types of secondary vocational education such as technical secondary schools; and professional technicians in administration, among others.

One of the highest scoring groups, the one scoring 85-88, made up only 0.5% of the entire employed population and includes managers; executives of banks, financial enterprises and securities companies; senior technical personnel such as doctors, professors, and lawyers; heads of state, Party, and popular organizations; and senior judicial officers like lawyers and judges.

Social scientific social theories must, however, be able to stand the test of social practice. Does the inverted "T-shaped" social structure mentioned above truly reflect actual conditions in China? Facts we hear and see every day should be able to provide persuasive evidence of this kind of bipartite reality of social structure. For example, the worrying issue of agriculture, rural areas, and the rural population is, in essence, a problem caused by the huge size of the lower class in the inverted "T-shaped" social structure. Furthermore, severe imbalances in regional development are further examples that reflect this structure.

When it comes to using survey data to provide evidence for social stratification structure, two models are internationally popular at the moment. One is the aforementioned Erik Olin Wright Model and the other is the John Goldthorpe Model. Lu Xueyi, an academic in the Chinese Academy of Sciences, recently divided Chinese society into ten strata in the book for which he was chief editor, *A Research Report on Social Stratification in Contemporary China.* Lu's method is actually a combination of the Erik Olin Wright Model and the John

Goldthorpe Model[91] and does not contain any original elements. How, then, does the approach of employing ISEI scores differ from and conform to Lu's approach? The Wright Model is based on three indexes: ownership, organizational assets, and technical assets; the John Goldthorpe Model is based on occupational structure, which is itself interwoven with ownership, organizational assets, and technical assets. ISEI scores are based on income, education, and occupational structure. So, in terms of what is analyzed, the approach of using ISEI scores overlaps with these aforementioned models. This method does, however, differ from the above models in that social structure can be depicted more accurately with IESI scores, since every single detail of a social structure can be shown. Additionally, this study directly presents the ISEI scores being examined, with the purpose of avoiding subjective interference and presenting a more objective picture of social structure. Other models, because artificial grouping is used, always use some subjective elements.

2. The Urban-Rural Relationship: The Core of the Inverted "T-shaped" Structure and China's Polarized Society

In previous research into social stratification structure, findings indicating that the inverted "T-shaped" structure was rare, indeed. Compared with the pyramid-shaped structure, it shows more radical hierarchical distinctions, forming a right angle rather than a sloped gradient. There are no buffering or transitional groups between the lower classes and the other classes, so it is an either-or bipartite structure. Why is that? Why does Chinese society present itself as an inverted "T-shaped" social structure?

There are certainly many reasons for this, among which the most obvious one is the separation imposed by the household registration system. Seen from this point alone, we can see that China's inverted "T-shaped" structure was not formed recently, but rather has existed for a long time. So far, as one can see, such a structure already existed before Deng Xiaoping's economic reforms. At that time, however, the dominant social structure was the "political stratification" structure[92] in which groups of lower socioeconomic status—poor and lower-middle

91 Lu Xueyi. *A Research Report on Social Stratification in Contemporary China.* Beijing: Social Sciences Academic Press (China), 2002. 7–23.
92 Li Qiang. "Political and Economic Stratification." *Sociological Studies* 4 (1997): 34–43.

peasants for example—possessed fairly higher political status. The inverted "T-shaped" social structure was thus temporarily concealed.

During the last thirty-plus years of economic reform, China's economy and economic structures have developed significantly, but the inverted "T-shaped" social structure has remained unchanged. Here, we must reflect on the urban and rural household registration system. I think that the most deeply rooted cause of the ever-enlarging gap between urban and rural areas is the slow pace of China's urbanization, which is constrained by the household registration system. Modernization and industrialization have invariably occurred in parallel with urbanization throughout the rest of the world. Is there any proportional relation between industrialization and urbanization? Of course there is. Samuel Preston investigated the relationship between industrialization and urbanization from 1950 to 1970 throughout much of the world, but not in China. Based on the large amount of data he collected, he calculated that there is a constant 1:2[93] ratio between industrialization and urbanization. This ratio means that a 1% increase in the size of the industrial labor force will normally coincide with a 2% increase in the size of the urban population.

Taking this as a reference, we can see that urbanization has radically lagged behind industrialization in the last thirty years. This is clearly demonstrated by the data from the years between 1978 and 2000. In 1978, Chinese industrial laborers accounted for 26.2% of the total number of laborers, while the urban population was 17.92% of the total population. In 2000, industrial (non-farming) laborers accounted for 50-60%, so an intermediate value of 55% of the laborer population will be used. The proportion of Chinese laborers working in industry had therefore increased by 28.2% from 1978 to 2000. According to Preston's ratio, the urban population should have correspondingly increased by 56.4%. However, the urban population in 2000 accounted for 36.22%, representing only an 18.3% increase from 1978. This clearly indicates that China's rate of urban development is half the speed of the international standard.

Why does this urbanization lag lead to a widening gap between urban and rural areas? The reason is obvious: compared with rural areas, cities act as political, economic (with respect to old and new

93 Samuel H. Preston. "Urban Growth in Developing Countries: A Demographic Reappraisal." *The Urbanization of the Third World.* Ed. Josef Gugler, Oxford: Oxford UP, 1988. 24–25.

industries as well as commerce), and cultural centers, occupying a position of preeminence, unavoidably attracting and agglomerating capital flow, a high quality labor force, and technology.

Conversely, compared with cities and industry, villages and agriculture are in an obviously disadvantageous position. Agricultural products have longer production cycles and agriculture is easily disrupted by natural disasters. Agricultural producers are dependent upon large volumes of low-value goods and the products themselves are hard to preserve and have low demand elasticity, which equates to low profits and high risks.[94] This is why countries all over the world have taken measures to protect domestic agriculture. Naturally, it is easier to execute such a policy when the agricultural population is relatively small. In China, however, the urban population being so small and the rural population so large decreases the value of per capita subsidies, making them an utterly inadequate measure in the face of such enormous population. In summary, the drawbacks of dependence on agriculture have put the rural population in a disadvantaged position. In Chinese social structure, the agricultural population tends to belong to low-income groups.

Furthermore, migrant workers in cities have to work in the secondary labor market. In other words, they are restricted by the household registration system and thus have to take unstable jobs with low pay and poor welfare. Although they earn a certain amount of money in cities, compared with urban residents, their incomes are still rather low.

Without readjustment or reform of the household registration system that separates China's urban and rural populations, which would allow unlimited free competition between cities and the countryside, development in rural areas will increasingly lag behind. The income and consumption gap between urban and rural areas reached its peak in the late 1990s and early twenty-first century. At this point, both the per capita incomes and per capita consumption rates of urban residents were four times as much as those of farmers. According to some experts' analyses, urban residents' incomes—if various forms of welfare and subsides are included—are as much as

94 Zhao Manhua, *A Research of the Income Gap between Urban and Rural Residents in China*. Beijing: Economy and Administration Press, 1997. 121.

seven times larger than those of farmers. This huge urban-rural separation makes the inverted "T-shaped" social structure very hard to change.

As was shown above, the inverted "T-shaped" social structure reflects the reality of the separation of cities and countryside. Rural residents (a huge group of lower-class people) stay at the bottom of Chinese society, forming the horizontal line of the inverted "T," while most of the urban residents form the vertical line. In the following paragraphs, an analysis will be made of the relevant data to see whether cities and countryside have totally different structures.

Let us begin with the social structure in rural areas. As measured by the ISEI (see table 11-2), the group of farmers scoring 23 accounted for 78.3% of the total sum of rural laborers. Other proportionally sized groups mostly scored below 40. Combined, the groups scoring below 40 accounted for 96.7% of the total population of rural laborers, with very few groups scoring above 40. Of the groups scoring under 40, several groups are particularly noteworthy. The 29-31 group accounted for 7.2% and was mainly comprised of migrant construction workers, miners, and those performing heavy manual labor, as well as workers in irregular township enterprises. The 33-34 group accounted for 2.4% and mainly included those in rural areas engaged in transport, catering services, manufacturing, and processing. The 37-38 group accounted for 3.7% and was principally made up of those in rural areas who performed mechanical device repair, electronic component and equipment manufacturing, assembly, instrument and meter repair, handicraft work, and textile work.

Besides the above listings of professions, there were no other proportionally large groups in the countryside, so we can make a safe conclusion that rural areas in China are typically populated by lower-status people and that the middle class simply does not exist in rural areas.

Table 11-2 "Standard International Socio-Economic Index of Occupational Status" (ISEI) of Rural Chinese Population Aged 16 to 64[95]

ISEI Score	Number of People	Percentage	Cumulative Percentage
16.00	797	0.2	0.2
20.00	4446	0.9	1.0
22.00	627	0.1	1.2
23.00	397105	78.3	79.4
24.00	2734	0.5	79.9
25.00	914	0.2	80.1
26.00	645	0.1	80.3
27.00	2429	0.5	80.7
28.00	3345	0.7	81.4
29.00	13444	2.6	84.0
30.00	15248	3.0	87.0
31.00	8346	1.6	88.7
32.00	3153	0.6	89.3
33.00	6190	1.2	90.5
34.00	6192	1.2	91.8
35.00	747	0.1	91.9
36.00	3683	0.7	92.6
37.00	13943	2.7	95.4
38.00	5203	1.0	96.4
39.00	788	0.2	96.6
40.00	991	0.2	96.7
41.00	1348	0.3	97.0
42.00	30	0.0	97.0
43.00	3971	0.8	97.8
44.00	449	0.1	97.9
45.00	644	0.1	98.0
46.00	65	0.0	98.0
47.00	1176	0.2	98.3
50.00	529	0.1	98.4
51.00	575	0.1	98.5
52.00	139	0.0	98.5
53.00	255	0.1	98.6
54.00	194	0.0	98.6
55.00	53	0.0	98.6
56.00	8	0.0	98.6
57.00	29	0.0	98.6
58.00	62	0.0	98.6
59.00	3	0.0	98.6
60.00	12	0.0	98.6
61.00	249	0.0	98.7
64.00	49	0.0	98.7
65.00	18	0.0	98.7
66.00	1818	0.4	99.1
67.00	6	0.0	99.1
68.00	1968	0.4	99.4
69.00	1467	0.3	99.7
70.00	24	0.0	99.7
71.00	37	0.0	99.7
73.00	273	0.1	99.8
74.00	54	0.0	99.8
77.00	63	0.0	99.8
78.00	3	0.0	99.8
79.00	3	0.0	99.8
82.00	6	0.0	99.8
83.00	78	0.0	99.8
85.00	146	0.0	99.8
88.00	685	0.1	100.0
90.00	4	0.0	100.0
Total	507463	100.0	

95 The data used in this chapter were extrapolated from the Fifth National Population Census in 2000 by systematic sampling technique. 1,180,111 samples were selected by a fraction of 0.95‰ from the national population, among whom those aged 16 to 64 were selected, making 641,547 in total. There were a few miss-samples after making the urban and rural division.

Among urban residents, the middle stratum accounts for a considerable proportion; low scoring groups are of moderate size and the highest scoring groups are very small. The distribution of the scores follow an olive-shaped curve, meaning there are few people at either ends and many in the middle. So, the distribution of urban strata, which can be understood to tend towards a middle stratum structure, is totally different from that of rural strata.

According to the ISEI scores for the urban population (see table 11-3), it could be evenly divided into three groups: high, middle, and low. The low-status group scored between 16 and 40 and accounted for 55.3% of all urban residents. This contrasts starkly with rural areas, where 96.7% of residents scored within this range. This gives us some idea of the radical differences in social structure between city and countryside. The middle status groups—meaning those with scores between 41 and 66—accounted for 26.5%, and those with scores between 67 and 90—the high status group—accounted for 18.2%. Accordingly, the general structure therefore had a larger proportion of low status groups occupying slightly over half of the total and the middle, while upper status groups occupied the other half. This type of structure is clearly distinct from the inverted "T-shaped" social structure of the countryside.

An attempt will now be made to analyze the urban high, middle, and low status groups. In the lower status group, which was proportionally large, the real situations of each subgroup were distinct and separate. In this group, those who were of the lowest status in a real sense constituted a very small proportion. For example, the typical 23 scoring subgroup (agricultural workers) accounted for only 6.2%. The 24-40 subgroups accounted for 83.9% of this lower status group—including 46.4% of its employees—and mainly comprised street vendors, slaughter house workers and meat processors, food processors, vegetable and fruit processors, construction workers, plumbers, earthwork construction workers, concrete preparation workers, bricklayers, carpenters, painters, barbers, beauty salon workers, porcelain and glass makers, metal processors, geological exploration workers, coal and metallurgical minerals miners, building material processors, metal processors, babysitters, cooks, storekeepers, accommodation superintendants, property management staff, all types of service staff, advertisement scaffolders, construction scaffolders, underwater workers, blacksmiths, tool makers, textile workers, dressmakers, electromechanical workers, electronic equipment makers, telephone installers, music instrument makers, wrist watch makers, exquisite craftsmanship makers, printers, audio recording producers, and assemblers. In summary, this subgroup was mainly composed of operational workers and constituted the bulk of the urban industrial laborers. They were not really of the low stratum, but rather of a middle-lower position. Though not very rich, they live stable lives.

The middle status groups, as was mentioned above, accounted for 26.5% of the population, and was composed of train drivers and ship pilots, electric equipment service personnel, electrical engineers, electro-communication engineers, chemical engineers, medical engineering technicians, surveying and mapping engineers, other kinds of engineers, computer technologist/engineers, engineers of all kinds of electronic equipment, ship and airplane technicians, salesmen, models, exhibition personnel, firefighters, policemen, client service staff, cashiers, ticket collectors, information service staff, travel agency staff, telephone operators, all types of performers, athletes, clerks, secretaries, word processors, data processors, accountants, statisticians, production and transport management personnel, commercial organization clerks, publicists, ordinary government management clerks, social workers, insurance salespeople, traditional Chinese medicine medical staff, staff in special education, employment agency staff, labor management staff, laboratory assistants, medical and health assistance staff, pharmacy assistants, nurses, translators, editors, kindergarten teachers, primary school teachers, some secondary school teachers and librarians, etc. To sum up, they were typical of the middle stratum of the Chinese society, which could be further subdivided into upper-middle groups and lower-middle groups. They could be called the middle group and middle-lower group of the middle class.

The high status group, which scored between 67 and 90, accounted for 18.2% of the population, and mainly consisted of judicial officers like lawyers, judges, and attorneys; types of social scientists like economists, personnel administration officials, and business professionals; types of professionals and administrators of higher education like university professors; types of medical professionals like doctors and pharmacists; senior professional and technical personnel like engineers and architects; natural scientists like physicists and chemists; computer engineers of software, hardware, and program design;

heads of state organizations and other institutions, heads of Party and popular organizations, principals of enterprises, managers of various businesses, and leaders of public institutions, etc.

From the analysis above, we can safely say that the following are characteristic of urban social structure. First, although the middle and lower strata are very large in urban society, it has a completely different social structure from rural society. The lowest status groups do not occupy too large a proportion—instead, the groups between the low and the middle strata are proportionally large. Second, the middle stratum clearly exists in urban society, though it is not as large as the lower-middle groups. This is, however, in contrast with the countryside. Third, there is no sharp division between the middle class and upper class, as many aspects of these groups overlap. The middle class accounts for a considerable portion of the high scoring groups.

Table 11-3 "Standard International Socio-Economic Index of Occupational Status" (ISEI) of Urban Chinese Population Aged 16 to 64

ISEI Score	Number of People	Percentage	Cumulative Percentage	ISEI Score	Number of People	Percentage	Cumulative Percentage
16.00	939	0.7	0.7	51.00	3955	3.0	73.1
20.00	2195	1.7	2.4	52.00	895	0.7	73.7
22.00	523	0.4	2.7	53.00	517	0.4	74.1
23.00	8185	6.2	8.9	54.00	1209	0.9	75.0
24.00	2579	1.9	10.8	55.00	159	0.1	75.1
25.00	508	0.4	11.2	56.00	66	0.0	75.2
26.00	304	0.2	11.5	57.00	62	0.0	75.2
27.00	681	0.5	12.0	58.00	258	0.2	75.4
28.00	1221	0.9	12.9	59.00	30	0.0	75.5
29.00	5240	3.9	16.8	60.00	7	0.0	75.5
30.00	14702	11.1	27.9	61.00	2311	1.7	77.2
31.00	1146	0.9	28.7	64.00	210	0.2	77.4
32.00	1278	1.0	29.7	65.00	606	0.5	77.8
33.00	1397	1.1	30.8	66.00	5327	4.0	81.8
34.00	4704	3.5	34.3	67.00	99	0.1	81.9
35.00	1780	1.3	35.6	68.00	5402	4.1	86.0
36.00	3258	2.4	38.1	69.00	12241	9.2	95.2
37.00	12083	9.1	47.2	70.00	71	0.1	95.2

ISEI Score	Number of People	Percentage	Cumulative Percentage	ISEI Score	Number of People	Percentage	Cumulative Percentage
38.00	6348	4.8	51.9	71.00	35	0.0	95.2
39.00	2658	2.0	53.9	73.00	687	0.5	95.8
40.00	1760	1.3	55.3	74.00	423	0.3	96.1
41.00	1236	0.9	56.2	77.00	2264	1.7	97.8
42.00	101	0.1	56.3	78.00	14	0.0	97.8
43.00	5960	4.5	60.7	79.00	42	0.0	97.8
44.00	304	0.2	60.0	82.00	135	0.1	97.9
45.00	7155	5.4	66.4	83.00	99	0.1	98.0
46.00	374	0.3	66.6	85.00	333	0.3	98.3
47.00	2063	1.6	68.2	88.00	2112	1.6	99.8
50.00	2521	1.9	70.1	90.00	213	0.2	100.0
				Total	132985	100.0	

Comparing urban and rural social structures makes it clear that the root cause of the inverted "T-shaped" social structure is the household registration system that separates urban and rural areas. Additionally, entirely different social structures can be observed in urban and rural areas. Can it therefore be said that there is only one system of social stratification in China? I uphold that there are in fact two—the social stratification system of urban society and that of rural society—which operate separately and have no overlap. There are indeed a considerable number of migrant workers moving between cities and the countryside, but most of them cannot assimilate into the cities where they work, so they eventually return to the countryside. This is largely because of the serious restrictions placed on them by the household registration system.

3. The Inverted "T-shaped" Social Structure and "Structural Strain"

What will be the social consequences of the inverted "T-shaped" social structure? Since the lowest group is very large, it connects to other groups in a bipolar (or rectangular) fashion, which in turn causes "structural strain" between social groups and even throughout the whole of society. It is necessary to introduce the concept of "structural strain." "Structural strain," or "social structural strain," means different social groups are in a state of opposition, contradiction, or conflict as a result of social structure. In other words, social relations are in a state of excess tension. Social contradictions in this state are more likely to become aggravated; social problems and crises are inclined to arise more easily.

Here arises a basic theoretical question of sociology regarding whether social structure is a kind of entity that exists beyond individuals or only a term without practical significance. According to Max Weber and other sociologists, social structure is not social reality, but only a term or a speculative abstraction with no basis in experience. Weber argued that social reality is built on individual social behavior and that social structure is merely a kind of possibility and cannot exist independently of people's behavior. In contrast, another school of sociologists holds the viewpoint that social structure is an entity that greatly influences every single person. Emile Durkheim, for example, holds that social structure—as an objective entity—exists beyond individuals and influences and constrains every single person's behavior. As a well-known saying by Emile Durkheim goes, "society comes prior to individual." On this point he was in accord with Karl Marx, who assumed that social structures are an objective entity independent of human will. This study is conducted on the basis of the latter standpoint. As is assumed in this chapter, "structural strain," or "social structure strain," is a kind of objective entity beyond individuals that is beyond social class structure. Once it comes into being, it will restrain the behavior of all members of society. "Structural strain" is therefore the fundamental cause of many social problems and conflicts.

In sociology, it was the American sociologist Robert Merton who originally proposed the concept of "structural strain," with which he attempted to explain situations in which a social structure would result in social problems. He asserted that "structural stain" is the imbalance between cultural goals and the means people have at their disposal to achieve those goals. For example, if the ruler of a society ideologically over-stresses the importance of money and wealth while the institutionalized means available to the public for making money are insufficient, this will lead to "social strain." As a consequence, strain is generated and produces an increasing number of social contradictions, crimes, and conflicts.

Thereafter, Neil J. Smelser further analyzes the consequences of "structural strain." Similarly to Merton, he assumes that strain is generated when public expectations of living conditions and social reality fail to coincide. In a state of strain, it is easy for people to develop radical beliefs or engage in extreme behavior such as rebellion, and people will use these radical beliefs to explain the social situation. When contingencies like racial violence occur, the public is very likely to mobilize to take collective action, which may deteriorate into collective behavior such as social movements if these actions are not constrained by a force for social control.[96]

Karl Marx held viewpoints on this issue similar to those of "strain theory," though he did not use the term directly when he studied the basic contradictions of Western societies. As he proposed in *The Communist Manifesto,* the social structure most likely to intensify social contradictions was that of a whole society split into "two great hostile camps," a bipartite social structure. "Structural strain" is therefore generally in accord with Marx's perspectives on society.

The inverted T-shaped *social* structure illustrated in this study and the structural strain it can cause can be used to explain the present social conflicts and problems in China, such as the widening gap between rich and poor, which is a direct consequence of the inverted T-shaped social structure. It was also demonstrated by the aforementioned social prestige survey, in which social values were found to have been split into conflicting value systems and disintegrating social assessment systems. This trend precisely reflects the existing gaps between social groups[97] and is a good example of the inverted T-shaped social structure. For example, the "melamine incident" and food safety issues have resulted in serious food safety scares among the upper and middle groups in the inverted T-shaped social structure, while most of the food producers and processors of the lower group neither received good educations nor were made aware of the enormous harm caused by auxin or additive abuse. Moreover, manhole cover theft has caused a number of serious injuries, in spite of repeated crackdowns. Thefts of public and semi-public property such as street lamps, fire hydrants, and even the window screens of buildings are also common. Those who are in the lowest social group have even been known to destroy public facilities, causing huge losses for very tiny benefits. Actually, they make almost no money from selling

96 Neil J. Smelser. *Theory of Collective Behavior*. New York, Free Press, 1962.
97 Li Qiang. "Occupational Prestige Measure in Transformation Period." *Social Sciences in China* 4 (2000): 100-11.

materials such as scrap metal obtained from public goods. The radical distinctions of the inverted T-shaped social structure even make it difficult to protect and operate social infrastructure. The barrier to improved Chinese social operation lies in the enormous demand discrepancies between different social groups. Social exchange is hard to conduct. The basic facilities the middle class demand for a decent living are regarded by the low groups as luxuries, and even the means by which they may earn their own livings. That said, explanations for almost all social problems—like social disorder, insecurity, poverty, AIDS, and prostitution—can be found in the inverted T-shaped social structure and resultant structural strain.

The inverted T-shaped social structure and structural strain are nothing new—they have existed for a long time. Over the past centuries, structural strain in China—resulting from discrepancies between social strata—has never been resolved and has taken different forms in different periods. Before the establishment of the People's Republic of China, it was illustrated by the conflicts between big landowners and peasants who lost their land and land-poor peasants. Since the foundation of the People's Republic of China, it has taken the form of criticism and struggles between social groups in large-scale political movements. Some people believe that the structural strain was the result of Chairman Mao Zedong's "class struggle" policies, but it was not. Individual policies may intensify or ease the situation, but they cannot create social structures. In the face of the huge discrepancies between social groups, Mao attempted to relieve the strain by improving the political status of poor and lower-middle peasants and manual laborers. His policies were generally in favor of the bottom portion of the inverted T-shaped social structure, which pacified the masses and stimulated their enthusiasm, but heavily impacted the well-being of the upper and middle classes.

To sum up, one of the important structural causes of long-lasting "social structural strain" in China, I believe, is the aforementioned inverted T-shaped social structure. In other words, structural transformation is the fundamental solution for a society in a state of social structural strain to ease that strain. How, then, can this transformation be achieved?

Many answers to this question have been proposed in Chinese sociological circles. As far as I am concerned, the very heart of the question is still the transformation of class structure or group structure, though the term "social structure" has more than one interpretation.

In order to achieve social structure transformation, we must first recognize that changing social stratum structure is a historical process that only occurs under certain socio-economic conditions; it cannot be induced by mere human action. Variables like significant events or social policies may influence social structure, but only in a very limited way. Second, it is a long-term historical process that cannot be accomplished quickly. Third, subjective interventions in social structure transformation that disregard the historical process of development will teach us further bitter lessons.

The best way to study the trend of structural transition is to analyze the past trajectory of structural change. Here I have attempted to compare the social stratum structure changes reflected in three censuses. Since there was no initial data for the 1982 and 1990 censuses, I did not use ISEI scores, which are also based on people's occupations. Instead, I compared the general changes in occupational categorization reflected in those three censuses. In this way, I was able to outline the major transitional trends. The two methods had no difference in essence.

In chapter 10, table 10-2 (Changes of the Percentages of the Main Occupational Groups in China from 1982 to 2000) shows that the occupational structure changes for nearly twenty years from 1982 to 2000. This happened to be a period of substantial reform, so the data also reflect the transitions of social occupation structure and stratum structure due to these reforms.

The following discussion attempts to probe into the occupational transitions shown in table 10-2, chapter 10.

During the period from 1982 to 1990, the reform (though intensified) did not bring about significant changes in social occupation structure, which indicates that there is a lag time between a policy and its influence on social structure. The agricultural laborer population (farming, forestry, herding, and fishing) decreased by only 1.4%, while the number of industrial workers remained roughly the same, even showing a slight decline. However, the structure of the industrial worker stratum did undergo internal change, in that fewer workers came from the cities and an increasing proportion came from the countryside. The number of commercial employees and workers in service trades increased by only 1.4% and there was very little change in the number of clerks, technical professionals, and administrators.

From 1990 to 2000, the pace of social structural change accelerated to some extent. The number of agricultural laborers decreased by 6.12%. Looking at the numbers, the overall percentage of industrial and transport workers underwent no obvious change, but there were, in fact, significant internal structural changes. The number of urban residents becoming workers decreased and these people were replaced by workers who had migrated from the countryside; the majority of those who made up the 6.12% decrease in the number of agricultural laborers became industrial and transport workers. The number of employees working in commerce and service trades almost doubled. But, there was only a very small change in the number of clerks, technical professionals, and administrators.

These changes did not, however, lead to any fundamental pattern changes in the inverted T-shaped or pyramid-shaped social structure. When will such changes take place? In my article summarizing China's thirty years of reform, it was predicted that the proportion of agricultural laborers would decrease to 10% of all employees in 2038 if China's current pace of development continues. If this prediction is to be realized, an earth-shaking change in China's social structure will have to take place in the next thirty years. At that time, the inverted T-shaped social structure will have undergone a thorough change.

One of the distinctive features and the key issue of China's social structure is that it has too small a middle stratum. Other countries' experiences indicate that the middle stratum mainly consists of four occupational groups: administrative and managerial personnel, technical professionals, clerks and employees in businesses, and service personnel. Due to worries over increasing administrative costs, the general principle of state policy has been to streamline rather than enlarge the ranks of administrators and cadres. But, it is beneficial for society (as well for the trend of modernization) to increase the proportional size of the other three groups—especially technical professionals, commerce traders, and service providers. There is currently a shortage of skilled technical operators, so preferential policies and greater training expenditure is necessary for this group. It is also necessary to open up certification channels for professional and technical operators and clear away the obstacles to their establishing social identities, such as residence registration. For commerce traders and service providers, it is important to improve their levels of education and professionalism.

Changing Chinese social structure will require a two-step strategy. The first step is to tangibly achieve the transference of many primary industry employees to secondary industries. This will require the formal employment of migrant workers who have entered industries in cities and workers in township enterprises. Formal employment means that the employers would have to sign labor contracts with their employees in accordance with China's labor laws and safeguard their rights, such as equal access to employment, salary, vacation time, health and safety protection, job training, social security, and welfare. Social security mainly includes such features as pensions, medical insurance, and unemployment benefits, which enable the creation of a stable workforce and prevent employers from forcing out their employees after enough profit had been obtained from their cheap labor. This practice leads to migrant workers returning to their rural homes and becoming heavy social burdens, an important cause of the slow worker-farmer transition in China. The first step of the strategy faces two major impediments—the residence registration system and the limited population capacity of cities—so it is necessary to conduct further research into household registration system reform, urbanization models, and urban population capacity. Now that a large number of farmers have become industrial workers, the task is to permit the transformation. If this proves too difficult for the first generation of farmers, we can achieve the farmer-to-worker transition in the second generation.

The second step is the transition from secondary to tertiary industries so that the middle stratum can be expanded. As is indicated in table 10-2, two of the main groups in the middle stratum, administrators and professional technicians, did not change much during a time span of twenty years, so these two groups should not be expected to grow significantly in a short time. The expansion of the middle stratum mainly depends on increasing the number of commercial traders, service providers, clerks, and other related groups. A social environment suitable for the development of these groups should therefore be created. A key element of this is improving these groups' levels of education and professionalism, as well as conducting research on how to develop the middle stratum in rural society.

At present, the focus is still on the first step and social policies mainly tend to look to what is required to accomplish this. For instance, in order to accomplish the first step, more citizens will have to receive at least nine-year or twelve-year educations that emphasize technical training for skilled operators. Educational expenditure is therefore mainly geared towards middle and primary school education.

To sum up, this chapter attempts to explain how social problems in China stem from the overall structure of the Chinese society. In order to ease the "strain," Chinese society must first achieve a fundamental adjustment of the inverted T-shaped social structure.

Chapter 12

Urban-Rural Stratification and the Relationship between Urban Residents and Migrant Workers

Previous chapters analyzed social stratification in contemporary China from the perspective of the upper, middle, and lower strata. At its beginning, this book pointed out that one prominent feature of Chinese social stratification is the differentiation and stratification between urban and rural areas. Therefore, this chapter will focus on an exploration of the relationship between urban residents and rural residents, particularly the relationship between urban residents and migrant rural workers.

1. Historical Changes in the Interaction between Farmers and Civic Society Since 1949

Under the urban and rural binary structure, farmers in China were fairly isolated from urban residents. However, farmers began to have greater contact with urban citizens under the social trinary structure, which was a significant change. To illustrate this change, it is necessary to review the history of Chinese farmers' interactions with urban residents since 1949, when the People's Republic of China was founded. This history can be divided into four phases.

1.1 The Interaction between Farmers and Urban Residents before the Establishment of a Strict Residence Registration System (1949-1958)

Opinions differ as to when the residence registration system was established. Some hold the opinion that it began in 1958, but others think it began in 1956. I believe that it began on June 6, 1955, when the State Council passed the *State Council's Directions on the Establishment of the Residence Registration System.* From then on, a strict residence registration system gradually developed. Therefore, I regard this as the starting point of the residence registration system. In 1958, the National People's Congress passed the *Regulations on Residence Registration in the People's Republic of China,* which legally authorized the resident registration policy. But, the evolution of social policy and systems usually has a lag phase. It was not until 1959 that a strict residence registration system began to influence the lives of Chinese citizens. Therefore, I choose the period between 1958 and 1959 to research the residence registration system in the real lives of urban and rural residents.

Since the People's Republic of China was founded, the government has been characterized by a centralized and unified administrative system. This system can be referred to as a planned economic system. As a basic system, it has influenced every aspect of social life in China. The influence of this system on the relationship between urban and rural residents lies in its separation of urban and rural areas into different management systems, thus reducing the free market-oriented interaction between farmers and urban residents in civic society. Therefore, since the founding of the PRC, the nature of interactions between farmers and urban residents has changed, though it has taken a relatively long period of time for this change to reach its final state. The most significant feature of this change was the establishment of a rigid residence registration system.

Prior to 1949, the interaction between farmers and urban residents was usually free and market oriented. After the founding of the PRC, however, this system of free interaction was gradually replaced by an organized system controlled by the government. It can be divided into two kinds of contact. One kind was between urban residents and farmers who moved from rural areas to cities; the other was between farmers and urban residents who moved from cities to rural areas. For example, at the beginning of this new regime, many soldiers who used to be farmers moved into cities, exemplifying the first type. Their direct contact with urban residents was quite limited and, after a while, many of them were required to return to their rural homelands because they lacked the skills needed to find work in a city. As for the second type of contact, during the land reform process, many urban students, intellectuals, and young cadres were sent to rural areas to carry out the land reform campaign. Their interactions with farmers had strong political significance as they passed on political ideas to farmers and were influenced by farmers and farmer-officials, who performed both roles simultaneously.

Apparently, immediately after the birth of the PRC, social life retained some features of the previous society. For instance, the interactions between farmers and urban residents had some features of free interaction. Farmers could look for jobs in cities without limitation. They could also become vendors or conduct small-scale business. Prosperous urban households could hire farmers as nannies. The first Constitution of the PRC (the Constitution of 1954) expressly stipulates in writing that the citizens of the PRC have the freedom of life and movement, which meant that farmers enjoyed the freedom to move to a city and gain legal identities as urban residents. However, this civil right was taken away in a subsequent version of the constitution.

1.2 Interactions between Farmers and Urban Residents after the Establishment of the Residence Registration System (1959-1967)

Although a relatively strict residence registration system was put in place after the National People's Congress passed the *Regulations on Residence Registration in the People's Republic of China* in 1958, rigid control over farmers' movement into cities began to be exercised in 1959, or even as late as 1960. The direct reason for this rigid control was the severe economic recession caused by the abortive Great Leap Forward, which was also referred to as the "three-year difficult period" ("三年困难时期"). From then on, urban residents and farmers lived in two totally different worlds. That was the dual structure under which there was basically no free interaction between urban residents and farmers. Occasional contact was limited to farmers coming to the cities to shop or visit relatives, or to urban intellectuals, cadres, and students moving to rural areas to strengthen themselves through manual labor. This kind of short-term interaction was very superficial and formed no deep interrelations.

During this period of organized interaction, it was very rare for farmers to move to cities. Urban residents moving to rural areas was more common. There were two relatively large-scale forms of interaction. The first was during the "three-year difficult period" in the late 1950s and early 1960s, when many city residents were sent to rural areas. The residents being sent to rural areas can be divided into two groups. The first was made up of cadres and intellectuals who were exiled to rural areas for political reasons. For example, some rightists were exiled to rural areas. The other was made up of urban workers who were persuaded to move to rural areas by food shortages in cities. These are both examples of long-term migration. The second occasion was a period of short-term migration during the Four Clean-Ups Movement ("四清运动"), which ran from 1963 to 1966. Large numbers of urban cadres, intellectuals, and students were required to participate in the Four Clean-Ups Movement in rural areas; while working on the projects related to the movement, they ate, lived, and worked with farmers. The nature of these organized interactions was entirely distinct.

Let us first examine the movement of urban residents to rural areas. Because of political or economic problems, the status of urban residents who moved to rural areas tended to degrade. Therefore, when they moved to rural areas, they did not feel superior to farmers, psychologically. In addition, for political and ideological reasons, farmers—especially poor peasants—were afforded a very high status. Therefore, these urban residents did not discriminate against farmers; on the contrary, they had great respect for farmers. However, from the farmers' point of view, these people were urban residents and some of them were even high-ranking officials. Even workers in cities had a higher status. Farmers always judged people according to economic standards rather than political or ideological ones. When they learned that the incomes of people who migrated to rural areas were much higher than theirs, farmers decided that these people had higher status. Therefore, farmers never treated these urban residents poorly, but showed a lot of respect. Because of this mutual respect, they enjoyed very harmonious relations during that period and, on the whole, they enjoyed equal status.

The second movement of urban residents to rural areas was different. These urban residents were ordered by their superiors to go there as members of work teams in the Four Clean-Ups Movement. They went there to lead and give instructions to farmers during the Socialist Education Campaign. Therefore, farmers were in awe of them, even though they also treated farmers well and tried to eat, live, and work together with them. But, apparently, urban residents enjoyed a higher social status than farmers.

1.3 Interactions between Farmers and Urban Residents during the Cultural Revolution: A Period of Sending People to the Countryside (1968-1979)

During the Cultural Revolution, a large number of cadres and intellectuals were sent to the May 7th Cadre Schools or rural areas to transform their bourgeois ideology through labor. Millions of educated urban youths were also sent to work in the countryside or mountainous areas. During this period, the number of urban residents moving from cities to the countryside greatly exceeded the previous period, making this a large-scale, national interaction between farmers and urban residents. Due to the large scale and length of this campaign, the urban residents who moved to the countryside had much more contact with farmers. Their interrelations were similar to the previous period in that they had mutual respect for each other. Sometimes, conflicts arose because of the shortages of resources caused by the overflow of urban residents. Generally speaking, however, the relations between farmers and urban residents were harmonious. This period ended with the return of these urban cadres, intellectuals, and educated youths to the cities.

1.4 Interactions between Farmers and Urban Residents after the Reform and Opening Up Began (1980-present)

Since the reform and opening up began, the nature of the interactions between farmers and urban residents has undergone some fundamental changes, which can be categorized as follows. The interactions changed from being organized to being dominated by free, market, and civic society-oriented interactions. Structural reform in China at this point was essentially a transition from a planned economy system to a market-oriented system. One important feature of the market is the free mobility of labor, which results in free interactions between farmers and urban residents. The result of the market transformation was the formation of a civic society. Therefore, the interactions between farmers and urban residents were free from political and administrative control.

Additionally, the direction of mobility shifted from urban-to-rural to rural-to-urban. China witnessed large-scale, organized migration into rural areas in the 1960s and 1970s in a very special political and economic climate, which had great influence on the interactions between farmers and urban residents. After 1979, the political and economic environment that had previously forced urban residents to move to rural areas was gone. Consequently, China came to develop the same trend of urbanization as other countries that were also in the process of modernizing.

Living standards also changed from being largely equal to a situation in which farmers had to keep up with urban residents. For a long time, narrowing the gap between urban and rural areas was ideologically set as a basic social objective. However, before the reform and opening up began, the approach was taken of narrowing the gap by adjusting the lives of urban residents to make them more like those of farmers. This included the frequent organization of urban residents to send them to labor in the countryside, characterized by the slogan of learning from poor peasants, sending urban students to the countryside to receive re-education from poor peasants, and so on. After the reforms started, the basic direction of social coordination changed such that farmers were now encouraged to keep up with urban residents. Urbanization and urban life became the basic target of social development. As a result, the relations between farmers and urban residents changed accordingly, from following the farmers' model to following the example of urban residents.

Finally, the social statuses of farmers and urban residents changed from being similar to being obviously different. As mentioned above, in the 1960s and the 1970s urban residents moved to rural areas because of either political or economic issues. Therefore, the status of urban residents was declining and had no apparent distinction from that of farmers. Since the 1980s, when farmers began to move to the city, they have encountered very few urban residents who were similar to them in status. Again, the differences in social status caused by the residence registration system became prominent.

Since the 1980s, increasingly extensive, substantive, and direct interactions between farmers and urban residents have taken place, but their relationship is not as harmonious as before. Conflicts between farmers and urban residents became increasingly open and widespread, which not only influences the stability of urban society, but also the process of urbanization. This ultimately influences the alliance of workers and farmers and the stability of the whole of society. Therefore, researchers must pay attention to the relationship between migrant workers and urban residents. I will attempt to conduct a preliminary exploration of this issue in this chapter.

2. Conflicts between Migrant Workers and Urban Residents

When a large number of farmers came to settle and work in cities, interactions with urban residents in cities were inevitable. These interactions have had two aspects; one is harmonious and smooth while the other is contradictory and conflicting. The contradictions and conflicts between migrant workers and urban residents could result in many negative consequences, which could jeopardize the process of urbanization in China. The following section will focus on the contradictions and conflicts, paying less attention to less problematic aspects of these interactions.

2.1 Urban Residents' Attitudes towards Migrant Workers

From the viewpoint of urban residents, the migration of farmers into cities has both advantages and disadvantages. The positive side is that it provides a large supply of energetic young laborers to cities. Nowadays, most physical work in cities is performed by migrant workers. Migrant workers have been making a great contribution to the urban construction of China for decades.

However, many problems are caused by the farmers' migration. The inflow of large numbers of farmers has increased the burden on urban infrastructure. Their demands for clothing, food, shelter, and transportation has exerted pressure on cities' material supplies. Because of the limitations of urban infrastructure, living space, and job opportunities, the inflow of a large number of migrant workers has put pressure on cities to provide sufficient shelter, transportation, water, electricity, and fossil fuels. At the same time, it has brought about disorder in social security, the city environment, and public hygiene. Social security plays an important role in the stability of cities. In recent years, as the floating population has increased in size, the crime rate has continuously increased. According to statistics collected in the early 1990s, crimes committed by temporary urban residents accounted for more than half of all criminal cases. As indicated in the statistics for Beijing in recent years, the crimes committed by people who are not permanent residents of Beijing accounted for about 80% of all Beijing's criminal cases. Moreover, some migrant farmers produce and sell counterfeit and low-quality products, resulting in heavy losses borne by urban residents.

In addition, the inflow of rural residents into cities and the ability of migrant workers to move all over the country puts great pressure on the limited transportation capacity in China, particularly railway transportation. Hundreds of millions of farmers move back and forth between cities and the countryside year after year. Before Spring Festival, they return to their hometowns and move back to cities after the holiday. This kind of mobility is quite different from that of most industrialized countries. The mass movement of migrant workers into cities occurs only once in many Western countries. So, there is no such problem as this annual migration. In comparison, the transportation pressures produced by the floating population in China are much higher than those of Western countries. Every year after the Spring Festival, farmers leave their homes to look for jobs in cities, which produces a vast amount of railway traffic. In some railway stations, passengers accumulate in the station as they wait for trains with free space. Sometimes, the overload rate reaches 100%. Some trains are so overloaded that the compartments are damaged and accidents become very common.

Objectively speaking, the flow of farmers into cities puts pressure on the material supplies, transportation, and social security in cities, which indeed causes damage to the living environment of some urban residents. This leads to complaints among some urban residents, which gives rise to contradictions between migrant workers and urban residents.

The opinions of urban residents on migrant workers varies greatly due to the positive and negative effects of migrant workers in cities. According to surveys of urban residents' evaluation of migrant workers—which were conducted in Beijing, Shanghai, and Wuhan in 1996—29.2% of urban residents held a positive attitude. Of the urban residents, 48.8% thought there were both advantages and disadvantages, 18.1% held a negative attitude, and 3.8% of the people found it hard to evaluate.[98] According to these statistics, the second attitude was the most common, indicating that this trend has both positive and negative sides. This kind of attitude was rather objective. Overall, it was obvious that a positive attitude prevailed.

These surveys also reveal the specific contents of positive and negative attitudes. The reasons for and percentage breakdown of positive attitude were as follows.[99]

98 Ling Yue. "Conflicts and Tolerance." *China Youth* Jan. 18, 1997.
99 ibid.

"They do most of the tough jobs that urban residents are unwilling to do."	52.8%
"They bring convenience to the daily lives of urban residents."	7.8%
"Farmers coming to cities have narrowed the gap between urban and rural areas."	11.4%
"They bring income to urban residents, such as rent."	2.9%

Besides the above, 5.1% of urban residents gave other positive opinions. From the positive answers listed above, the chief reason is that farmers shoulder the toughest jobs in cities, a fact commonly acknowledged by urban residents.

The evaluation and percentage of negative attitude from the same surveys are as follows.[100]

"They have caused a rise in criminal offence and crime rates."	48.8%
"They don't care about personal hygiene and make places dirty."	14.0%
"They encroach on the public resources of urban residents."	13.4%
"They take jobs and development opportunities away from urban residents."	12.3%
"They objectively make urban residents lazier."	2.9%
"They make money from what belongs to urban residents."	2.1%

It is clear that the biggest problem that urban residents complain about is the rise of the criminal offence and crime rates. Psychologically, offences and crimes pose a threat to the lives and security of people, so it is understandable that most people choose this answer. The offence and crime rates have risen rapidly in cities recently, and migrant workers

100 Ibid.

account for a large proportion of convicted criminals. As a result, the mentality that migrant workers have led to a rise in offence and crime rates is very common among urban residents. However, out of all the migrant workers, the percentage of criminals is very small. So, migrant workers feel wronged and think it is unfair to blame the problem of crime on them. In an individual case study I carried out, many migrant workers argued that they never do anything against the law and are opposed to people linking crime with migrant workers.

As illustrated in the first part of this chapter, great differences exist between the migrant workers moving into cities and urban residents in terms of employment and living style. Difference means conflict. If we closely observe the relationship between rural migrant workers and urban residents, we will find many problems in their relationship. This does not mean that there are no good sides to their relationship. Some migrant workers enjoy a harmonious relationship with urban residents. However, this chapter will focus on an exploration of the problematic side from the perspective of research and analysis.

Regarding this issue, most current media are concerned with the opinions of other groups towards migrant workers rather than this group's self-evaluation. In recent years, assisted by students from the People's University of China, I have conducted investigations into the individual cases of more than 150 migrant workers. According to the analysis of these cases, it has been found that most of the migrants surveyed mentioned the problem of being looked down upon and discriminated against, believing this to be the most dissatisfactory and unbearable part of their lives, while poor material conditions came second. I believe this issue is very worthy of attention because the two groups will have to coexist in the long term in most cities, and so must adapt and cooperate with each other in the following decades.

2.2 The Discontent of Migrant Workers with Urban Residents

Case studies reveal that many migrant workers are rather discontented with urban residents, especially urban residents' attitudes. This can be found in most cases. The complaints they make will be listed as follows.

Case 16. Male, thirty years old, from Kaifeng, Henan Province. Education level: elementary school. Occupation: waiter in a restaurant. When asked what is most dissatisfactory to him, he said, "There are so many things I am dissatisfied with. The most unbearable thing is the urban residents' disdain towards us migrant workers. It is infuriating that all the people, from officials to dustmen, look down upon us."

Case 9. Female, twenty-two years old, from Ningjing, Hebei Province. Education level: junior high school. Occupation: waitress in a cafeteria. She said, "Although Beijing is a nice place, it is not suitable for us to live here for a long time. People in Beijing seem to be condescending, always looking down upon us, especially on those from the countryside. Their attitude is so apparent that I cannot stand it. Of course, I will tolerate it as much as I can. Once I've earned enough money, I will definitely go home and never go out again as a migrant worker."

Case 34. Male, twenty years old, from Zhe County, Henan Province. Education level: junior high school. Occupation: construction worker. He says, "Generally speaking, I find Beijing a very nice city; the environment and transportation are good here. But I always feel like Beijing residents look down upon nonlocals, especially migrant workers like us. Actually, we are the same as them. We make a living on our own. They have absolutely no reason to look down upon us."

Case 48. Male, twenty-six years old, from Liulin town, Yueyang City, Hunan Province. Education level: high school. Occupation: clerk in a company. He said, "Beijing residents are xenophobic and look down upon nonlocals. I find them lazy and bad tempered like the spoilt sons of a rich family. They speak much while doing little. When they talk, they can speak much more than anyone else, but when it comes to real business, they are not as good as us southerners."

Migrant workers feel dissatisfied because of psychological discrimination, which leads to them developing an aversion to urban residents. The discontent reflected in the above cases is quite common among the more than 150 cases I analyzed. Other similar surveys also have produced the same result. According to surveys conducted in Beijing, Shanghai, and Wuhan from 1994 to 1996, 45% of those engaged in business with migrant workers suffer some discrimination from urban residents, and 18% have felt strong discrimination from urban residents. If these two groups were combined, that would mean that 63% suffered discrimination from urban residents. The same survey also shows that two-thirds of those engaged in business with migrants did not dare, or even want, to make friends with urban residents.[101] Therefore, both the case study and statistical surveys show that it is very common for migrant workers to face discrimination.

However, ironically, many urban residents do not realize that their behaviors show discrimination. As indicated before, only 16.5% of urban residents were aware that they looked down on migrant workers, meaning 83.5% people were not aware of their discrimination against migrant workers. One party believed that they were discriminated against while the other party did not believe this was the case. How could this happen? Sociological theory holds that what people genuinely feel is a social fact. How could such two opposite social facts appear? I hold it involves symbolic interaction. The communication between people cannot be carried out directly; it depends on symbols. The different interpretation of two people, or two groups, of these same symbols will often result in conflict. How could the same symbol have two opposite interpretations? This is a result of the different cultural and social backgrounds of the two people or groups. Urban residents and migrant workers live in different social situations, which inevitably results in them developing different concepts. Consequently, this indicates that this gap between urban residents and migrant workers cannot be closed in the short term.

The exploration of the problems reflected in the analysis of different cases will be continued. In the case study, migrant workers also give many examples of the discrimination they experience from Beijing residents.

Case 29. Female, eighteen years old. Education level: junior high school. Occupation: saleswoman. She said, "On the buses, the conductors look down upon us. When we ask about stops, they are indifferent. They always check our tickets just because we dress poorly. What's more, whenever something is missing, the first to be suspected are migrant workers. I never steal things. The most unbearable thing is that people from the industrial and commercial bureau are very mean to nonlocal vendors who sell their goods on the streets and who come from the country to sell vegetables. Sometimes stands are knocked over and goods are smashed to pieces, which is really infuriating. Sometimes we don't know much about the policy for migrant workers; they should explain it to us instead of being so rude. After all, they look down upon us."

101 ibid.

Case 34. Male, twenty-eight years old, from Huaining, Anhui Province. Education level: dropout from junior high school. Occupation: salesman. He said, "When I take a bus, some conductors are unreasonable. When we come back to the city with a lot of luggage, sometimes they charge us extra money, much more than one or two tickets as it says in their rules. We feel angry but have no choice but to pay."

Case 11. Male, forty-two years old, from Ningsha County, Hebei Province. Education level: junior high school. Occupation: service person in a private clinic. He said, "To be honest, People in Beijing are rather high handed. On the bus, I see conductors bully people who are not locals. I consider myself lucky not to have run into something like that, but I still feel angry as a watcher."

Case 21. Female, nineteen years old, from Sichuan province. Education level: elementary school. Occupation: nanny, later working as a cook for a construction team. She said, "At the end of 1992, a friend of my brother decided to look for a job in Beijing, so I came with her. People from Sichuan are very united. As soon as I came to Beijing, I was introduced to work as a nanny. The family I worked for had an apartment with three bedrooms and a living room. I slept on the balcony. During the daytime, they locked all the three bedrooms, so I could only go into the living room, kitchen, restroom and balcony. At night, I could not go into their rooms without calling, which meant I seldom had a chance to watch TV. I had to clean the bedrooms only when they were there. They gave me twenty or thirty yuan for preparing one day's meals in the morning and I had to note down all the things I bought in a notebook. The family was well-educated and wealthy. They paid me 200 yuan a month and sometimes gave me a little bit more or gave me some used clothes and daily necessities. They never scolded me. I got up at 5 o'clock every morning and wasn't able to go to bed until 12 at night. I didn't care about all this. But what made me uncomfortable was that they looked down on me and were suspicious of me. One day, some money was lost. They suspected that I had taken it. I couldn't stand it any longer and had no choice but to leave. Most Beijing residents treat us nicely, but sometimes some people say some rude things to us and look down upon us because we are poor or dressed shabbily. On the buses, whenever we get on, all people on the bus are not happy and stay away from us. The conductors always keep an eye on us and yell at us to buy tickets. In the department stores, salespeople tend to ignore us. If we want to look at or try on something, it takes a long time before they come to us. What scares me most are the people in various uniforms. It is said that people from the police station, the Industrial and Commercial Bureau, the Tax Bureau, the Urban Administrative Bureau and security are very powerful and they often yell at us. Sometimes, the foremen are very rude, too. Our salary is not paid on time every month and they act shamelessly. They make difficulties for us on purpose. But no one dares to say anything. What can we do about it? We make a living from them. Sometimes when I recall all these experiences, I feel that being a migrant worker is really miserable."

Case 57: Female, twenty-one years old. Education level: dropout from elementary school. Occupation: carrying vegetables to Beijing and selling them there. She said, "When I came to Beijing, I found it very clean and beautiful. It was after I came to Beijing that I realized that there are many new things in the world. However, Beijing residents are not friendly towards us and are high handed. On the buses, the conductors always stare at us and yell at us. People in uniforms such as policemen, people from the urban administrative bureau, the health epidemic prevention station, environmental protection, tax departments and neighborhood committees, etc., try to control us. They yell at us. I am so afraid of them."

Case 49. Female, twenty-one years old, from Xinyang, Henan province. Education level: high school graduate. Occupation: salesperson, nanny. She said, "I think Beijing residents tend to bully nonlocals, especially people like us from rural areas. The son and daughter-in-law in the family I work for as a nanny believe that people in rural areas are ignorant. What I dislike most is that they often show disdain and contempt."

From these cases, it can be seen that discrimination occurs in two different situations: in public and in residents' families. Public discrimination is described by migrant workers as often taking place on buses or in department stores. Are these facts? It should be admitted that these problems do exist. As newcomers in an unfamiliar environment, migrant workers face great pressure, which—compounded by unfriendly attitudes from service personnel—likely produces a strong sense of discrimination. Although attitudes towards customers among public service people in Beijing have improved, there is still a lot of room for improvement. Actually, not only the migrant workers but also urban residents in Beijing often criticize this tendency. However, as Beijing residents and migrant workers are from different backgrounds, they feel differently about it. Beijing residents notice these unfriendly attitudes towards customers without feeling any sense of discrimination. Some urban residents even attribute such bad attitudes to the flow of migrant workers into Beijing and show sympathy and understanding towards these rude service personnel. The same experience makes migrant workers feel discriminated against and oppressed. Of course, I do not deny the fact that some service personnel do discriminate against migrant workers.

Another form of discrimination occurs within residents' families, which is a problem mostly encountered by migrant workers who are hired as nannies. This kind of contradiction is more covert. Nowadays, it is common for migrant workers to work as nannies in urban residents' families. Most of them are young women. It should be noted, however, that most migrant workers enjoy a harmonious relationship with their employer's family. Those experiencing tense relationships and discrimination are few in number. In cases of conflict, some responsibility lies with urban residents as some do indeed discriminate against migrant nannies. Since most nannies are young women, there are also problems of sexual harassment. As for harassment, I once carried out an investigation and published papers especially devoted to this problem; therefore, it will not be re-discussed here.

2.3 Conflicts between Migrant Workers and Urban Residents

The accumulation of discord and distrust in interpersonal relationships usually produces conflicts, ranging from squabbling to violence. As for the conflicts between migrant workers and urban residents, the media has tended to take the standpoint of urban residents and focus on the threat and conflicts caused by migrants. In my case studies, another voice can be heard: the complaints of migrant workers, who mostly talk about conflicts caused by urban residents.

Case 26: Female, twenty-three years old, from Jiangxi province. Education level: junior high school. Occupation: waitress. She said, "Some urban residents look down on us. The other day, when I went out shopping, a middle-aged woman forced me to make way for her. She stared at me rudely and mumbled something which I didn't catch clearly. I thought to myself: who do you think you are? You guys would starve without the vegetables that farmers grow and sell."

Case 7: Male, thirty-five years old, from Wenling, Zhejiang province. Education level: dropout of junior high school. Occupation: street vendor. He said, "People in Beijing are very brusque. For instance, when they ride on their bicycles on the road and run into somebody's bicycle accidentally, they immediately curse at you, no matter whose fault it was. They are really good at cursing. Take another example; after buying some stuff, even if they break the stuff themselves and the problem isn't the quality of my goods, they often insist on returning the goods or demand a refund. If I don't agree, they quarrel or yell at me. What can I do? Quarrelling has a very bad effect, so I give in. Of course, many people in Beijing are nice, friendly; for example, my acquaintances."

Case 8: Female, twenty-eight years old, from Wuwei, Anhui province. Education level: junior high school. Occupation: clothing salesperson. She said, "Sometimes, people in Beijing bully nonlocals. Just take us, for example. If they buy clothes from us, they force us to let them return the clothes when they find a tiny part unsatisfactory. But if they buy clothes from Beijing locals, they would not dare to return the goods, even if they find them unsuitable."

Case 37: Female, twenty years old, from Wenzhou, Zhejiang province. Education level: elementary school. Occupation: salesperson in a market. She said, "People in Beijing are very mean to us and condescending. I don't like it. I know a lot of nonlocals who are cursed at or even beaten up. Although these beatings didn't happen to me, sometimes I hear rude remarks. At first, I felt extremely angry, but got used to it later. After all, I am not from Beijing."

Case 45: Male, thirty-five years old, from Jiangfang county, Rumu, Jiangsu province. Education level: junior high school. Occupation: private owner of a clothes shop. He has lived in Beijing for more than ten years. He said, "Most people in Beijing are nice but some guys are very arrogant. Last year before Spring Festival, a bicycle repairman and I got into a fight. He regarded himself as a Beijing local and supposed himself to be superior to me. But I thought to myself that he is no more than a repairman. Later, he called some gangsters to seek revenge. I went to the police station and found a policeman I knew. I asked the policemen to stop them. The policeman told the repairman not to make a scene. The repairman immediately obeyed the order and did not dare to confront me."

Case 64: Male, twenty-one years old, from Anqing, Anhui province. Education level: elementary school. Occupation: an interior decorator. He said, "Beijing locals are tigers in the mountains and we nonlocals are tigers that are climbing into the mountains. We don't want to or dare to wake up sleeping tigers. If we can tolerate something, we do. The most dissatisfying thing for me is that they don't pay migrant workers after we finish our work. These kinds of things haven't happened to me, but to my fellow townsman. He told me that they made a deal at the beginning. However, after the work, they put a knife to his throat and forced him to leave without any wages. What could he do? Nothing. He had to leave. What if this kind of thing happens to me? Just accept my bad luck. We won't go to the police station."

Case 24: Male, twenty-two years old, from Anhui province. Education level: junior high school. Occupation: first a carpenter, later a fruit vendor. He said, "Beijing is a great place that offers an opportunity for you to see the world. But, Beijing locals are arrogant and look down upon us. Besides, some gangsters or rascals often pick on us nonlocals, especially vendors like me. They come here for money. If we refuse to give it to them, they will attack our stalls, smash things and raid our businesses. Fortunately, three of my brothers-in-law are armed policemen and they help me to keep away from these gangsters. Last year after the Strike-Hard-Action, doing business became much easier with fewer troubles from these gangsters."

Case 22: Female, thirty-four years old, from Jiangxi province. Education level: elementary school. Occupation: vegetable vendor. She said, "Generally speaking, customers are polite and we are happy. But some Beijing locals are not that polite and bargain with us as if they are joking. Words such as 'hillbilly' and 'dirty' are often found in their conversation. We are human beings, too. If you want to buy things, just buy them. If you don't like them, you can leave. Why look down upon us like that? The other day, a young man riding a bicycle knocked over my stall. He refused to pay for my losses and even blamed me for making his tire dirty. He was so unreasonable. But they are local villains. I can do nothing about them."

According to a survey of 1,200 migrant workers in Beijing and Shanghai, carried out by the Horizon Research Consultancy Company in 1994, 24.3% of migrant workers admitted that they had gotten into a fight after moving to these cities. The people they fought were local urban residents in 43.4% of cases, fellow townsmen in 5.4% of cases, and unidentified people in 15.5%.[102] It is clear that migrant workers mostly get into fights with urban residents more often than with members of other groups.

In the past, when it has come to conflicts between migrants and urban residents, most media and research have taken the side of urban residents and claimed that these conflicts are caused by the migrants. It is worthy of note that I took the side of the migrant workers in my case studies and found that urban residents are usually the cause of conflicts. From the standpoint of migrant workers, the conflicts mentioned in the above cases can be categorized into two types. The first type involves conflicts between ordinary migrants and ordinary urban residents, mostly induced by trivial disagreements in daily life or business. The barrier and social distance between the two groups tends to aggravate such disagreements, which both sides should be held accountable for. Cases 7, 8, and 26 are of this type. The other kind involves conflicts between ordinary migrant workers and "aggressive urban residents." The term "aggressive urban residents" refers to bellicose urban riff-raff who are usually responsible for the conflicts. Cases 45, 64, and 24 are of this kind. Another type of conflict, which is only very sparsely reflected in these cases because of the subjects' biases, involves "aggressive migrant workers" and ordinary urban residents. "Aggressive migrant workers" means violent, or even criminal, migrant workers who obviously hold the responsibility in this type of conflict. Urban residents complain a great deal about this kind of conflict and some even grow hostile towards migrant workers.

3. How to Properly Handle the Relationship between Migrant Workers and Urban Residents

In order to handle properly the relationship between migrant workers and urban residents, the causes of conflicts between the two groups must first be analyzed.

3.1 Analysis of the Causes of Conflicts between Migrant Workers and Urban Residents

Discrimination against migrant workers and the social conflicts discussed here are very complex.

102 Yuan Yue, et al. "Naked Man." *The International Symposium of Mobility of Rural Labors in China.* Beijing: *Horizon Research Consultancy Group*, 1996. 121.

Firstly, it should be stated that some urban residents do discriminate against migrant workers, though they are few in number. However, the negative influence of discrimination cannot be underestimated because it is very easy for migrant workers to interpret the discrimination they face as being the general attitude of urban residents.

Although most urban residents do not actively discriminate, many are sometimes disrespectful towards migrants, which inevitably leads to their resenting urban residents.

Finally, it should be noted that due to their different cultures, backgrounds, and statuses, etc., it is inevitable that there will be some misunderstandings between migrant workers and urban residents. For example, some words or phrases may have no negative connotation in the minds of urban residents but might sound discriminatory to migrant workers. Just as the old saying goes, "The speaker has no particular intention in saying something, but the listener reads his own meaning into it" ("言者无意,闻者有心"). It is obvious that the two groups do not communicate effectively.

Besides this, many other problems have centered on migrant workers in recent years. Although most migrant workers are law-abiding citizens, a small number of them do violate the law or even commit crimes, tarnishing the reputation of migrant workers. Migrant workers have mentioned this kind of problem in some of my case studies. For example, in Case 8 (the one from Wuwei, Anhui province), the interviewee said, "The attitudes of Beijing locals are worse than before. Several years ago, they were very nice. It is not Beijing locals who are to blame. Some nonlocals do all kinds of bad things such as stealing, robbing, or murder, etc. in Beijing; the nonlocals are degrading themselves." In recent years, a tiny number of nonlocal criminals have committed crimes in Beijing, disturbing the social order and bringing a sense of insecurity to urban residents. This is a very important reasons for the distrust of, and discrimination against, migrant workers.

Of course, an exploration of the deep-rooted causes of this problem reveals the influence of the long-term system of separating urban and rural areas. The dual-structure system of separating urban and rural areas was first implemented in the 1950s and has endured to the present day. Before the reform and opening up began, although there were huge differences between urban and rural areas, most farmers did not have the opportunity to live in cities because of the strict restrictions of the residence registration system, not to mention the conflicts between farmers and city residents. After the reform and opening up began, farmers were allowed to move to cities and find work, so they flowed into cities in a large numbers. For the first time, these two groups with obvious social differences lived in the same urban community. The two groups lack the ability to communicate and identify with each other, as they were very different in terms of social norms, culture, and backgrounds. Such differences easily lead to contradictions and conflicts.

Over the past decade and more, a large number of rural residents have moved into cities, but the two groups are still sharply divided. As mentioned above, migrant workers differ from urban residents not only in their registered residence identities, but also in the worlds they live in. Migrant workers live in "villages within cities." Their occupations, lives, daily interactions, behaviors, marriage patterns, and living environment are significantly different from those of urban residents.

At a deeper level, the conflicts between urban residents and migrant workers are conflicts between the modern age and traditional times, between interest monopolization and egalitarianism. There is unfair competition for employment and conflicts occur between white-collar and blue-collar workers, between an "ego group" and a "foreign group."

To be more specific, migrants normally come from relatively backward communities. Their way of thinking and their behaviors carry traditional features, whereas the urban residents, who are born into a modern civilization, show more modern traits in terms of thinking and behavior. Therefore, it is inevitable that there will be conflicts between them.

For a long time, city residents enjoyed too many privileges and social welfare, so they came to take it for granted. When farmers come to cities and use the opportunities there to earn a living through their own hard work and improve their living standards, urban residents think migrant workers are taking advantage of them, as if the cities belong to them. In fact, urban prosperity and development and the increasing opportunities to become rich cannot be separated from the contributions made by all migrant workers in the last decade and more.

Although most migrant workers do jobs that urban residents are unwilling to do, some urban residents—especially the unemployed—still believe that migrants are taking their jobs. When some migrants' incomes rise higher than those of urban residents, some urban residents feel psychologically disturbed, as if their income has to be higher than that of "farmers." Given the limited number of available posts and insufficient employment among urban residents, it is understandable that urban residents feel anxious and complain. However, it is unreasonable that this anxiety does not encourage urban residents to improve themselves and is not directed at unsuitable policies and systems, but is vented instead on migrant workers.

With the adjustment of the industrial and occupational structures, many urban residents have entered the "white-collar" stratum, so they enjoy a new sense of superiority. In the meantime, migrant workers have continued to do the dirtiest and most exhausting jobs and have remained part of the "blue-collar" stratum. Thus, some conflicts between urban residents and migrants can be regarded as conflicts between different occupations and statuses.

Finally, because migrant workers have not been allowed to register as urban residents, they are a "foreign group" in the eyes of urban residents. This "foreign group" does not have a sense of urban identity, so they also have no sense of responsibility. Therefore, some migrants lack self-control in cities and feel no guilt or regret for behavior that obstructs and disturbs social order, and even for crimes that they commit. Some even take it for granted, since they knew that they will not live in the city permanently. Consequently, such behaviors harm the interests of permanent urban residents and give rise to conflicts between the two groups.

All the factors mentioned above cause or aggravate conflicts between urban residents and migrant workers. Compared with other countries, it can be observed that in developed countries, it is through the process of urbanization that class differentiation and class conflicts arise due to migration. This has remained one of the most important causes of conflicts in the cities of those countries up to the present day. Therefore, the importance of properly handling the relationship between these groups cannot be overemphasized.

3.2 Strategies for Regulating the Relationship between Migrant Workers and Urban Residents

These contradictions are generated by the interaction of two groups with different social statuses and interests, as well as by a lack of communication and recognition. Therefore, the basic principles of problem management dictate that a solution must involve reducing the differences and strengthening the communication between the two groups.

It is necessary to guarantee the legitimacy of migrant workers. In the current situation, though the time for the reform of the residence registration system in large and medium-sized cities has not yet been reached, it is feasible in small towns. In addition, all migrant workers should be offered official quasi-urban identities, providing them a certain status and reinforcing their sense of urban identity.

It would also be constructive to encourage urban residents to treat migrant workers fairly. Both migrants and urban residents are citizens of the People's Republic of China and therefore have the legal right to work and live in cities. They are equal in status. Migrant workers have made a great contribution to the economic and social development of cities. Besides, many good qualities can be learned from migrants, such as diligence and frugality. Urban residents who create difficulties for and bully migrant workers should be criticized and punished, and those who violate the law should be strictly punished.

Additionally, concern should be directed towards migrant workers and they should be made to feel welcome in cities, reducing their sense of alienation. In the meantime, it is necessary to encourage migrants to be disciplined and law abiding and to improve the personal qualities of all citizens. Even more important is to improve the cultural and technical qualities of migrant workers through different means and narrow the gap between migrants and urban residents. When the personal qualities of migrants improve, so will their confidence. Their sense of inferiority will disappear, which will be beneficial to their normal communications with urban residents.

Establishing a system capable of fairly and efficiently regulating the relationship between urban residents and migrant workers would also be constructive. The management of migrant workers should be improved, as should the system for responding to the needs of migrants, satisfying their reasonable needs. At the same time, the state must strengthen the supervision of management departments and their behavior, preventing the bullying and exploitation of migrant workers.

Finally, more opportunities for interaction, communication, and understanding between the two groups should be created. For instance, in Chinese cities, the "work unit" is the core of social activity for many workers. At present, most "work units" employ migrants but neglect them in social activities and inter-unit communications. There is a great deal of room for further exploration in this field of communication.

In the above case studies, some migrants mentioned that urban residents should treat them equally. Just as some migrants have said, discrimination is far harder to bear than material hardship. Recalling the guidance of public opinion and the mass media, little importance has been attached to migrant workers and the inadequate treatment they have received. In the past, migrants have been regarded only as a source of labor, not as a group with complex social needs. It could even be said that they are one of the two most important groups living in cities in contemporary China.

In summary, it is very clear that migrant workers have made a tremendous contribution to the construction and development of cities in recent years. In other words, migrant workers have shouldered great responsibilities and obligations in urban construction. However, responsibility, obligations, and rights are inseparable. One cannot enjoy rights without responsibilities and obligations. Those without rights will not be willing to shoulder responsibility and obligations. Even if they are forced to shoulder them, they will not do their best. At present, it is very common for people to not shoulder their responsibilities and obligations in urban public life, which damages the living environment in cities. Some problems indeed have much to do with the floating population of migrant workers. Therefore, it is very important that certain rights should be granted to migrants, as great responsibilities have been assigned to them in policy making.

Great importance should be attached to the contradiction and conflicts that exist between urban residents and migrant workers. Both the experience of modernization in all countries and the special situation and development trajectory of China indicate that migrant workers and urban residents will be the two major groups living in cities in future decades. Only through the reinforcement of communication, understanding, and cooperation between them—and the elimination of their estrangement, misunderstandings, and conflicts—can the long-term development of cities and the consolidation of the alliance of workers and farmers be achieved and the long-term and stable development of Chinese society be guaranteed.

Chapter 13

Housing Status Groups in the Transitional Period

The primary function of housing is to provide a place of shelter, warmth, and stability. However, in a stratified society, housing also symbolizes social status.

Since the market-oriented reform of the housing system in China, the biggest shift in housing status for the vast majority of urban citizens was that of becoming a homeowner rather than a tenant. Houses and property have become important assets for urban residents. Renting or selling houses has become an important means of making profit. Therefore, the significance of the economic status of homeowners has increasingly come under scrutiny.

A theory of the housing class was proposed by sociologists John Rex and Robert Moore in 1967.[103] It holds that people can be divided into different classes according to their housing. The number of housing classes was determined by the degree of variation in access to housing resources. However, this classification of class merely based on housing seems to lack sufficient scientific rigor. As a result, this theory has been widely criticized. The concept of a "housing status group," as it is used in this chapter, intends to avoid the confusion associated with the general concept of a housing class.

When a society functions normally, one's housing status is relatively stable, as is economic status. For most people, the chances of a sudden change in housing status are pretty slim. However, this is not the case in China. In the last thirty years, China has witnessed significant systemic reforms, particularly in its housing system. State-owned housing suddenly became privately owned overnight. The real estate market developed in leaps and bounds. Thus, forms of group status related to housing changed drastically as well. The status of some groups soared and that of some groups slumped. From the perspective of social stratification, this was also an important period for shaping social status. Therefore, the study of housing status groups in the transitional period is of great significance. Many social problems can be better understood through researching housing status groups.

---

103 J. Rex and R. Moore. *Race, Community and Conflict.* Oxford: Oxford UP, 1967.

1. How to Understand Housing Status Groups in Present China

1.1 What Are Housing Status Groups?

As mentioned above, social status refers to the particular position of an individual or group in a social structure. Status group refers to a group of people who occupy the same position in the same social structure. An individual can have various statuses—such as income status, educational status, and professional status, etc. Thus, people can form different status groups. The English equivalent of *di wei qun ti* ("地位群体") is "status groups," or groups with similar social position. The term "status groups" emphasizes that a group of people who are in a particular position in a social structure share the "same identity." When examined from a different perspective, or when a different position is reached, people will fall into different groups. So, status groups are flexible, loosely defined, and not as strict as the group boundaries of class. Therefore, housing status groups are one of a number of status groups. This group is composed of a group of people who share a similar social position through the influence of the housing they own or live in. The factors influencing their housing status are ownership of housing, price, location, rent differential, community environment, community cultural characteristics, and so on.

Housing status is just one of the various social statuses everyone has, so it only reflects one's overall social status from one perspective. People who share the same educational status may hold different income statuses, just as people who share the same housing status may be differentiated by other forms of status. Why, then, is housing status particularly emphasized here? Because, housing is the most important "material entity" without which people cannot live. Of all the indispensable material entities in people's lives, houses take up the most space, have the highest value, provide the strongest protection, and have the greatest durability. They can even provide lifelong protection. From the perspective of consumption stratification, clothing, apparel, jewelry, and appliances (including cars) also have consumption status, social status, and symbolic meanings. But, they are all of secondary importance compared to housing. Of all material entities associated with humans, the symbolic significance of housing-related social status is second to none.

In fact, there is a long history of housing symbolizing one's social identity and status. In ancient China, people emphasized the symbolic meaning of housing as an indicator of one's social status. According to the etiquette of Chinese imperial society, the rules related to the social status of housing were very complex, involving the size and modeling of the main gate, the steps of porches, the doornails in the main gate, the size and number of door stops, the design and specification of the gate pier, the pattern of a front door's engravings, the size and patterns on screen walls, the size of the courtyard, the type and number of carvings of beasts on the roof and ridges of houses, and so on. All of these reflect the precedence of determining the hierarchy, gentility, power, and wealth of different strata according to housing. Obviously, housing acting as an indicator of one's social status is not unique in China; it is a very common worldwide trend.

1.2 Housing Status Reflects Relatively Stable Social Relationships

People's lives involve many social factors, but not all these factors can be regarded as relatively stable status symbols. We will take the four basic necessities of life—clothing, food, shelter, and transportation—as examples. In recent years, some social groups and social strata have increasingly sought to possess famous brand-name clothes, a trend which has also caused a glut of knock-offs. This clearly indicates that the association between clothing and social status is not as stable as it appears. As for eating, luxurious and delicious meals are not exclusive to wealthy people; middle or lower income individuals occasionally indulge themselves with such meals. Of course, in comparison with other commodities, the significance of cars as a material entity is more evident. Therefore, wealthy people are seen frequently rushing to buy fancy cars at luxurious car exhibitions. For these wealthy people, cars worth tens of millions of yuan embody their social status in a way that is unaffordable for ordinary people. Additionally, cars, used for driving around, can show their income status at any time. However, in terms of their significance as status symbols, fancy cars cannot be compared to luxury housing.

Housing occupies a comparatively large space and area, putting it beyond comparison with other commodities. Firstly, space and area are associated with land prices and are granted important economic significance by rent differentials. Secondly, space and area are associated with the surrounding environment, thus establishing differences between an advantageous environment and a disadvantageous environment. So, the space taken up by housing refers not only to the space the housing directly occupies, but also to the many "available resources" related to the space. For instance, a house in the CBD of Beijing or Shanghai, or a not-very-spacious room in the financial district of Manhattan, will usually be associated with the apparent advantages of social and financial resources. The available resources in these locations are enormous. Thus, housing gains the significance of status as a symbol of class and stratum because of the importance of space. People of a similar class status often live in the same areas—meaning the area's residents gain the distinct characteristics of a group. So-called affluent and poor neighborhoods thus come into being, giving rise to further differentiation through social factors—the quality of schools, hospitals, sports facilities, service facilities, consumption differences, cultural patterns, and symbols, etc. Adjacent housing forms a neighborhood and its residents become neighbors. It bears the significance of social interaction between residents and leads them to form different social interaction groups. Therefore, living area is one of the factors contributing to the formation of class.

Put briefly, housing makes people enter a relatively stable social network. Neighbors, the communities they live in, the school their children attend, nearby hospitals, stores, and the availability of sports and leisure facilities all bestow social meanings on the residents in the neighborhood. Thus, relatively stable living patterns form, attaining symbolic meanings for both social economy and social status.

1.3 The Great Significance of Studying Housing Status Groups in Contemporary China

Since the late 1990s and the twenty-first century, China's urban housing system has undergone a number of fundamental changes. The reforms made housing status groups in urban China become more prominent as the differentiation of housing status became one of the most important elements of social stratification and social differentiation. Thus, they are a very worthwhile object of study.

Before the reforms began, state-owned housing was the dominant system and there was no such thing as housing status groups in

China. After 1979, although the goal of housing system reform—that of changing the domination of state-owned housing and establishing a market-oriented mechanism of real estate—was established at the beginning, the process was very slow until the mid-1990s. In the late 1990s, the government's policies on housing system reform became rather more radical. After July 1998, the regulations of the State Council banned the distribution of welfare-oriented public housing and implemented a policy of selling formerly publicly owned housing to work unit employees. Within a few years, urban China witnessed a shift from a society dominated by public housing to one in which private housing prevailed. Since the beginning of the twenty-first century, the marketization of real estate has witnessed enormous development, which has resulted in two outcomes. On the one hand, the crazy "enclosure movement" took place throughout China, leading to the fad of property development. On the other hand, the blocking of the public housing distribution system and the new mortgage loan system caused the real estate market to gradually expand. The general public has suddenly become aware that the later they buy a house, the more money they will lose. Consequently, real estate prices have increased drastically. In some big cities, three-room apartments are almost universally priced at more than one million yuan. In most cities, housing has abruptly become people's most important asset. Of course, housing prices vary greatly from city to city and in different areas. Details aside, the overall outcome of the reform has been that most urban residents have changed from being tenants to the owners of their houses. In the past, televisions, refrigerators, and cameras were the durable consumer goods that most prominently expressed the economic status of a family. Nowadays, housing is the first and foremost indicator of social status difference. Therefore, housing status groups have become the subject of widespread attention.

Before the mid-1990s, the housing status of most urban residents was very similar. This similarity existed for two main reasons. All houses belonged to the state, so regardless of whether a resident was a leader or an ordinary worker, he was still just a tenant. Also, housing was built and distributed by work units. Although there were some differences—such as the privileges enjoyed by officials—the gap could not be too wide, as all employees lived in the same neighborhood. There were indeed some special houses, such as the ones occupied by a few top officials, but this type of housing was exceptionally uncommon. By the late 1990s, however, more general housing status groups had appeared as a result of urban housing system reform and the marketization of real estate.

This chapter attempts to explore the phenomenon of housing status groups. The research method adopted in this chapter is an analysis of the features of residents' housing status by means of a survey of some communities in Beijing. The survey found six significantly differentiated housing groups: commercial housing households, households displaced by building demolition, households that purchased publicly owned houses from work units during the housing reforms, households living in economy buildings, low-rent housing households, and traditional private housing households. Although there are subdivisions within the six groups—for example, the commercial housing households can be divided into superior-quality and lower-grade commercial housing—the results of this survey prove that "inter-group distinctions" are more significant than "intra-group distinctions." There are some limitations to this survey. It was carried out in only one neighborhood in the Chongwen district of Beijing. Although this is a diverse community in terms of household types, it does not include all types of households. For instance, there are no very luxurious communities, villa communities, or communities for foreigners and celebrities. In short, the results of this survey cannot be generalized to all Chinese cities.

2. Pluralistic Communities and Six Housing Status Groups

This research was conducted on East Huashi street in Chongwen district of Beijing, a typical and unexceptional district. Its sub-district office has eight neighborhood committees with 15,557 households and 41,279 residents. There used to be 117 old buildings and 206 bungalows. Forty-four new apartment buildings were built after the 1990s. All eight of the neighborhood committees had their own distinctive features, which will be introduced in turn. Due to the principle of academic anonymity, each community will be referred to with an English letter.

Community A is a commercial housing community that is classified as being of high quality. The houses there were constructed in 2004. When this research was conducted, the average price in this area was around 18,000 yuan per square meter.[104] It was run by new real estate management service and was separated from the outside by brick walls. Thus, a community has formed within this enclosed space.

Community B is also a commercial housing community, but is classified as having both medium and high-grade housing. The buildings in this community were constructed at different times, mostly between 2003 and 2006. It is also run by a new real estate management service. In contrast to Community A, it consists of two housing status groups—commercial housing households and households displaced by building demolition. Initially, the intention was to accommodate both types within the same community but later, due to some conflicts, the two housing status groups were geographically separated and permitted different levels of access to community resources. Finally, the two apartment buildings were separated by walls, despite the fact that they belonged to the same community and were run by the same real estate management company.

The situation of Community C is more complex. There are commercial housing households, households displaced by demolition, traditional unit-housing households, and bungalow blocks where subsistence allowance receivers with relatively low living standards congregated. The commercial apartments in this community were built in 2004. At the time, houses here were priced at around 12,000 yuan per square meter.

Community D and Community E have a lot in common. Most residents live in commercialized work unit-buildings that used to be office buildings and dormitories belonging to factories and institutions. They belonged to more than thirty property owners (work units). By the time of this study, most of the dormitories had been sold to their residents. The quality of these dormitories varies greatly. The prices range from 14,000 to 16,000 yuan per square meter.

Community F is characterized by multiple household types, but most are bungalows. There is also the potential for further renovation. The housing price in this community is above 10,000 yuan per square meter. The level of population mobility in this community is rather high, which has led to severe security problems.

Community G contains two subcommunities: a top-quality commercial housing community and a low-rent housing community

104 From this chapter, all the prices for commercial housing refer to the pricing for the construction area.

for income support recipients. The top-quality housing consists of serviced apartments built in 2005. The price is about 13,000 yuan per square meter. The other subsection is a commercial housing community built in 2004. The price is also about 13,000 yuan per square meter.

The situation of Community H is also complex. There is a subcommunity of commercial housing built in different years. The price of those houses ranges from 11,000 to 12,000 yuan. There were also some poorly appointed economy buildings constructed during the Cultural Revolution. During 2005 and 2006, when this research was carried out, they had not yet been renovated, so information was obtained on the residents of the economy buildings. When I carried out research in the autumn of 2008, the economy buildings had been dismantled and new buildings were under construction. The new residents were divided into two types: households displaced by demolition and commercial housing households. These buildings were arranged densely and were lower and middle grade.

From the above brief introductions, it can be seen that the housing situations in these eight communities is very complicated. Clearly, they are very suitable research subjects because, in these communities, many different housing status groups related in complex ways can be found. Between 2005 and 2008, I enlisted students from Tsinghua University to carry out research on this community many times and completed a number of questionnaire surveys and case studies. All the data used in this chapter is drawn from questionnaire surveys of the residents of these eight communities in 2005. Samples were selected randomly by house number. However, the sampling was conducted by clerks in the neighborhood committee, so the sampling was not carried out very strictly. This questionnaire involved 981 valid samples; the distribution of samples in every community was as follows: 50 samples in Community A, 218 in Community B, 191 in Community C, 100 in Community D, 74 in Community E, 147 in Community F, 62 in Community G, and 139 in Community H.

Through analysis of these 981 samples, it was found that, although the housing types were diverse and complicated, there are six notably distinct housing types: households that had purchased commercial housing (hereafter referred to as commercial housing households); households that purchased publicly owned houses from work units during the housing reforms (former work unit-housing households);

households displaced by building demolition (referred to as replacement housing households); low-rent housing households; privately owned housing for self-use households (privately owned housing households), and economy building households. Obviously, there are other types of housing besides those named in this group of six basic types. Sample distribution of all the housing groups can be seen in table 13-1.

Table 13-1 The Distribution of Sample Housing Types and Residents in the Chongwen District of Beijing

Housing Type	Household	Percentage
Commercial Housing	218	22.6
Former Work-Unit Housing	210	21.8
Replacement Housing	307	31.9
Low-Rent Housing	29	3.0
Privately Owned Housing	33	3.4
Economy Buildings	52	5.4
Rented Privately Owned Housing	21	2.2
Work-Unit Housing (Publicly Owned, Un-Purchased)	33	3.4
Directly Managed Public Housing	25	2.6
Self-Built Housing	13	1.3
Other	22	2.3
Total	963	100.0

Notes: missing samples: 18; total number of samples: 981.

As shown in table 13-1, there are many housing types in East Huashi District, where this research was carried out. Besides the aforementioned six housing groups, there are other housing groups such as directly managed public housing households—meaning households occupying public housing directly managed by a neighborhood administration office. Housing of this type used to be very common, but it has greatly declined in number since the housing reforms of the 1990s. Take work-unit housing households as another example. The housing status of these residents was not stable. It was possible that these houses would be sold to them, enabling them to become households that had purchased formerly public housing. It was also possible for them to leave work-unit housing and purchase commercial apartments instead, thus becoming commercial housing households.

Households that rented apartments are another example of households that are certainly not a very stable housing group. There is also self-built housing, which is often temporary or constructed without authorization. Additionally, in figure 13-1, there are twenty-two households in the category of "other." According to the survey, the housing types of these twenty-two households were disparate. Some individuals were living in their parents' houses, some in other relatives' houses, some had borrowed other people's houses, or were housed in temporary dormitories, etc. Compared to other housing groups, the above-mentioned six housing groups are most significant out of all the housing groups, in number and percentage. Their housing status is relatively stable. Therefore, these six housing groups are typical.

3. Differences between the Six Housing Status Groups in Terms of Income, Living Standards, and Education

Admittedly, social status as it relates to housing is very complex. This chapter has described the existence of six significant housing status groups found in a survey of East Huashi District. So, will the empirical research data support the hypothesis that there are great differences between the six housing status groups? In the following section, an attempt will be made to determine whether there are significant social differences between the six housing status groups in terms of income, education, total living space, and per capita living space, etc.

Table 13-2 lists the index of three variables: each family's per capita monthly income, a self-evaluation of their living standard, and householders' educational level. In the following discussion, these indexes will be analyzed.

Table 13-2 Economic Status and Education Level of the Six Housing Status Groups

	Family Per Capita Income (yuan)	Self-Evaluation of Living Conditions (Average of Grade Index)	Average Educational Level (Index)
Commercial Housing Households	3138.15	3.0469	3.63
Former Work-Unit Housing Households	1694.99	2.6202	3.10
Replacement Housing Households	1086.76	2.1051	2.86
Low-Rent Housing Households	873.33	2.0000	2.41
Privately Owned Housing Households	753.91	1.7879	2.76
Economy Building Households	849.50	1.7115	2.50
ANOVA (Analysis of Variance) for All Groups	F = 44.370 Sig.= 0.000	F = 51.112 Sig.= 0.000	F = 22.441 Sig.= 0.000

Firstly, we will explore the significance of the index for per capita monthly family income.

This table shows that the per capita income of commercial housing households (3,138.15 yuan) is noticeably higher than that of other groups. The per capita monthly income of the other five groups, if their mean value is calculated, is only 1,052 yuan. Therefore, commercial housing households demonstrably tend to have greater economic power and higher economic status. In recent years, housing prices in Beijing have risen wildly. East Huashi District—the object of study for this research—can be regarded as a central area of Beijing, such that being able to afford commercial housing in this area shows the strong purchasing power of this group.

The second highest income in this group was that of households that had purchased former public housing. However, the situation of this group is a little bit more complicated. After the housing reforms of the late 1990s, most employees bought the apartments (formerly the property rights belonged to work their units) that they lived in. The living standards of these people are closely related to the economic profits of the work units they work for. Basically, they are traditional employees and are still able to enjoy traditional welfare. In this respect alone, they share great similarities with each other. The third group is replacement housing households. The per capita monthly family income for this group is 1,086.76 yuan, which is a medium level income according to the overall standards of these groups. But, this is a special group. Generally, people hold the opinion that the economic status of this group is not very high. However, this depends on where the demolition of their former homes took place. If they have been relocated from a downtown area, they are normally residents who have lived in this area for a long time and have incomes that are not high, but are also not the lowest. Therefore, they hold medium income status, according to this survey, which is consistent with the characteristics of replacement housing households in the old town of Beijing.

The lowest three groups in the income ranking are low-rent housing households, privately owned housing households, and economy building households; the per capita monthly family income for these groups is 873.33 yuan, 753.91 yuan, and 849.50 yuan respectively. The incomes of these three groups are similar—all belong to low-income groups. As for the low incomes of low-rent housing households and economy building households, they are understandable since these two housing status groups are evidently of a low-income nature. However, people may ask why the incomes of privately owned housing households are so low. Generally, it is believed that privately owned housing households belong to the high-income group. Why, then, does this research indicate a contrary result? There are two reasons for this. Firstly, the number of traditional householders in the old town of Beijing became very small after the 1956 housing reforms. In accordance with government policy, the big houses of rich householders were handed over to the state, leaving only a few small private houses. In the 1990s, these small private houses were usually shabby, poorly appointed, and had horrible living environments. The residents were usually senior citizens with low incomes. I found a similar situation when conducting research in other old town areas in Beijing. Although it cannot be stated with certainty that there are few high-income traditional private householders, most sold their houses and moved to other places. By the late 1990s and early 2000s, those still living in this kind of old private house were usually poor. Secondly, this research was carried out in the Chongwen district of Beijing. "Poor Chongwen," "shabby Xuanwu"—these are old sayings among common people in Beijing. Traditionally, most members of low-income groups live in the south of Beijing. Therefore, the data collected in this research reflects the situations of traditional private householders in this area.

The ANOVA for the income of these six housing status groups supports the hypothesis that the differentiation of the six economic status groups is significant. Test value F for per capita monthly family income was 44.370, significantly higher than the test value F (4.10) with a condition for significance of 0.001. Therefore, the mean difference of the six housing status groups in terms of family monthly income was of valid significance.

Secondly, the index for the respondents' self-evaluations of their living standards will be analyzed.

In order to facilitate a comparison with economic income, this questionnaire involved some subjective questions. The original question was: "Compared to other residents in Beijing, which level do you think your family's living standard belongs to?" This question had five possible answers: upper, upper middle, middle, lower middle, and low. This feedback, arranged from high to low, correlated with the data for the households' incomes. Commercial housing households gave themselves an average rating of 3.0; former work-unit housing households, 2.1; replacement housing households, 2.1; low-rent housing households, 2.0; privately owned housing households, 1.8; and economy building households, 1.7. Therefore, the economic status, objective economic income, and subjective self-evaluations were all reasonably consistent.

The ANOVA for the self-evaluations of living standards for the six housing status groups show that test value F was 51.112 and Sig. was 0.000, which is clearly significant. Therefore, the families' self-evaluations of their living standards were significantly different for each of the six housing status groups.

Thirdly, we will examine the average educational levels of the six housing status groups.

Education levels were divided into five discrete categories for the purposes of the questionnaire: (1) elementary school and below, (2) junior high school, (3) high school or secondary technical school, (4) university or college, (5) postgraduate and above. From the results of the examination of the education index, the educational levels for each of the six housing status groups related closely to the income index.

The highest average educational level was that of commercial housing households (3.63), followed by former work-unit housing households (3.10). There were differences between the four other groups, but all of them were lower than 3.0: 2.86 for replacement housing households, 2.41 for low-rent housing households, 2.76 for privately owned housing households, and 2.50 for economy building households. The test value F for the analysis of variance of the six groups' educational levels was 22.441 and Sig. was 0.000. Therefore, the observed differentiation of the six housing groups' educational levels was statistically significant.

4. Differences between the Six Housing Groups in Terms of Living Conditions

As the term "housing status group" suggests, an important factor in this discussion is the significant status difference related to housing, to which a part of this chapter is devoted. The results of the survey under discussion proved that, in terms of living conditions, there were significant differences between the six housing status groups. However, the hidden message behind the data is the difference in housing pattern. This can be seen in table 13-3.

Table 13-3 Differences in Living Area among the Six Housing Status Groups

	Building Area Per Household	Usable Area Per Household	Usable Area Per Capita
Commercial Housing Households	116.84	93.67	31.82
Former Work-Unit Housing Households	91.06	69.98	23.12
Replacement Housing Households	73.64	55.07	18.54
Low-Rent Housing Households	60.29	42.82	12.06
Privately Owned Housing Households	29.56	22.52	8.75
Economy Building Households	11.54	9.47	4.38
ANOVA for All Groups	F = 86.417Sig. = 0.000	F = 99.293Sig. = 0.000	F = 64.337Sig. = 0.000

(Unit: Square Meter)

Firstly, the commercial housing group will be examined. This group tends to have the largest living area, with an average building area of 116.84 square meters per household and a usable area of 93.67 square meters per household. Domiciles with more than 100 square meters of building area per household emerged and were developed with the rapid marketization of real estate in Beijing. The advent of this kind of spacious housing was the result of housing commercialization and marketization. I believe that this trend is not only reflected by an expansion of space, but also shows a new type of housing pattern and structure. The most noteworthy features of this new type of housing are the ubiquity of relatively large living rooms and having more than one washroom. Living rooms are used for family get-togethers or for entertaining guests. The structures of this kind of housing include "two bedrooms and one parlor," "two bedrooms and two parlors (one living room and one dining room)," or "three bedrooms and two parlors." The role of the living room is very important. It is not used for sleeping, so there are no beds. There were no houses of this type before the housing reforms because, back then, houses were in short supply. It would have been very luxurious to have a room specifically for entertaining guests. Therefore, the advent of the living room is the result of roomy and spacious housing. It is common to have two parlors: one for dining and the other for entertaining guests. This is considered luxurious compared to old, traditional housing. As a matter of fact, this kind of structure is very practical for family life as it provides a special place for dining and a living room where people can watch TV. For example, in the Chinese TV serial *I Love My Family,* most scenes are set in these two parlors. Apart from these two areas, more spacious commercial housing includes other rooms, such as halls or vestibules. This shows that housing with a building area of more than 100 square meters does not just mean an expansion of area, but also a diversification of function. A more luxurious type of housing is the duplex, a two-story domicile with a living room and dining room downstairs, multiple bedrooms upstairs, and usually more than one bathroom—indicating the growing importance of personal privacy. Therefore, the purchase of commercial housing has demonstrated the emergence of a new lifestyle. From the development of commercial housing, a trend of area expansion can be observed. Under these circumstances, policies such as *Eight Measures on the Regulation and Control of the Real*

Estate Market and *Six Measures on the Regulation and Control of the Real Estate Market* were issued by the State Council in an attempt to control the area expansion of commercial housing. It was proposed that the area per apartment should not exceed ninety square meters in at least 70% of newly built commercial housing.

The second housing group is made up of those occupying former work-unit housing. The average building area for this type of household is 91.06 square meters with a usable area of 69.98 square meters. On the surface, it seems that only the housing area is smaller than that of commercial housing. The average building area was 25 square meters smaller than that of commercial housing and the usable area tended to be around 24 square meters smaller. However, this revealed differences in housing structure, and even lifestyle. Most former work-unit housing is characterized by a traditional structure with a very small living room, or even no living room at all. This kind of housing primarily serves as a place for sleeping. Traditional one-room, two-room, and three-room housing all belong to this type. Because the living room is so small, entertaining guests or other social activities have to take place in bedrooms. So, the typical scenario for entertaining guests or hosting family gatherings is based around the bedroom. The social effect of this kind of housing is that individuals have very little personal privacy or private space. Since the mid-1990s, some newly built work-unit housing has employed new design structures characterized by more spacious living rooms or two parlors. Therefore, domiciles with 91.06-square-meter building areas and 69.98-square-meter usable areas are the results of this new type of design profile. If only traditional work-unit housing is counted, the building and usable areas would be even smaller. However, this type of work-unit housing was only constructed for a very short period because the work-unit housing distribution system was scrapped in 2000. So, most former work-unit housing is of the traditional structural type.

The third group is that of the replacement housing households. For this group, the average building area per household was 73.64 square meters, with a usable area of 55.07 square meters per household. The replacement housing households surveyed often live in the same community as commercial housing households. But, because the purchasing power of replacement housing households tends to be much lower, small housing has been designed specifically for them. This kind of small housing cannot facilitate the new lifestyle associated with bigger commercial housing; some of it is hastily and ineptly designed, causing many inconveniences. There are replacement housing households and commercial housing households in the same neighborhood within Community B. From the outside of the buildings, there is not much to distinguish these two kinds of households. However, if you walk inside their apartments, you will notice the tremendous differences between these two types of households. In this study, some replacement housing households complained that there is a so-called "black eight households" ("黑八户") problem. That means that they cannot afford the bigger apartments originally designed for commercial housing, so they have to settle in smaller, temporarily modified apartments. Eight apartments on every story have been modified into small, alley-like apartments. Once you enter this kind of apartment building, you remain in darkness until you walk all the way to the windows. The whole structure is like a dark alley. The average usable area of replacement housing households is 55.07 square meters, which reflects not only a difference in area, but also in lifestyle.

The fourth group is made up of low-rent housing households. The low-rent housing examined in this survey was built using special funds provided by the Beijing municipal government between 2002 and 2003. This housing covers a total area of 30,000 square meters and more than four hundred apartments. This is the only community in Beijing offering large quantities of low-rent housing at the time of the survey. The data collected in this survey shows that the average building area, usable area per household, and usable area per capita are 60.29 square meters, 42.82 square meters, and 12.06 square meters, respectively. Apparently, compared with the aforementioned types of housing, this type tends to be much smaller; this is the feature of housing built through reliance upon government subsidies. The rules governing who is eligible to apply for this type of housing will be explained later in this chapter.

The fifth group is that of privately owned housing households. The average building area per household is 29.56 square meters with only 8.75 square meters of usable space available per capita. In terms of building area per household, usable area per household, and usable area per capita, this type of housing is very limited. This is a very typical housing condition for bungalow residents in the old town of

Beijing. I used to conduct research in this area and found that most old bungalows are characterized by narrow living spaces. The rooms in traditional bungalows are defined by the number of beams under the roof. So, a so-called multiroom bungalow is actually one big room without partition walls. For example, "a three-room bungalow" is actually one big room where all the residents' activities take place, regardless of whether they are having dinner, entertaining guests, or sleeping. Additionally, the heating installations and bathrooms are in a terrible condition; sometimes a coal stove is needed in winter and placed in the middle of the bungalow. People even cook their meals in the bungalows, while some others cook in self-built sheds. After the 1976 Tangshan Earthquake, residents in Beijing began to build sheds in their yards, with the intention of preparing for possible earthquakes. Later, such self-built housing became popular. Yards, which used to be spacious, became occupied by poorly constructed huts, leaving only a narrow passageway. The living conditions in these areas deteriorated, and even posed some potential dangers. These dilapidated bungalows are often accompanied by other problems such as obsolete infrastructure, neglected water and sewage systems, aging circuitry, and a lack of central heating equipment. Many residents believe that someday these residences will be demolished and rebuilt, so they do not bother to expend much effort on renovation and repair work. This kind of small private housing householder will not benefit much, even if his or her house is torn down and rebuilt. The compensation paid for the demolition will not be a large amount of money because of the small size of the buildings. Therefore, these people will be unable to afford to buy new houses. These problems tend to encourage conflicts in these bungalow areas in the process of urban renewal.

The final group is made up of economy building households. Bungalow householders are able to enlarge the amount of usable space they have access to by building in their small yards, but economy building householders are deprived of any such opportunity. This survey found that the average building area per household is 11.54 square meters for domiciles of this type. The average usable area is 9.47 square meters per household and 4.38 square meters per capita. This amount of living space is really too little. Economy buildings of this type were built during a particular period of time when the principles of hard work and living simply were advocated. They are so crudely constructed and poorly equipped that there are no support facilities whatsoever: no heating devices, no bathroom, no balcony, and no gas line. Some economy buildings are built with bricks incredibly laid side-on-side in order to save on materials (size of normal brick: length: 240mm, width: 115mm, and thickness: 53mm). These buildings have quickly become dilapidated and are widely acknowledged as the worst rank of housing.

5. Differing Access to Housing Resources among the Six Housing Status Groups

Different living conditions are only one aspect of housing status group differentiation. More important is the path by which people gain their housing status. Status acquisition is referred to as "status attainment" by sociologists and has a long history of research. The exploration of the intergenerational transfer using the path diagram method by P. Blau and O. D. Duncan in the U.S. is regarded by many people as the beginning of status attainment research.[105] However, before them, S. M. Lipset and R. Bendix[106] and S. M. Miller[107] had already explored the factors influencing intergenerational status attainment. Status attainment as it is examined in this chapter does not focus on the inheritance relationship between generations, but rather the methods of acquisition and factors influencing how people obtain their housing or economic status.

It was not until the late 1990s, or even early in the twenty-first century, that significant numbers of Chinese people really started to buy commercial houses. According to the data released by the National Bureau of Statistics, calculated on the basis of the total area occupied by residential commercial housing, only 33.73% of the residential commercial housing sold in 1991 was purchased by private individuals. This index rose to 49.27% in 1996, 66.56% in 1998, 80.08% in 2000, and 96.43% in 2003. This is because, before the 1990s, the majority of residential houses were purchased by the state or by enterprises; very few were bought by private households. However, after 2003 private purchasing became the norm and far fewer commercial properties were

105 Peter M. Blau, and Otis Dudley Duncan. *The American Occupational Structure.* New York: John Wiley & Sons, 1976. 163–205.
106 M. Lipset and R. Bendix. *Social Mobility in Industrial Society.* London: Heinemann, 1959.
107 S. M. Miller. "Comparative Social Mobility." *Current Sociology* 9 (1961): 1–89.

purchased by the state or by enterprises. The fundamental motivation for private purchase was the housing system reform in the late 1990s, which abolished the welfare-oriented public housing distribution system. People no longer had any choice but to obtain housing through the market. The ability to purchase housing depended on two factors: having enough money and being in need of housing. Obviously, of these two factors, the first is the more important. Combined with the marketization of housing, this new demand changed housing from a living requirement to a potential source of profit—a trend that was fueled by a seemingly endless demand. A recent social survey found that wealthy families often bought numerous commercial properties; some of them were found to own more than ten houses or apartments (see chapter 8). In brief, a sudden surge of ordinary people with purchasing power into the housing market caused urban commercial housing prices to rise after 2003 and 2004. The housing prices mentioned in the survey conducted in Donghuashi Street, Chongwen District, were typical of prices at that time: even second-hand commercial housing cost more than 10,000 yuan per square meter. This means that an apartment would have cost more than one million yuan at a time when the monthly income of most ordinary people in Beijing was 2,000–3,000 yuan. The high price of commercial housing has become a kind of "social selection mechanism" that has eliminated a lot of low and middle income earners. The households that are able to bear these costs are usually those with stronger economic power, so they have entered into the higher stratum of housing status.

Former work-unit housing households are a big group in urban China. At present, nearly 70% of homeowners have obtained their houses through the reform of work-unit housing. We call them "house from housing reform" ("房改房") householders. Most of them are long-serving employees of state-owned enterprises, public institutions, collective enterprise, or are civil servants. They have already served many years in their work units and so obtained houses or apartments when they were distributed by their work units. They were generally the tenants of these properties before they purchased them. Against the background of national housing reform in the late 1990s, these properties were sold to employees at subsidized prices with the level of subsidization according to the provisions in governmental policy, such as the employee's terms of service. The prices were naturally rather lower than today's housing prices. In the Chongwen district survey, the total transaction price of the work-unit houses sold to work-unit employees was found to have usually ranged from 30,000 to 50,000 yuan per apartment. The present market value has reached one million yuan for many of these properties. Therefore, this group has been a beneficiary of housing reform. Of course, this is a very special group in terms of access requirements, as these workers had contributed to their units for many years for very low wages. This cheap housing is a kind of reward for their past contributions. In other words, without a long history of contribution, employees such as new workers are unable to enjoy such advantages.

Of course, housing conditions vary greatly within this group. The standards for the homes of senior cadre members are much higher. The survey under analysis showed that there are high quality "houses from housing reform," but the houses of low-ranking enterprises and collective enterprises are of a much lower standard. However, since the former work-unit houses under survey are located in Donghuashi street, Chongwen district, they have the advantage of being located in the city center—meaning that the market price of these houses is not low. Therefore, it is reasonable to treat those who have bought work-unit houses as belonging to the same housing status group. Additionally, people in this group are generally long-term work-unit employees. They tend to have very similar ways of living and even of thinking, and, generally speaking, all enjoy the formerly common social welfare benefits of medical insurance, pensions, etc. Because of these advantages, some of the above data also shows that the difference between this group and other housing groups is much greater than the differences within this group.

Replacement housing householders are basically the original tenants in this area. Most of them used to live in single-story housing in quadrangles in Donghuashi street. This kind of residential group can be divided into four subgroups. The first are old private householders, who have been referred to as privately owned households in this chapter. Many of them were the owners of their homes before the founding of new China. According to the survey, some households had even lived there for more than one hundred years. Some preferred to retain their old houses rather than let them be demolished because they already had independent property rights over their own courtyards. At the same time, when demolition occurred, they tended to have higher expectations of their compensation. The second group was the most complicated. After the housing system reforms enacted in 1956 and during the Cultural Revolution, the government or sub-district offices arranged for some residents to live in the houses or yards of the "old private households" ("老私房户"). They can be considered "old lessees" ("老承租户") because they are long-term residents in the area and have held "official" housing status. After the Cultural Revolution, policies were once enacted in an attempt to resettle the people who had been moved into private houses during the Cultural Revolution. However, a lack of housing made it difficult to solve this problem. Some of these people were not resettled until their homes were eventually demolished. These "old lessees" also obtained economic compensation and were granted the right to return after their houses were demolished and rebuilt. The third group is also made up of "old lessees," but these individuals rent public housing owned by the district housing management office (directly managed public housing) rather than private houses. This was the main method by which ordinary people rented houses after the 1956 housing reforms. The fourth group is that of employees who rent public work-unit houses (self-managed public houses). Urban renovation, renewal, and demolition often involve public work-unit houses, so compensation and the right to return have to be given to the occupants of public housing. Of course, these issues are often resolved through negotiation between local government officials, the construction managers, and relevant units. In short, a common characteristic of replacement housing households is long-term residence in the area and officially registered urban residency; some of them have property rights, some of them have just been placed there by the government. They have gained their homes through legal and formal means of access. Because of this, they are able to gain compensation and the right to return to their homes. Therefore, the right to return is a very special kind of "historical status right" ("历史身份权利").

The policy basis for allowing people to return to their homes and providing compensation is largely drawn from two documents: *Implementation Measurements for Accelerating the Transformation of Dilapidated Houses in Beijing No.19*, which came into force on March 23, 2000, and *Management Measures for Urban House Demolition and Relocation in Beijing (Municipal Government Decree No. 87)* which took effect on November 1, 2001. The primary difference between the two documents is that the first allowed for the provision of replacement housing while the second recommended monetary compensation. In fact, replacement housing became less common after 2002 and monetary compensation has been issued in the majority of cases. During the years between 2002 and 2004, the standard for demolition compensation was 8,000–10,000 yuan per square meter, in accordance with the policy of the time. However, due to soaring real estate prices, the compensation standard has doubled, sometimes several times over. Recently, the compensation standard for housing in central Beijing reached more than 40,000 yuan per square meter, which has caused a lot of jealousy among former replacement householders.

Low-rent housing is a kind of welfare housing leased to urban residents with low incomes and disadvantaged urban groups by the government. There are two documents regulating the admittance mechanism for low-rent housing. One is *Management Measures for Low-Rent Housing for the Lowest Income Urban Households*, which was jointly issued in 2003 by the Ministry of Construction, the Ministry of Finance, and the Ministry of Civil Affairs. The other is *Security Mechanisms for Low-Rent Housing*, a document produced in September 2007 by the Ministry of Construction, the Development and Reform Commission, the Ministry of Supervision, the Ministry of Civil Affairs, the Ministry of Finance, and the Ministry of Land and Resources. Those documents describe the admittance mechanism for low-rent housing in detail, but these minutiae will not be repeated here. The research under discussion found that there is a significant disparity between the admittance mechanism set out in these documents and the methods which are actually employed. According to government policy, the distribution of low-rent housing should be based on *Management Measures for Low-Rent Housing for the Lowest Income Urban Households*, which was published in 2003. However, the actual situation is much more complex. The survey found that Dahuashi Street held the only concentration of low-rent housing construction in Beijing at that time. The low-rent housing there was mostly good quality high-rise buildings. They could not be distinguished from commercial housing on the basis of outward appearance alone. Low-rent housing householders were selected from Beijing's entire population. Serious problems occurred during the selection of low-rent housing residents, caused by a lack of housing and an overabundance of applicants. Therefore, it proved difficult to distribute low-rent housing. Under such circumstances, there is no alternative but to create another set of very strict standards for admittance. Beijing's municipal government therefore applied "The Low-Rent Housing Renting Scheme" to the problem of housing distribution, which clearly states that applicants must be recipients of welfare support with valid certification. The acceptable forms of certification include: Certificates of Revolutionary Martyrdom, Certificates for Revolutionary Soldiers Killed in Action, Certificates for Revolutionary Soldiers Killed by Illness, Certificates for Disabled Revolutionary Servicemen, Certificates for Fixed-Term Pension, and Certificates of Periodical Allowance. In the actual process of distribution however, some low-income "Model Workers" ("劳动模范") award winners with housing difficulties (who also held certificates) were also admitted. So, the most critical admittance criteria abruptly changed from housing difficulties to the political standards of one's contributions to the Chinese revolution and reconstruction. Of course, besides such political standards, economic conditions were also proposed as criteria with the intention of including households whose per capita living space was less than ten square meters. In practice however, standards were relaxed to include households with 15 square meters under certain circumstances. This happened because it was a burden that all stakeholders could accept. No one complained because those individuals had made great contributions to the country. Moreover, there were no fake certificates of this kind. All low-rent housing householders were actual certificate holders. In sum, because of a shortage of low-rent housing, other standards were insufficient for selecting applicants to receive housing, and the distribution of low-rent housing was characterized by "social closure."[108]

A total of 271 households lived in low-rent housing in the residential district under survey as of June 2005, of whom 59.8% were disabled veterans, 22.5% were family members of revolutionary martyrs and servicemen, and 10.3% were "Model Worker" award winners. These three politically selected groups made up 92.6% of the total recipients of low-rent housing in this area. This explains why the monthly per capita income of low-rent housing households is higher than those of

108 Li Qiang. *Ten Lectures on Social Stratification*. Beijing: Social Sciences Academic Press, 2008. 110–11.

households living in privately owned housing and economy buildings. However, with the expansion of government investment in low-rent housing, the admittance conditions for this kind of houses are likely to change in the future.

The proportion of privately owned housing households in the survey is very low—only 3.4% across all samples. As mentioned in this chapter, the households of this type that were surveyed occupy a kind of small, traditional private housing left over from the 1956 reforms. From the perspective of status attainment, they are owners of their houses but have experienced a lot of changes during the popularization of private housing. The residents of this kind of housing are often the original residents, and so are very old. As mentioned above, Donghuashi Street was traditionally a poor area; these households' inheritance was certainly not luxury curtilages and courtyards. The survey also shows that the living standards of these residents are comparatively low. So, it would be inappropriate to judge too quickly and believe that "private housing owners" must be high-income earners. Many similar cases have been found in investigations in Beijing's old town. Such houses are often dilapidated and in a state of disrepair because of long neglect. Some owners hope that the government will rebuild or renovate their houses, but this involves the problems of demolition and compensation.

Economy buildings in Beijing were mostly built during the Cultural Revolution. Essentially, they are a product of the public housing system. Residents obtained their housing through the housing distribution system. Superficially, this kind of housing appears to be similar to the self-managed houses distributed by work units and the directly managed houses distributed by management offices in the traditional fashion. However, economy buildings were designed as simply as possible in order to save money. These houses were often built by low-ranking work units and distributed to common employees. In any case, most people's housing conditions have improved a lot in the last thirty years and many people who once lived in these economy buildings have moved out. By 2005 and 2006, when this survey was conducted, those still living in economy buildings had very low living standards. Through the survey, we found that those residents include laid-off workers whose former work units had gone bankrupt or been disbanded and those out of work due to long-time illness. Most of them are old and unable to secure new housing either through the pre-housing reform housing distribution system, or through the commercial housing market. So, their only hope lies with the government. Most of them expect that the government will renovate and rebuild their houses. When conducting this survey, I saw that these economy buildings are indeed very shabby, very crowded, and very dark. During the survey, the residents voiced their various problems with their living conditions and expressed the expectation that the government would transform their homes. In fact, the government has managed to rebuild these houses in addition to an adjacent commercial housing project. However, this project had failed because the developer proved unable to meet high costs. To our pleasure, when the survey was repeated in 2008, the renovation of these housing had been accomplished. New buildings were under construction and the relocation ratio of original residents was very high. So, from the perspective of housing status groups, this is a case in which economy building households became replacement housing households. From the outside, these new buildings still appeared to be of a relatively low grade; however, there is a world of difference between them and the old economy buildings.

The above analysis of housing status attainment by different groups can be confirmed in terms of time. Housing status groups are the product of different governmental policies enacted at different times, since the housing policies of different periods produced different selection criteria for housing status attainment. Commercial housing is distributed by money and the market; work-unit housing by length of service, position, and qualifications; replacement and low-rent housing on the basis of all sorts of titles and qualifications stipulated by a number of policies. The following table (table 13-4) lists the average index of when the households surveyed began living in the accommodations they occupied at the time of the survey.

Table13-4 Different Residence Times of the Six Housing Status Groups

	Year Residence Commenced (Average Index) (Year)
Commercial Housing Households	2001
Former Work-Unit Housing Households	1988
Replacement Housing Households	1981
Low-Rent Housing Households	1994
Privately Owned Housing Households	1950
Economy Building Households	1971
ANOVA (Analysis of Variance) for six groups	F = 43.399 Sig. = 0.000

In the variance analysis of "The Year Residence Commenced," the following results are obtained: the F value is 43.399 and Sig. is 0.000, indicating that the differences between these six groups are statistically significant. If we scrutinize the average year in which residence commenced for each, it is not difficult to see their unique characteristics. The year residence commenced is most recent in the case of commercial housing households, since most of these residents moved in at the start of the twenty-first century. This coincides with the year when welfare housing distribution stopped in China and many residents were forced to enter the real estate market. The average year in which residence commenced for former work-unit housing households was 1988. If we expand this period to stretch from 1978 to 1998, it will include the period when work-unit housing distribution in China was most widespread. Although the welfare housing distribution system was established before the reforms, it was an empty policy with no housing to distribute. The vigorous construction of housing by work units did not take place until after the reforms. Therefore, the actual implementation of welfare housing distribution took place twenty years after the reforms. Welfare housing distribution was ended in 1998, theoretically because the State Council issued documents ending welfare housing distribution in that year. Table 13-4 also shows that most economy building households began to live in their houses in 1971, at the climax of the Cultural Revolution. Most of the economy buildings in Beijing were built during the Cultural Revolution. The year residence commenced is of significance to the history of this period. The average year residence commenced was 1950 for privately owned housing households. This means that most of these households bought their houses immediately before and after the establishment of the PRC. Legally speaking, the sale of private housing was not permitted after the 1956 housing reforms, but data shows that these deals continued secretly in quite a few cases after 1956. The data for replacement households is more complex. For psychological reasons, some people tend to emphasize their original residence status. They completed the questionnaires using the date when they, or their families, first began living in the area, rather than the date when they were relocated. According to the data, four people said 1800, two people said 1920, seventeen people said the period between 1925 and 1949, and twelve people said 1950. Of course, most people completed the questionnaire using the year they were relocated, as was requested. In addition, 99 people chose the period from 2002 to 2005. Regardless, due to the ambiguity in the answers, the average index for the data of this group was 1981, making it meaningless for analysis.

6. Conclusion

An attempt will be made to summarize the above discussion as follows.

6.1. Housing Status Differentiation and Social Division

Based on the research done in Donghuashi Street, Chongwen district, this chapter has proposed that housing status and housing status groups have become an obvious feature of social stratification. China experienced a period of housing unification. After the urban housing system reform in 1956, many private houses became public houses. Additionally, the social and work unit welfare housing distribution system started at that time and was conducted on the basis of "equalization." The differences in housing became very small among the vast majority of urban residents. Although there were a few better houses in the cadre compounds, only a very small number of people could live there. Public houses were also not inheritable; it could not be said that there were stable housing status groups in cities until the 1990s; the universal housing status groups discussed in this chapter did not emerge until after the mid-1990s. The documents that triggered the development of housing status include: *Decision on Deepening the Reform of the Urban Housing System*, which was issued by the State Council in July 1994 and *Notice On Further Deepening Urban Housing System Reform and Accelerating Housing Construction*, which was issued by the State Council in July 1998. More importantly, the State Council also decreed that the whole country should stop welfare-based housing distribution in the latter half of 1998. By 2000, the traditional welfare housing distribution system had been entirely eliminated. At the same time, the sale of public houses to employees or original residents and the reform of the real estate market reform were also thoroughly enacted. So, the privatization and marketization of urban housing is the most striking aspect of city management system reform in the early twenty-first century.

When work-unit employees or households became homeowners, their statuses changed. In the past, the maintenance of housing had nothing to do with the residents because the houses were publicly owned. But now, they have more responsibility for their houses. Of course, they can also profit from housing. The leasing of housing has become an important source of income for many urban residents. Leasing purchased work-unit housing or apartments obtained through the housing reforms has also contributed to the prosperity of the second-hand housing market in cities.

Many studies have proved the extent to which economic divisions have developed since China's economic reforms began. In the last thirty years, these divisions have manifested themselves in different material ways at different stages. In the 1980s, for example, the division was manifested in the possession, or lack thereof, of large, durable consumer products such as furniture, black-and-white or color TV sets, single- or double-door refrigerators, and single- or double-cylinder washing machines. Whether or not one had a video camera or cell phone became the markers of consumption differences among people who had just become rich. In the 1990s and the early twenty-first century, the key factor was whether or not a household had a car. Only when housing status groups emerged did relatively stabilized standards for social status differentiation begin to take shape.

Housing as an important vehicle for indicating social division was illustrated at the beginning of this chapter. In addition, a number of factors have strengthened the function of housing as a vehicle for showing social division. Crucially, housing has the characteristic of being overt. Although a variety of material objects have indicated economic differentiation since the economic reforms, the overt nature of housing—such as the difference between a mansion and a shack—make divisions more conspicuous. Additionally, due to the selection of households by rent price differentials and the variation of property prices in different cities, people at different economic levels are configured into different classes of housing, and housing at different economic levels is gradually separated into zones. This is referred to as "zone-separation" in cities. This has once again strengthened the explicit nature of social division. Finally, dense population and insufficient land are unique features of cities—especially big cities. Land in any area in these cities is unique. The price of houses located in the city center area or the CBD has increased hugely in this period. Therefore, a multiplier effect on the price of these houses has become the difference between these properties and houses in other areas.

Of course, the assessment of China's housing groups expounded in this chapter is still limited. Only six housing status groups are analyzed here, but this does not mean that there are only these six groups in China. The intention has been to emphasize the important function of housing status differentiation in social division through those six groups. There are many other kinds of housing status groups beyond the six mentioned above. For example, commercial housing can be differentiated internally into: high-grade villas, high-level residential houses, ordinary commercial houses or apartments, and limited price commercial houses or apartments. Grades can be classified solely with relation to the residential property management of commercial housing. There is a big difference between the property management fees for high-grade residential communities and those of ordinary ones.

In addition, of the housing status groups listed in this chapter, some will disappear due to urban renovation, such as economy building households. However, some new groups will come into being. Therefore, the existence of housing status groups is not a temporary phenomenon, but one that will last for a long time. This is especially so in China now that, after the privatization of the housing, houses and apartments have become inheritable, bestowing the characteristics of social continuity upon housing status groups.

6.2. Sudden Changes in Housing Statuses and their Consequences

As early as 1978, Deng Xiaoping advocated the marketization path for urban housing reform and started the experiment of housing commercialization in the early 1980s. However, because of the inertia of the former system, it was not until the late 1990s that the reform of state-owned enterprises was completed and positive measures were taken by the central government to stop welfare-based housing distribution that this system was finally dismantled. It was not until then that the vast majority of urban residents came to understand what the coming of housing marketization meant for them. Formerly public houses suddenly became private ones, and housing prices rose very quickly in a very short time. Housing prices had doubled before their new importance began to dawn on many people. As a result of the sudden surge in housing prices, two kinds of residents emerged in cities. One group was made up of those who were fortunate enough to have purchased commercial houses or apartments. These people suddenly owned huge properties and their wealth increased one hundred times over, but the ones who did not buy commercial property would have felt like they had been thrown into the lower classes and been made obsolete by history. However, since only a few people bought commercial properties by chance, the majority of residents were psychologically disturbed by these events.

When public houses became privately owned, the sudden rise of housing prices and the formation of housing status groups were very sudden changes; the accidental decisions of residents caused sudden changes in their social statuses, which naturally resulted in the maladjustment and psychological disturbance of many people. This maladjustment is always explained by "relative deprivation theory" ("相对剥夺") in sociology. That means that, although the economic conditions of all people improved, the widening prosperity gap became a main source of dissatisfaction. Marx once gave an incisive explanation for such developments. According to him, when residents are all surrounded by small houses, they are satisfied. But, when the differentiation of houses is suddenly exaggerated, maladjustment and dissatisfaction will increase among those who live in the comparatively smaller houses, even if the conditions of all the houses have been improved.[109] Therefore, housing differentiation and status group differentiation led to popular dissatisfaction and a concern with social fairness and justice. Since the government conducted housing reform, status groups derived from this reform had their own distinct integration characteristics. All these led to the phenomenon of some groups becoming wealthy because policies favored such groups.

Of all the sources of dissatisfaction about housing injustice, the most prominent has been discontent with housing corruption. Government policies have played a very important role in the shaping of housing statuses in a short period of time. Some people have profited hugely from housing by exploiting the advantageous aspects of the government's policies, causing the prevalence of all kinds of speculation and corruption during this period. There have been many loopholes in these housing policies and systems as they have been hurriedly formed in just a short time. For example, power-for-money deals have occurred frequently between the government and real estate developers because the acquisition of land was an essential step in the process of building new housing, which had to be approved by the government. It is an open secret that real estate developers have often given houses or apartments to officials in charge in various ways during the development of their real estate. In residential areas under survey, many people bluntly criticized the corrupt way real estate developers presented houses to officials. They angrily denounced some of the buildings as "corrupt buildings." As ever, it is undeniable that "the masses have sharp eyes" ("群众的眼睛是雪亮的"). So, both people's strong sense of relative deprivation and the corruption endemic to housing differentiation have become the most prominent social problems in the formation of housing status groups.

6.3. Special Mechanisms for Generating Housing Status Groups

From the perspective of social stratification research, a special mechanism is needed to deliver status groups to their strata in the formation of stable housing status groups. It has been argued in this chapter that housing status groups of various ranks appeared under the influence of a number of government policies during the period of system

109 Karl Marx. "Lohnarbeit und Kapital." *Selected Works of Marx and Engels.* Vol.1. Beijing: People's Publishing House, 1972. 5, 367.

transformation. The question now is, will the housing status groups endure when transformation is ongoing and the system is still in flux?

The mechanisms that generated the housing groups discussed in this chapter were hybrids. With the progress of the reforms, three kinds of mechanism will continue to have an effect. First, the real estate market will undoubtedly continue to be important. Also, replacement households will continue to exist for a long time. China is in a state of rapid urbanization. The basic pattern of urbanization is city renovation and expansion into the areas surrounding cities. So, a stable mechanism exists for generating replacement households. Finally, as an essential supplement to the real estate market, low-rent housing plays an important role in the Chinese welfare system. The entire Chinese government has promised to continuously expand its investment in low-rent housing. In Western countries, low-rent housing is for members of relatively low social classes. Many scholars in the West have criticized the low-rent housing mechanism, believing that it forms an underclass. However, the situation in China is different; the low-rent housing households investigated are a special political group. Nevertheless, it has proven difficult to evaluate the situation of the low-rent housing households throughout the country in this chapter, as their situation tends to depend on how low-rent housing operations are implemented at a local level. However, it is certain that low-rent housing households hold a special housing status in China.

Furthermore, it is likely that the other three mechanisms of housing status acquisition detailed here will decline in importance over time. The proportion of former work-unit housing households among property owning urban residents has been large up until now; in my estimation, it is as high as 70%. But, with the termination of the welfare-oriented public housing distribution system and the continuing expansion of commercial housing, the number of these households are likely to fall dramatically. In the long run, most former work-unit housing households will eventually enter the market. For example, the number of former work-unit housing properties on sale in the second-hand market has grown consistently in recent years. Therefore, this form of housing will be integrated into the market system and the residents' status will thereafter be decided by the market, for these houses differ greatly from each other in quality. The mechanisms governing privately owned housing will also change. Traditional privately owned properties are bungalow courtyards leftover from the 1956 housing reforms. They are small in number, and will gradually disappear from existence through the process of urban renewal. Although a few large, modern courtyard bungalow properties have appeared in the center of Beijing, such housing cannot be compared to traditional homes of this type. They should instead be listed as a form of luxury commercial housing. Finally, the economy buildings constructed during the Cultural Revolution will also gradually disappear as a result of urban renewal. Of course, the original residents will become replacement housing households.

All in all, according to the analysis above, commercial housing households are likely to become the mainstream of future urban housing status groups. However, not all commercial housing residents are of the same status group. There are great internal differences within the group. Commercial housing households are generated by the mechanism of the real estate market, but the Chinese real estate market is still underdeveloped, so it has a lot a room to change. In general, Chinese people have a strong desire to own property. Therefore, it can be expected that most people will become part of the commercial housing household group. This research did not distinguish between the levels or grades of commercial housing because the district under survey is in the center of Beijing where the prices of commercial housing are relatively high. Actually, the grade of commercial housing in Beijing is closely related to which district it is located in. Economically affordable houses and limited price houses are often far from the downtown area, in places where land prices are lower. So, if the future of the real estate market mechanism is considered, it can be predicted that a variety of housing status groups will be sorted out, such as a luxurious mansion status group, a high-grade commercial housing status group, a middle-grade commercial housing status group, and a low-grade commercial housing status group, among others.

The real estate market, which has generated the new commercial housing status groups, has become very powerful. In other words, the Chinese real estate market has a strong driving force in terms of purchasing power in the development of the commercial housing market in China. There are four factors driving people into the new commercial housing market in cities. First, there is "newlywed power." Every year, millions of newlyweds begin looking for housing and more than six million graduates start work each year. Although not all of them are able to afford a house or an apartment, their demand for housing is very strong. Second, there is immigration power. The migration of registered rural residents in search of work into cities and towns is also an important driving factor. Although few of them can afford to buy a house, they could attain a housing status linked to various housing statuses as long as they live in a city or town. Third, the migration of workers from small cities to big ones, or between different cities, has a similar influence. In order to find a good job, people may have to buy houses in the city. Finally, the power of changing lifestyles has proved a crucial factor. New commercial housing symbolizes a new way of life and has become very attractive. Based on international experience, when a country's per capita GDP is lower than $8,000–$10,000, the demand for real estate becomes very intense. China has now reached this point and the potential purchasing power of its vast population is enormous. It can therefore be predicted that this housing status group generation mechanism will continue to play an important role.

6.4. Lifestyle as It Is Shaped by Different Housing Statuses

As mentioned above, new commercial housing is not only larger in area, but represents a brand-new lifestyle. By contrast, the lack of differentiation in housing patterns prior to the reforms resulted in a single, monolithic way of life. The diversification of housing status groups has allowed different ways of life to develop. In my stratification study, I argued that the main characteristic of social structural change in contemporary China is that there is a tendency towards the solidification of stratum structure. That is to say, the boundaries between strata have formed, the internal identities of each stratum have been strengthened, and a shared culture and lifestyle has formed within each stratum. The different lifestyles of different housing status groups are an important aspect of status solidification. How should China deal with this phenomenon? Economic differentiation and status stratification are unavoidable in any market economy society. The formation of the different lifestyles associated with different housing status groups have the function of maintaining social order, but there is no denying that this stability comes at a price. For example, "zone-separation" is a big problem. The "zone-separation" of urban housing into rich and poor areas, and the living space differential between different strata, are synergistic with the formation of so-called lifestyles. Although this is a common phenomenon in the economically developed countries of Europe and North America, China has experienced a period of communalized living. It must, therefore, be decided whether to allow the market to develop and shape housing segregation or to try to create another model that is different from that of Europe and North America. This is a real challenge facing Chinese society.

Chapter 14

Housing Reform and the Construction of a Ranking Order of Social Stratification

In the previous chapter, analysis revealed that housing status groups have formed because of the vast diversity in housing conditions in Chinese cities. In this chapter, further analysis will be made of the institutional foundation of such formations. I believe that the levels of social stratification revealed by housing status groups are merely external symptoms of stratification which is, at a high level, a manifestation of the construction of a new ranking order of social stratification.

Of China's housing system reforms, the reform of state-owned enterprises is the most representative because it converts publicly owned housing into employee-owned housing. This means it is a process of changing a housing distribution system into a commercialized private housing property system. In this chapter, the reform of state-owned enterprises will be the main focus of our attention.

The general thrust of this chapter is that the reform of state-owned enterprises has solidified, in the form of property rights, a housing inequality that was based on power and status under the previous housing distribution system. This has not only maintained and reproduced the old system of housing distribution, but also constructed a new ranking order of housing stratification based on property.

The reform of state-owned enterprises has not only changed the nature of house ownership, but also influenced the whole structure of social stratification as well. This chapter will therefore deal with how the transformation of China's housing system from a distribution system dominated by public ownership to a commercialized system dominated by private ownership has affected the pattern of people's economic statuses and interests, as well as how new forms of social differentiation and inequality have emerged.

China's urban housing system reform was implemented at two levels: the state level and the work unit[110] level. The state developed

110 All the public and collectively owned entities in China before the reforms were referred to as "work units," including government organizations, schools, hospitals, companies, enterprises, grocers, etc.

policies and directives while work units managed their implementation. Before the reform, work units independently provided social security for urban residents; housing supply and management were an important part of work units' duties. When the reforms took place, units in cities became the key loci where transformation took place and the main entities responsible for the implementation of housing reform policies in urban areas. Compared with work units involved in other forms of ownership, state-owned enterprises (because of their large number and high status among the enterprises in China at that time) became pioneers and the most important area of urban housing reform. Employees throughout all levels of these state-owned enterprises were the individuals greatly and directly affected by these urban housing reforms. A field study of housing system reform in one large state-owned enterprise will be utilized in this chapter to investigate the influence of changes in the housing system on the pattern of people's interests and on social stratification structure.

The case that will be discussed in this chapter is that of a large state-owned enterprise. It is a relatively old, competitive enterprise in Shandong province. Similar to other state-owned enterprises, the enterprise had been implementing a welfare housing policy before the mid-1990s. After people received these distributed houses, they were granted rights of residence and occupation and a very low rent was deducted directly from their monthly salary by their work units. Since houses were provided by enterprises and were built close to the enterprises' production areas in a relatively tight concentration, typical work-unit communities developed. Since the housing system reforms, this Shandong enterprise has experienced and implemented the policies and programs of Chinese urban housing reform, witnessing a transformation from a distribution system under a welfare housing policy to a monetized and commercialized housing system.

1. Inequality in the Housing Distribution System

Less than a decade after 1949, a system of public ownership was set up and a planned economic system was established in China. In response, an urban housing system dominated by public ownership was set up as part of a full-scale socialist transformation of urban housing to alleviate the severe housing shortage confronting urban residents. Under this system, a housing distribution system based on the principles of welfare was put into effect in cities. This means that the majority of housing expenditures were deducted by the state or by work units according to a unified standard coupled with a system of issuing universally low salaries. Houses were built by the state or by work units and housing distribution was, at the point of delivery, almost completely free for the recipients of new homes. The houses were treated as a form of nonmonetary income, a kind of welfare benefit and right of all urban employees. Employees just needed to pay a nominal rent, while the state or work unit took care of housing management and maintenance. The publicly owned housing distribution system that was established and implemented immediately after the foundation of the People's Republic of China solved China's urban housing problems at that time and was able to meet the basic needs of urban residents and guarantee their right to housing.

Although "meeting the basic needs of workers" ("满足职工基本生活的需要") was regarded as the most important principle of the process of distributing housing, the allocation of housing expenditures—which were monthly deducted by the state or by work units—was not actually redistributed evenly to employees in these work units. Large and high quality houses—a scarce resource at that time—were usually used as rewards for those occupying managerial positions, or for senior and skilled employees, and others with abundant social capital. Therefore, inequality and stratification in housing distribution were inevitably produced. Under the housing redistribution system, the most obvious and prominent inequality was that between the majority of ordinary workers and the minority of the elite. The most important differences were in floor space, quality, construction, and facilities of the houses being allocated. The houses allocated to workers tended to be comparatively small and equipped with minimal facilities, while elites enjoyed special favors—usually larger, better quality homes. In 1988, China's social investigation department conducted a door-to-door questionnaire survey that collected data from 1,000 Beijing households. The results showed that group diversity based on political status and administrative grade existed in housing distribution. The most prominent was the differentiation and inequality of housing benefits between cadres and workers. Scholars like Yanjie Bian conducted investigations into Chinese urban housing redistribution, discovering that the political capacity of individuals was the most prominent factor in determining the quality of housing that they were allocated. The term "political capacity" refers to the extent to which an individual is able to influence the distribution of housing. According to the known indicators by which an individual's political capacity can be measured, cadres had the most significant influence on housing distribution. This means individuals with more political capital were prone to obtain better houses and conditions.[111] Wang Yaping and Murie have also discussed the disposition of the mechanism of redistributing urban housing in China in their research. They have pointed out that one's official status—cadre or worker—was an important factor in determining how one was prioritized: workers had almost no right to precedence.[112] All this research indicates that significant housing stratification based on political power and official status existed in the housing distribution system.

In Chinese society, power and status inequalities were not only reflected in urban housing allocation, but were actually a common feature of Chinese social stratification. Before the 1979 reforms, urban employees could basically be divided into two social identity groups: cadres and workers. This division of social identities within urban areas and the household registration system, which separated urban and rural areas, acted together as social barriers to resource allocation and became two major social barrier mechanisms in China. In urban areas, identifying people as cadre members or workers was not only an aspect of record keeping, but an indication of immense differences in material benefits, including political power, pay grade, working conditions, travel allowance for official business, housing conditions, medical treatment, and pension, among others. Generally speaking, cadres had a clear advantage over workers in these respects. The division of identity between cadres and workers maintained social stratification and social order after social stratification according to assets was destroyed. As for housing, cadres were often given priority in housing allocation, receiving large, high-quality, and well-equipped houses.

In the case of the enterprise under discussion, obvious differences and disparities between cadres and workers can be found in housing distribution. Field research has revealed that the factors influencing

111 John R. Logan, et al. "Housing Inequality in Urban China in the 1990s." *International Journal of Urban and Regional Research* 23 (1999): 7–25.
112 Wang Yaping and Alan Murie. "Social and Spatial Implications of Housing Reform in China." *International Journal of Urban and Regional Research* 24.2 (2000): 397–417.

housing distribution can be arranged in order of importance as follows: length of service, administrative position, professional title, and household population. The available evidence indicates that people with long service records, power, and skills had preferential access to housing resources. Other factors such as household composition, which was a factor symbolic of equalitarianism, also played a critical role. Preferential treatment on the basis of administrative position and status, which are largely symbolic grades, was also regarded as reasonable. In addition, it was found that this enterprise would design and build special "houses for cadres" every time a new building project began in order to ensure the housing of the enterprise's leaders and managers. These "houses for cadres" were consistently of the best quality and equipped with the best facilities of all the houses built in this period, clearly demonstrating the advantages held by cadres and their differences from ordinary people. Compared with their advantageous housing conditions, the housing conditions of ordinary workers were of an intermediate or low level. In some workers' apartments, lack of space and the poorly designed internal structure of the houses made them rather uncomfortable and inconvenient for daily living. These worker households, which had between 30 and 40 square meters of living space, were not the worst houses in the enterprise. At least these workers owned their own apartments. Some worker households even lived in low-quality bungalows and economy buildings. The bungalows usually had no more than 20 square meters of living space and their inhabitants were forced to build sheds to act as kitchens or storerooms in front of or behind their houses. Most of the economy buildings took up about 20 square meters, but, in their allotted service period of fifty years, they became very dilapidated.

In the enterprise under study, inequality between cadres and workers in housing distribution was a direct consequence of a distribution system in which political power and status were considered important criteria for housing distribution. Although the aim of distribution in this housing distribution system dominated by public ownership was to achieve demand-oriented equalitarianism, in the actual process of distributing houses as a form of nonmonetary income, there is a return to the role of political power and status. Therefore, a ranking order of housing stratification based on power and status was established.

2. Housing Reform and the Reproduction of Social Stratification

Power and status playing a key role in the allocation of social resource was a fundamental characteristic of social stratification and an important foundation of China's ranking order prior to market transition. Whether these characteristics and the ranking order changed after the transition and what the changes were, if any, are the focus of dispute and a basic starting point of research for Chinese scholars concerned about social transformation in China. In *A Theory of Market Transition,* Ni Zhiwei asserted that the relationship between market transition and social stratification in state socialism is that the transition from a redistribution to market system in state socialism will fundamentally change the stratification mechanism, which is based on political power in a redistribution system. Power can never play as important a role in a market economy as in a redistribution economy because a transition of power and privilege from the political classes to producers occurs.[113] However, critics of this theory of market transition note the continuity and path dependence in such social transformations, believing that market-oriented reform has not decreased the efficiency of power, but rather enabled redistribution and the market to coexist. Therefore, the influence of power on social stratification and inequality persists.[114] Ni Zhiwei's assumption that market-oriented reform in state socialism has had an "effect of equalization" is based on research conducted on earlier market transitions. He believes that the beneficiaries of the process of market transition are those who were in disadvantageous positions in the system existing prior to the transition. However, in the middle and later periods of the reform, expansions of social inequality in state socialism became widely admitted facts. Since the 1990s, with the deepening of reform in China, scholars researching state implementation of market transitions have confirmed that market-oriented economic reform and social transformation tend to be more beneficial for redistributors than for producers, leading to divisions forming within society. These social consequences are revealed by social structure, through which pre-transition inequalities between redistributors and producers persist and become consolidated.

113 Nee Victor. "A Theory of Market Transition: From Redistribution to Market in State Socialism." *American Sociology Review* 54 (1989): 663–81.
114 Yanjie Bian and John R Logan. "Market Transition and the Persistence of Power: The Changing Stratification System in Urban China." *American Sociological Review* 61.51 (1996): 739–58.

Urban housing reforms confirmed that China's market-oriented reform brought about the reproduction and consolidation of social stratification. The transformation from a planned economic system to a market economic system began in China in the late 1970s and early 1980s. Urban housing reform was carried out against this wider background and was regarded as an important measure towards fulfilling the goal of economic reform. After a dozen years of exploration and transition, in July 1998, the State Council of China stated in the *Notice on Further Deepening Urban Housing System Reform and Accelerating Housing Construction (No. 23 [1998] of the State Council)* that housing allocation would be halted and that housing distribution would be gradually commercialized in the second half of 1998. This was regarded as the end of the welfare housing system and the beginning of a new housing system, namely that of a market-oriented housing system and the establishment of commercial residential housing as a market entity. Implementing urban housing reform and establishing the new housing system were intended to alleviate housing shortages and prevent uneven housing distribution. The reforms did indeed reduce this shortage and housing conditions were improved. Nonetheless, inequality and injustice in housing distribution were not resolved according to the original intention. Scholars have pointed out that, although the reform in 1979 has made citizens' living conditions better, the serious misdistribution it invited has been ignored.[115] After the mid-1990s, China's urban housing reform proceeded apace; research into housing reform has concentrated on the relationship between housing reform and social inequality. Researchers have tried to find out how housing reform—as an important aspect of the transition from a redistribution system to a market system—has influenced social stratification and inequality, as well as whether the stratification system based on redistribution persisted after this transition. Researchers have reached a basic consensus that these housing reforms did not reduce or degrade the inequality observed under the housing redistribution system but in fact, to a certain extent, continued or even strengthened housing inequality.

In the case under discussion, the state-owned enterprise experienced the transition from welfare housing to monetized and

115 James Lee. "From Welfare Housing to Home Ownership: The Dilemma of China's Housing Reform." *Housing Studies* 15.1 (2000): 61–76.

commercial housing. One of the most important programs of the reform was to sell to the employees the publicly owned houses that they already occupied. This was one of the most important aspects of the reforms in all state-owned enterprises, including the one under discussion, which owned lots of public houses. It was also the key step in initiating housing reform. Public houses—also known as publicly owned or state-owned houses—are houses constructed, managed, and distributed by government (central or local) and state-owned enterprises or institutions. In terms of property rights, before they were sold, they were all publicly owned. The public houses built by local or municipal governments, referred to as "directly managed public houses" ("直管公房"), were managed directly by Housing Management Bureaus, which were responsible for renting and selling housing to urban residents. Public houses constructed by state-owned enterprises and institutions, referred to as "work-unit-managed public houses" ("单位自管工房"), were managed by enterprises and institutions and were rented and sold to their employees. In the early 1990s, urban housing in China was proportioned as follows: 9% were directly managed public housing, *75%* were unit-managed public housing and only 16% were private housing.[116] It is apparent that the majority of urban publicly owned houses were owned by work units. Accordingly, the sale of work-unit-managed public houses was one of the most important tasks in the implementation of housing reform. In the early stages, the priority of housing reform for state-owned enterprises—the most important arena of urban housing reform—was the selling of work-unit-managed houses, which was regarded as having a leading role and being a breakthrough in housing reform. By selling work-unit-managed houses, the enterprise discussed converted all its publicly owned houses into private houses with attendant property rights. This accomplished the goal of commercializing housing, freed the enterprise's employees from dependence on the enterprise for housing, and established a new housing allocation mechanism.

Field research on the process and outcome of the sale of public houses in the enterprise studied found that the housing inequalities between managerial staff and employees that appeared during the housing distribution period were reproduced and strengthened after the reforms. This was caused by structural disparities between these

116 Wang Jiafeng. "Critics on Housing System Reform in China." *Journal of Zhongnan University of Finance & Economics* 5 (1991): 37–40.

groups. Structural disparities developed from the systematic differences in the rights and social benefits enjoyed by managerial staff and workers. These disparities were the products of housing distribution under the redistribution system, which restrained the adjustment of housing benefits after the housing reforms. In the enterprise studied, after housing reform was implemented through the sale of public houses, the reproduction and consolidation of housing stratification between cadres and workers could be observed in two respects.

The first was the priority received by managerial staff in the allocation of housing resources. Though employees in the same enterprise (both cadres and workers) possessed their own houses after the sale of public houses, their houses differed in living space and value. As has been stated by some scholars, since the preferential policies of the sale of public housing enterprises ignored the different housing conditions of employees, they produced a situation in which those who lived in large houses "gained more benefits" ("多得利") and those who lived in small houses "gained fewer benefits" ("少得利"). Those who had no allocated housing "gained nothing" ("不得利").[117] With the deepening of market-oriented reform, the segregation caused by the sale of public houses led to the housing disparities between cadres and workers becoming wider in terms of the value of houses rather than their living space. Thus, the housing inequality between the two groups under the former housing distribution system was enhanced by the reform.

Second, in the process of selling off public houses, the houses offered at cost to the enterprise's managerial staff and cadres were much larger than those ordinary employees were entitled to. This means that the actual prices they paid to purchase these houses were much lower than those paid by ordinary employees. According to the rules governing the sale of public housing in the city of the studied enterprise, the area of the housing that employees could purchase depended on the employees' grades: workers and grassroots-level cadres could purchase houses of 65 square meters; section-level cadres, 75 square meters; cadres at county level and department level, 95 square meters; and cadres at the provincial department level and bureau level, 120 square meters. This preferential policy obviously enabled the cadres of enterprises to obtain better houses at lower cost, meaning that those who had

117 Li Bing. *Housing Interests Differentiation and Transition of Social Stratification Mechanism.* Changsha: Zhongnan UP, 2004. 301.

benefited from the former housing distribution system were guaranteed to continue to benefit from the new housing system.

Empirical research conducted by Yanjie Bian and John Logan indicates that a key strategy of urban housing reform in China was selling public houses to the inhabitants at preferential prices. Though this encouraged people to ignore or forget housing inequality within the work unit, it could not erase the huge differences in housing benefits between people. As a consequence of the housing reforms, existing inequalities persisted after the process of housing commoditization and marketization, and social inequality molded by both power and the market emerged. Beneficiaries were those employees who held higher political status and better socio-economic conditions, as well as employees with power and organizational resources, which deepened the inequalities caused by the old welfare housing system.[118] In their research on Chinese housing reform in state-owned enterprises and its impact, Wang Yaping, et al. pointed out that huge internal inequalities between managerial staff, professional staff, and ordinary workers in housing distribution had always existed in every state-owned enterprise, a trend that greatly increased during the reforms.[119] In the enterprise under study, as is demonstrated by the above research, housing inequalities caused by the distribution of publicly owned housing persisted and were reproduced among cadres, managerial staff (who were regarded as elites in redistribution), and productive workers during the change in the housing system.

3. Setting up a Ranking Order of Social Stratification Based on Property

In the city where the enterprise in question is located, almost all private urban houses were nationalized in a socialist transformation of private housing before the marketization reforms, allowing an urban housing system dominated by public ownership to take shape. By 1956, 95% of all leased houses and urban land had

111 John R. Logan, et al. "Housing Inequality in Urban China in the 1990s." International Journal of Urban and Regional Research 23 (1999): 7–25.
112 Yaping Wang, et al. "Chinese Housing Reform in State-Owned Enterprises and Its Impact on Different Social Groups." *Urban Studies* 42.10 (2005): 1859–78.

been nationalized.[120] Under this highly centralized public housing system, the right to privately own houses was nonexistent; houses, as public property, were allocated by the government in a top-down fashion. This was the only method by which the city's residents could legally obtain houses. The actual process of housing distribution was implemented by work units according to the statuses of cadres and workers. The result of the distribution was housing differentiation and inequality between cadres and workers from the perspective of social stratification, which set up a ranking order of social stratification based on political power or status.

Urban housing reform in China is a process of constantly promoting "houses for those who need it" ("居者有其屋") and providing as much support as possible. As one scholar put it, along with the marketization of the Chinese economy, the privatization of houses was not only a requirement of the socialist economy, but also a significant part, or even a basic target, of housing reform. By means of housing reform, great progress was made in improving property rights and the privatization of housing. As a result, a housing ownership system dominated by private ownership, in which various forms of housing ownership co-existed, took shape.[121] From the state-owned houses of the redistribution system to the privately owned houses of the market economy, the transition of urban housing in China is not only a change of ownership but also a transformation of housing from a form of remuneration for labor into a private asset in the market economy system. In the case under study, a core change in housing reform was the transition from public ownership to private ownership. It was through the sale of public housing—an important aspect of the reforms—that houses stopped being remuneration offered by the state or work units to individuals and became private property in a market economy. As for employees in state-owned enterprises who lived through the transition from a redistribution system to a market economy, the houses that they obtained from work units in the past were houses to which they only had the right of use and residence. After the reforms, however, they were permitted to purchase these public houses that they lived

113 Xiangming Chen. "China's Urban Housing Development in the Shift from Redistribution to Decentralization." *Social Problems* 40.1 (1993): 266–83.

114 Jin Jian. "Progress of Housing Privatization in China and Its Impact on Society." *Chinese Journal of Law* 3 (2004): 107–11.

in and thereby owned their houses, which later became their most important private assets. Housing was therefore redefined as a form of private property after the housing reforms.

In the late 1990s, Deborah S. Davis, professor of sociology at Yale University, turned her attention to the impact of urban housing privatization on Chinese society and studied the housing reforms and the transition between property rights systems. She pointed out that the majority of houses in urban China had become fully capitalized and transferable individual assets. Urban housing reform in China, therefore, brought forth a new urban property rights system. In her view, the effect of urban housing privatization on society was the furtherance of a social differentiation that originally resulted from housing distribution. It was because of the very process of turning the supply of houses, which were formerly public welfare benefits, into privately owned property that the gap between the households of officials and managerial staff and the households of productive workers and immigrants became wider.[122] Xueguang Zhou once mentioned in his research that, by the late 1990s, most of the houses in urban China had been sold to citizens as part of the experiment in housing privatization, marking the formation of a housing market and the legalization of long-term housing inequality.[123] In recent research, scholars such as Yanjie Bian have included housing property rights in the index system for social stratification. For them, the reform of the housing distribution system in China brought about the recognition of housing property rights. As a result, the possession of housing property rights has become an important indicator of stratum disparity and differentiation. By analyzing data from the Fifth National Population Census, Yanjie Bian, et al. found that householders who are administrative and professional elites make up a larger proportion of homeowners in possession of housing property rights. Compared to productive workers, the property ownership rate for leaders of Party and social organizations, government organizations, public institutions, and enterprises are 2 times, 1.8 times, 1.7 times, 1.7 times, and 1.4 times higher than those of workers, respectively. The advantages held by Chinese administrative elites lie in their comparatively high economic

122 Deborah S. Davis. "From Welfare Benefits to Capitalized Assets: The Recommodification of Residential Space in Urban China." Housing and Social Change: East-West Perspectives. Eds. Ray Forrest and James Lee. London: Routledge, 2003. 183–98.

123 Xueguang Zhou, *The State and Life Chances in Urban China: Redistribution and Stratification, 1949-1994*. Cambridge: Cambridge UP, 2004. 244.

capability in purchasing commoditized apartments. Besides this, they also enjoy the advantage of the power bestowed upon them by the redistribution system through which most of them purchased "public houses they lived in" ("现住公房") or "economical and functional houses" ("经济适用房")[124] at preferential prices.

The legal approval of the private ownership of houses as property brought about by the housing reforms has had a significant impact on the interest patterns of housing, and even the entire pre-reform social stratification structure. In the housing distribution system dominated by public ownership, houses could not be possessed as private assets. Although there were still privately owned houses in cities, such as the business premises of individual businesses and the self-owned tenements of citizens, they were not protected by any solid legal provisions. The housing property rights system dominated by private ownership established after the reforms made houses legal assets that could be privately owned. As a result, the realization of transferring the ownership of houses to their residents through the sale of public houses, in a sense, solidified housing stratification under the distribution system of physical housing in the form of property rights, making property a new symbol of social stratification. As for its impact on inequality, I believe that changing the housing system in urban China not only maintained or strengthened housing inequality, as many scholars have claimed, but also constructed a new ranking order—a property-based ranking order. In other words, housing reform not only changed the degree of inequality but also the foundation of that inequality, constructing a new ranking order in the market economy—stratification based on assets. During the housing transformation in enterprises, the housing inequality that previously existed between cadres, managerial staff, and ordinary workers was reproduced and strengthened. This occurred because, after the enterprise-controlled housing became privately owned assets that could flow through the market, housing assets possessed by cadres and managerial staff were much greater than those of ordinary workers. This enabled the beneficiaries of the former system to gain more while low-ranking employees and ordinary workers gained much less. The impact of housing reform on people's interest patterns led to the former housing inequality producing

124 Yanjie Bian and Liu Yongli. "Social Stratification, Housing Property and Housing Quality – An Analysis on the Fifth Census Data." *Sociological Studies* 3 (2005): 82–98, 243.

a cumulative effect under the new system, causing a "Matthew Effect" in urban housing benefit distribution.

Urban housing reform, a significant systemic transformation, has greatly influenced China's social stratification structure in the past thirty-plus years. It has made property increasingly important in social stratification in China. Before the reforms, Chinese society was a society dominated by public property; it was also a proletarian society with regard to the ownership of the means of production. In this social system, private property was not effectively protected while the ownership of personal property was weakened, and social stratification was not based upon property ownership. This means that the standard for classifying social status was not property or income, but a special index of identities including political position, registered residence, and work position like cadre or worker, among others. It was the establishment of such an identity system that facilitated the mechanisms of resource redistribution and social order maintenance that were formed all over the country. The great significance of the housing reforms for social stratification structure lay in facilitating the proliferation of privately owned houses within the market economy. Housing commercialization dominated by occupant ownership was established and houses became the most important assets most people possessed, symbolizing their social identity and status. On March 16, 2007, *The Property Law of the People's Republic of China*, which proposed equal protection for both private and public property, was approved by the National People's Congress. On October 15th of the same year, Chairman Hu Jintao proposed in his report to the Seventeenth National Congress of the Communist Party of China **that "better conditions should be created to enable more citizens to have property income."** With all this rhetoric, it is to be expected that property would gradually become a major indicator of economic and social status and social stratification in China. A system of property ownership rights has become increasingly important, tending to become the new primary system of maintaining social order. The housing status groups described in the previous chapter have emerged precisely because of this system.

Chapter 15

A Review of Social Stratification Theories in China

Most of the previous chapters have employed empirical analysis. From the perspective of theoretical study, what theoretical research has been carried out on social stratification in China? What kinds of problems have been encountered? What arguments have been made? This chapter will simply disentangle the existing theoretical research on social stratification in China.

As previously mentioned, social stratification essentially reflects the unequal relationships between different social groups. Inequality is usually hidden deep in the internal structure of a society. Theoretical research into social stratification attempts to analyze the vertical social divisions, integrations, and mobility that may reveal the roots of social inequality. Generally speaking, research into stratification in China is usually done within Karl Marx's framework of class theory and Max Weber's theory of multielement stratification.[125] Although it is commonly understood that these two theories fundamentally disagree on the causation, process, results, and even decisive factors of stratification, they arrive at the same end by different means. Both of them reveal that social stratification is closely related to political and economic systems. The features of social stratification cannot be discussed without considering system environment, economic system, and political power.

This chapter divides the development of social stratification theories in China into five stages by reviewing different theories and discourse models produced since the founding of the People's Republic of China. The following will analyze the features of social stratification theories and practices at different stages.

1. Class Stratification Theory and Practice (from the Founding of the People's Republic of China to 1957)

The first stage is from 1949 (the year in which the People's Republic of China was founded) to 1957. In this stage, China was a new

125 Let us, for the moment, consider functionalism and Weberism as representing the same ideas.

democratic society. The intention at the time was to "gradually eliminate exploitation and achieve Great Harmony" ("消灭剥削和实现大同"), then transition to socialism. Because the economy and culture were backward before the founding of the People's Republic of China, "efforts should be made to develop the economy" ("努力发展经济") and unite all the people of different classes. Chairman Mao Zedong had once said: "At the present stage, people are referred to as the working class, the peasant class, the urban petty bourgeoisie, and the national bourgeoisie." In the meantime, the economic form of a new democratic society must allow the "five economic components" ("五种经济成分") of state economy, cooperative economy, private capitalism, individual economy, and state capitalism to exist and develop together for a long time. Chairman Mao Zedong also believed that allowing the coexistence of the five economic forms was the only way for a backward China to develop productivity and build a prosperous and powerful country after the success of the revolution. After the Second Plenary Session of the Seventh Central Committee of the Communist Party, Mao Zedong proposed the economic policy of "taking both public and private interests into account, benefiting both labor and capital, and enhancing mutual assistance between urban and rural areas, as well as internal and external communication," so as to satisfy different interests and achieve the goal of developing productivity and improving the economy.[126]

In order to achieve the dual goals of a new democratic society, the Communist Party of China (CPC), under the leadership of Mao Zedong, employed the theory of Marxism to analyze Chinese society and, on the basis of a vast amount of investigation and research, proposed many innovative ideas. Mass movements were launched in rural and urban areas and the old class system was dismantled, which promoted the development of social productivity. In rural areas, the National Administration Council of the Central People's Government passed *The Decision on the Classification of Class Status in Rural Areas* and *Agrarian Reform Law of the People's Republic of China*. In terms of land reform, the land of landlords was seized and divided equally amongst peasants. Therefore, "feudal land ownership was abolished in rural areas and farmers obtained ownership of the land." In urban

126 Yang Xiancai, ed. *A History of China. The Volume on the People's Republic of China*. Beijing: Higher Education Press, 2001. 3.

areas, the "socialist transformation of capitalist industry and commerce" movement was launched, in which the Chinese bourgeoisie was transformed by means of the government purchasing their property at "low interest rates." Private enterprises were transformed through "state-private joint ownership" and then state and collective ownership. Through the official process of "redemption" ("赎买") of privately owned assets, capitalists were deprived of their rights.

The classification of class and stratum during this period had the following features. There was only one criterion specified for identifying classes—that is, whether or not one held the means of production (mainly the land in rural areas), how much one possessed, and the relations of production generated from that ownership (exploitative relationships). Landlords and traditional rich peasants were ranked as targets of the revolution, and new rich peasants (peasants who became rich through hard work after the land reform), middle income peasants, poor peasants, and farmhands constituted the peasant class.[127] The timeline for identifying class status was made clear, as it was expressly stated in the provision for determining the class status and land ownership status of intellectuals, soldiers in the Red Army who were born to landlord and rich peasant families, landlords, rich peasants, capitalists, workers, peasants, and poor peasants who married into a different class. The alteration of the timeline for classifying landlords and rich peasants was adjusted. For instance, "exploitation should be calculated from the time of local liberation rather than from any other time." Using facts from before liberation to determine one's class status was not right. It was expressly stated in the provision that "intellectuals should not be regarded as a class status, and that their class origin should be determined by their family class status and their class status by their means of living." All these concrete land reform measures linked class status closely with economic status, ensuring that the social class system objectively reflected the social structure of the day. Because such documents as the *Land Reform Law* and *The Decision on the Classification of Class Status in Rural Areas* had established detailed provisions for different class statuses, quantity limits, the timeline, and accounted for some special situations—while clearly answering questions encountered during the process—a systematic rural class and stratum classification theory was formed.

127 Xu Gongxi. "The Outline of Mao Zedong's Standards of Identifying Rural Class Statuses from 1925 to 1956." *Journal of Shangrao Teachers' College* 2 (1994): 1–6.

To sum up, in this period, Marxist class theory prevailed in the field of Chinese social stratification studies, the core point of which concerns the exploitative economic relationship between the ruling class and the servile class. By employing this theory, the Communist Party of China broke the class system of Old China,[128] "nationalized almost all the private property in urban areas and collectivized almost all the private property in rural areas,"[129] laying a legal foundation for the new government and legitimizing the legal status of the leading class, which seized and consolidated political power.

2. Political Stratification Theory and Practice (from 1957 to 1978)

The second stage lasted from 1957 to 1978, about twenty years. Deng Xiaoping once regarded this period as a stage of "twenty years of leftist error" ("20年左倾错误").[130] He said that the "leftist" ideas originated in 1957 and prevailed gradually. For instance, the "Great Leap Forward" ("大跃进") of 1958, a rushed mass effort to form rural people's communes—which strongly emphasized being "large in size and collective in nature" ("一大二公") and "everybody eating from the same big pot" ("吃大锅饭")[131]—led to a catastrophic famine, not to mention the Great Cultural Revolution. After the defeat of the Gang of Four, the government lingered on for another two years, basically following these "leftist" errors until 1978. From 1958 to 1978, the incomes of peasants and workers increased little, resulting in low living standards and underdeveloped productivity. In this period, the "leftist" error was in the incorrect judgment that the principal social contradiction in contemporary Chinese society was undoubtedly "the contradiction between the proletariat and the bourgeoisie and between socialism and capitalism." This judgment led to a poorly conceived political path geared towards enlarging class struggle. This naturally resulted in a failure to develop the economy, which

128 "Old China" usually refers to China before the founding of the People's Republic of China whereas "New China" refers to China after the founding of the People's Republic of China.
129 Huang Zongzhi. "The Rural Class Struggle in China's Revolution." *Foreign Sociology* 6 (1998): 37.
130 Zhang Zhirong. "Correctly Understanding Deng Xiaoping's Judgment of Twenty-Year Left-Leaning Mistakes." *Journal of Shangrao Teachers" College* 8 (1994): 22.
131 "Everybody eating from the same big pot" was an egalitarian practice common during the movement of the so-called "Great Leap Forward." It dictated that everyone should labor for the public and should not keep food in their own houses. Instead, people should dine in public canteens where the same portion of food is provided to everyone regardless of the amount of work they have done. This practice was later referred to as indiscriminate egalitarianism.

was organized according to the methods of class struggle. Political and ideological standards were, therefore, unavoidably brought into the system of social class stratification. In the political movements that followed, the importance of these political and ideological standards were exaggerated in social stratification by assigning political labels, such as "the rightist" ("右派"), "the class with serious historical problems" ("有严重历史问题"), and "the bad man" ("坏分子"), which changed the more objective class stratification system in the first stage.

Class stratification in this period actually had nothing to do with economics because a landlord was no longer a person who possessed land, but one who had possessed land prior to the land reform movement in 1949; a capitalist was no longer a person who possessed an enterprise, but one who had possessed an enterprise before the socialist transformation of capitalist industry and commerce in 1956. These political labels were pinned on them according to a subjective ideological classification. Therefore, class stratification in this period was actually political stratification: people were stratified into different hierarchical social groups according to their family status, political status, political positions, and political views and attitudes. Regarding my political stratification theory, see also chapters 4 and 5 of this book.

Political stratification had the following characteristics. Firstly, political stratification had nothing to do with economic interests. One's political status did not reflect one's economic condition. For instance, the political status of a capitalist was quite low, but his economic status and living standard might have been rather high. Those who were labeled as "rightists" suffered a sudden decline in political status, but their economic status and living standards may have remained almost unchanged. Secondly, the political stratification system was largely a closed one. Political stratification in this period identified one's family status by the class status identified during the first stage. Although the class status identified during the first stage had something to do with one's economic status, in the following twenty years, the original landowners and industrialists no longer possessed their property and their economic status had fundamentally changed, but their class status remained unchanged. Thirdly, the social mobility that allowed one to change one's political status usually took place through political movements. The main forms of political status mobility in this period were "being sent to the countryside" ("下放"), "being assigned a political label" ("戴帽子"), "a political label being removed" ("摘帽子"), "being criticized while being provided for" ("一批、二养"), "rehabilitation" ("平反"), and "redressing cadres of a grievance" ("解放干部"). In these surging tides of political movements, a number of political classes were created in order to meet the needs of class struggle, causing a great deal of injustice. In short, the leading catchphrase in this period was "never forget class struggle" ("千万不要忘记阶级斗争"), which emphasized that class, class contradiction, and class struggle must be mentioned every year, every month, and every day. Thereafter, class and class struggle theory became increasingly emphasized, which led to the policy guidance of enlarging political struggle. As Mr. Huang Zongzhi said, "Such terms as class, class identity, class struggle, class enemy, targets of attack infiltrated into standard everyday language. In this complicated situation, the meticulous standards for analyzing relations of production no longer existed. All the enemies in the revolution had been combined into one: the class enemy, who represented all the evils in the old society."[132] The class structure stated in this state power discourse was disjointed and entirely distinct from the objective social structure. Class had gradually become an identity and a label of ideology.

3. Returning to "Objective Class Structure" (from the End of 1978 to the Mid-1980s)

The third stage began with the Third Plenary Session of the Eleventh Central Committee of the Communist Party of China in December, 1978 and continued up until the mid-1980s. The Third Plenary Session proposed that the focus of the whole party should be shifted, which achieved the strategic transformation from class struggle to economic construction for the CPC and the whole country. This was a fundamental correction. In 1979, the Central Committee of the Communist Party of China made a major decision and proposed that landlords and rich peasants should be re-identified. "Those labeled as landlords, rich peasants, reactionaries and bad men who have obeyed all laws and regulations, labored hard, and done nothing evil shall be unpinned from their class status and treated as rural people's commune

132 Huang Zongzhi. "The Rural Class Struggle in China's Revolution." *Foreign Sociology* 6 (1998): 47.

members. The class status of those born to the families of landlords and rich peasants shall be commune members.... Henceforth, their political performance shall be mainly considered without any prejudice when they register in a school, take a job, join the army, join the Youth League, join the Party or are assigned to a job."[133] As the strategic focus on economic construction was established, political stratification reflecting political attitudes and ideological standards gradually lost their meaning. The political stratification system began to collapse and the differences in political status between people began to shrink dramatically. Simultaneously, the economic hierarchy marked by income difference and property difference was gradually re-established.

During that period, in order to promote the reform and opening-up policy, China put forward the slogan of "letting some people grow rich first through hard work." This proposal was immediately put into practice and became a social policy. In a short period of several years, a group of wealthy people appeared in China and a new class system with comparatively high economic incomes gradually took form.[134] For instance, statistics show that from 1978 to 1984, the number of self-employed laborers increased from 150,000 to 3,930,000; among all urban workers, the proportion of self-employed laborers rose from 0.2% to 2.8%, while the proportion of workers in state-owned enterprises declined from 78.3% to 70.6%. Take another example: by the end of 1984 there were 4,260,000 rural households producing special products besides crops, accounting for 2.3% of all rural households; the labor force was 9,890,000 (or 10,580,000 if apprentices are included), accounting for 2.9% of the whole labor force. Their total incomes amounted to 19.68 billion yuan, accounting for 4% of the total value of the rural economy. The average household income was 4,625 yuan, which was 79% higher than the national average household income for farmers; the commodity rate was 69%, which was 53% higher than the national average. All that means is that at this stage, rural households producing a special product besides crops actually became wealthy. On July 25, 1985, the *Economic Information Daily*, sponsored by China's official *Xinhua News Agency*, published a report entitled *The Characteristics of Rural Wenzhou, the Leading Role*

133 Shen Chong and Xiang Xiyang, eds. *Theories, Policies and Practice within Ten Years*. Vol. 1. Beijing: Qiushi Press, 1988. 200.
134 Lu Xueyi, ed. *The Vicissitudes of Social Structure*. Beijing: China Social Sciences Press, 1997. 101.

of Individual-Run Economic Undertakings.[135] The social development trend showed that, at this stage, China's ownership structure began to change and "diversified economic sectors and modes of operation emerged, which changed the existing ownership structure and class and stratum structure, and the power and benefits to which different classes were entitled also changed."[136] Based on the articles published at the time, all these changes may be summarized into the following three points. "The first is the reform of the state-owned economic system and the change of the relationships within the working class; the second is the reform of the rural collective-owned management system and the change of the internal relationships within the peasant class; the third is the recovery and development of the rural individual economy and the formation of the new petty bourgeoisie and their similarities and differences from the former petty bourgeoisie."[137] How best to understand these changes of class has been fiercely debated in the field of social stratification. Take, for example, these articles published in the *People's Daily* in 1979 such as *How to Correctly Understand the Fundamental Changes of Classes in China, The Theory Training Class of Shanghai's Municipal Party Committee at the Party School Studies the Report on the Work of the Government, Discusses Class and Class Struggle, Makes Clear the Principal Contradiction at the Present Stage and Determines to Focus on Realizing the Four Modernizations, To Make Clear the Principal Contradiction in Discussing the Criteria for Truth, Comrade Xi Zhongxun at the Meeting of arty Secretaries of Guangdong Province-Prefecture-City Cadres at All Levels Requires that the Changes of Class Relationship in Our Country be Correctly Understood, and that We Consciously Transfer the Focus of Our Work to the Building of the Four Modernizations on the Basis of Their Years of Practical Experience.* These kinds of articles were very common. All these articles express one core concept: "over the past three decades, social class in China has experienced fundamental changes in that the class position of the working class in our country had been greatly enhanced, peasants have been collectivized for more than twenty years, and the mass of intellectuals have become a part of working

135 The data is cited from He Jianzhang's "The Adjustment of China's Ownership Structure and Changes of Social Class Structure." *Sociological Studies* 3 (1987): 2–9.
136 Zhu Qingfang. "Analysis on the Transformation of China's Class and Stratum Structure during Recent Ten Years." *Sociological Studies* 3 (1990): 1–8.
137 He Jianzhang. "On 'Middle Class' in China," *Sociological Studies* 2 (1990): 1–7.

class. All the ethnic groups in China have developed new socialist ethnic relations based on unity, friendliness, mutual assistance, and cooperation. The majority of capitalists who were able to work have become laborers and live on their own labor. The non-communist parties have developed into political unions representing their group of socialist workers and patriots supporting socialism."[138] Under the guidance of this ideology, rich peasants were reconsidered according to Marx's theory of social class. On the basis of investigation, various kinds of false and incorrect cases were addressed, which eased the tension between different social groups caused by political stratification. For instance, according to the *People's Daily*, the Party committee of Liya commune in Zheng'an county, Guizhou province was praised for the way in which they addressed the case of the poor peasant Huang Shaojun, who was re-classified as a "new rich peasant." In May 1977, an unknown individual in the Xinming production team of Liya commune reported that poor peasant Huang Shaojun was a new rich peasant and *nouveau riche* because his family income was much higher than other community members. Some cadre members in that commune had perniciously adhered to the pseudo-left (but actually right wing) thought advocated by Lin Biao and the Gang of Four, according to which a rich life was revisionism and a good life was capitalism. They forcibly confiscated Huang's savings of 897 yuan, a sewing machine, and some of his furniture. At the same time, he was regarded as one of the main targets of attack and was delivered over to every brigade in his commune to be publicly denounced. However, the investigation showed that his family, which lived simply and worked hard, worked actively in their production team and developed household sideline activities. He bred a large sow that produced broods twice a year and brought more than 20,000 yuan to this family. Over the previous decade, he had sold heavy pigs to the country every year and also grew tobacco and pepper on his private plots. His children drove their cattle over the hill to feed them while they dug out raw herbal materials for medicines in the mountain and sold them to the state. Huang was therefore not practicing capitalism and he was not a class enemy, but was, in fact, a poor peasant.[139] Reports like this were numerous in the newspapers at that time. All these facts suggest that,

138 *People's Daily* Jun. 16, 1979.
139 "Uncompromisingly Rehabilitate Those Commune Members Who Have Been Persecuted." *People's Daily* Jan. 18, 1979.

in this period, efforts were made to restudy the social division that had emerged in Chinese society under the framework of Marx's theory of social class, which politically ensured that individuals and intellectuals were not assigned labels—meaning that "the class structure stated in the governmental discourse" in political stratification returned to an "objective class structure."

4. Free Discussion of the Gap between the Rich and the Poor (from the Mid-1980s to the Early 1990s)

The fourth stage ran from the mid-1980s to the early 1990s. During this period, a tremendous change in China was the introduction of a market mechanism. The market transition inevitably caused changes in the social stratification mechanism; the coexistence of diversified forms of ownership produced differences in the possession of the means of production. The property stratification system—in which payment distribution was based on one's possession of property—led to the division and reconstruction of social class and stratum in China. This was most clearly manifested in the polarization of incomes and the widening gap between rich and poor. Therefore, the gap between rich and poor in mainland China and related social equity became the focus of social discussion during this period.

According to international practice, the Gini coefficient is an important indicator of the gap between rich and poor. The policy of allowing some people to become rich first caused the rapid increase of China's Gini coefficient, reflecting the personal income gap between urban and rural residents in China after the mid-1980s. For instance, the personal income Gini index of urban residents was 0.19 in 1986—but by 1987, it had increased to 0.20, to 0.23 by 1990, and 0.37 by 1994. The Gini index of personal income of rural residents was 0.30 in 1985, and it increased to 0.31 by 1986, 0.338 by 1988, and 0.411 by 1994. China's total personal income Gini coefficient, for all urban and rural residents taken together, was estimated to be 0.382 in 1988 and 0.434 in 1994. These figures showed that the Gini index, which reflects the gap between rich and poor, increased too rapidly and caused a remarkable structural division in China's society. This was a regional division that was reflected in significant variation in per capita income in different regions. For example, supposing that Zhejiang province, Hubei province, and Gansu province represent the eastern region, the middle region, and the western region separately, the annual per capita income of farmers in these three regions would have been 721 yuan, 467 yuan, and 296 yuan, respectively. The yearly income difference between Zhejiang province and Gansu province was 425 yuan.[140] These regional divisions also manifested themselves in the great differences in social structure and behavioral patterns in different regions, with a clear-cut line of demarcation that further exacerbated regional differences. The division was also organizational. Organizational difference was mainly manifested in the decentralized economic management of central government and local governments, which was characterized by "local governments being financially responsible for themselves." Horizontally, within the government, those departments with the function of economic administration were separated from administrative departments. Additionally, various levels of governmental organizations and professional institutions were also differentiated by "separating government functions from enterprise management, delegating power, and allowing enterprises to keep a bigger share of profits." However, great heterogeneity developed within professional institutions and between enterprises, due to the different resources available to them after decentralization and the way in which economic function increased more for rural households than urban households. There were also individual divisions. Individual division meant the emergence of new social groups and the decomposition of old ones, as well as the growing plurality of individual roles. It also meant changes in the social positions of different members of society and changes in the meanings and types of social position. For instance, the individual division in urban areas mainly manifested itself in five respects. First, there was a gap between fixed income earners and non-fixed income earners. Non-fixed income earners gained various kinds of "pay-offs," "commissions," and "extra income" by illegal means, which were more profitable than a fixed income and became a typical example of corruption in society. Second, there was an income gap between workers in different kinds of ownership enterprises. The income of workers in private enterprises, foreign enterprises, and joint ventures was double that of workers in state-owned and collective enterprises. Third, there was an income gap between different occupations. Some common

140 See *Social Sciences in China* 4 (1990): 124.

occupations, such as taxi driver and tour guide, had extremely high salaries. Fourth, there was an income gap between workers in different industries. In 1987, employees with the highest incomes in the construction industry had salaries 90% higher than those with the lowest incomes in the textile industry. Fifth, there was an income gap between mental workers and manual workers, which could be summarized as "the reversal of the wage differential between mental labor and physical labor."[141]

These sharp social divisions provoked a storm of social reaction. As Zhang Wanli said in her 1990 paper named entitled *A Review of China's Social Classes and Strata in Recent Periods,* "Since 1987, social classes and strata in the present phase have become a research hotspot in academia... On the one hand, since the reforms, not every class is content with the status quo and sometimes they blame each other. On the other hand, there are indeed some people who became rich first and social wealth has become more tightly concentrated in the hands of a minority of people in our society. This state of affairs has provoked the division of the original classes and strata to varying degrees, which has had significant effects on social interest groups. Determining how to acknowledge and adjust the interest relations of different groups and how to achieve the overall goals of China's reform has become a tough, urgent, and important task."[142] Newspapers of the day published articles with such hot topic titles as *On Unfair Distribution, Shanghai: Ten Types of People with High Incomes,* On *Seven Unjustified Practices of Distribution,* and *Three Questions on Unfair Social Distribution.* The gap between rich and poor and unfair distribution has gained wider and wider attention. Ai Feng, a journalist from the *People's Daily,* published an article entitled *A Debate on Social Equity,* in which he summarized different views on the gap between rich and poor and unfair distribution as four main topics of debate. Topic one: If we allow some people to become rich first, who should these "some people" be? Topic two: How do we measure the equity between income gaps? Topic three: Which should be stressed at present—social equity or social efficiency? Topic four: What measures should be taken to achieve social equity?

141 Research Team of Social Division of Peking University. "The Division and Integration of Social Structure in China at the Present Stage." *China Social Sciences* 4 (1990): 126–28.
142 Zhang Wanli. "A Review of Research on China's Class and Stratum in Recent Period." *China Social Sciences* 5 (1990): 173.

In spite of this heated discussion, the unavoidable fact is that, along with the progress of the economic transformation and modernization at that time, the original social structure of two classes and one stratum (working class, peasant class, and the intellectual stratum) were separated by remarkable divisions. New social strata gradually formed, such as the stratum of individual laborers, the stratum of private entrepreneurs, the stratum of white-collar workers engaged in social management, and the stratum of entrepreneurs in urban areas. In rural areas, the peasant class was no longer solid, but in a state of constant decomposition and recomposition. Many intellectuals made the appeal that "the peasant class should be redefined again." All the above indicates that social economic change led to the emergence of a new social stratum structure and a new social stratification mechanism based on the gap between rich and poor. Income stratification was replacing the old political stratification mechanism based on political identity, residence registration, and administrative identity.

Theoretically speaking, how should the new stratification mechanism be defined? Should stratification studies continue to be carried out according to the framework of Marxist class theory, or should Weber's multielement stratification model be introduced? Many seminars on stratification theory were held to figure this out. For instance, in April 1988, the Institute of Sociology of the Chinese Academy of Social Sciences, the Hebei Academy of Social Sciences, and Cangzhou's prefectural Party committee of Hebei province held "A Seminar on China's Present Social Class and Strata Structure," at which social stratification theories and the status quo of China's present stratification were explored. At the seminar, three representative theoretical approaches to stratification were advanced. The first approach held that Marxist class analysis should be applied when classes are analyzed, but that some reasonable elements should be absorbed when strata were analyzed. The second believed that the traditional class analysis method was not feasible any more or was limited, while Western stratification theory was more realistically meaningful. The third approach held that social group differentiation caused by inequality in interests and power distribution existed in all societies and that China was no exception. Therefore, social stratification studies should take interest groups as the basic unit of research and refer to Western social stratification theory.[143] In the meantime, in order to reconcile those two kinds of social stratification theories, another viewpoint was that the concept of class was larger than that of stratum and that stratum could be divided into smaller "interest groups." According to this viewpoint, class was the highest level and so included strata composed of different interest groups. From the articles published at that time, it can be seen that some theoretical studies of stratification were indeed conducted and papers were written in accordance with this viewpoint. Actually, people became entangled in distinguishing the concepts of class and stratum simply because, in the face of the new social reality, efforts were made to introduce Weberian stratification theory while fears of betraying the class analysis approach endured. At that time, using the term "class" held the connotation of accepting Marxism while using the word "stratum" seemed to have the connotation of non-Marxism. So, when there was no obvious social trend to apply Weberian multielement analysis to stratification studies, many scholars declared their stances with the following words, which may demonstrate the practical characteristics of social stratification in this period: "Our studies are based on various theories; abstracting the concept and theoretical framework of China's social stratification from Chinese reality is neither limited to Marxist class theory nor simply copies Western stratification theory, but seeks new theories from the Chinese reality of social structural changes by modifying the existing related theories and creating new theories in our research practice."[144]

5. The Trend of Diversified Analytical Models in Stratification Study (from the 1990s until Now)

It has become increasingly obvious in this period that the domestic analysis models of stratification research in China are diversifying. As many scholars who engage in stratification studies have pointed out, "At present, China's social structure is complicated and obviously characterized by diversity, which means that it is insufficient to explain social stratum division only according to the possession of the means of production. In contemporary Chinese society, material possession is not the only standard for measuring people's social

143 "A Review of Seminars on Social Class and Stratum Structure in China at Present Stage." *Sociological Studies* 4 (1988): 141.
144 Li Lulu. "Theoretical Research on China's Social Stratification." *Sociological Studies* 5 (1989): 35.

stratum or social position. In order to clearly understand the features of China's present social stratum structure, it is necessary to restudy the traditional class analytical theory and develop it scientifically and form a more realistic, diversified standard that conforms to new social changes." Furthermore, the purposes of the research in this period are definite: "to unite and encourage more social forces to realize the aim of socioeconomic development and to build a modern socialist country."[145] This is in line with the arrangement of the CPC central committee. For instance, the Chairman of the CPC Central Committee Jiang Zemin stated confidently in his "July 1st Speech" that, "At present, it is improper to judge whether people are politically progressive or backward simply based on whether they own property or how much property they own. They should instead be judged mainly by their political ideology and actual performance, the way they have obtained their property, and how they distribute and use their property, as well as their contributions to the cause of socialism with Chinese characteristics through their labor."[146]

Under the guidance of the aforementioned diversified stratification concepts, social stratification research in China since the 1990s has achieved some success by analyzing national conditions in China in depth from different perspectives. The first perspective is that of dynamic research on status acquisition mechanisms and social stratification structure in China as they relate to social resource allocation at a given time in an attempt to identify the special laws dictating social mobility in China. The second advocates the analysis of the gains and losses of different social interest groups in China by looking at interest groups, believing that in modern Chinese society interest structures change so rapidly that social interest groups continually divide and reorganize. This would mean that "it is not suitable for China to employ the relative stable concepts of class and stratum," and that inflexible and simple classification and stratification of China's social structure should not be advocated. Researchers with this view proposed substituting "interest group structure" for "stratification structure" with a concentric circle model and different types of interest groups. The third perspective attaches importance to state and political mechanisms in

145 Lu Xueyi, ed. *A Report on Contemporary China's Social Class and Stratum*. Beijing: Social Science Academic Press, 2002. 6–7.

146 Cited from Zheng Bijian's "Chinese Communist Party Comprehensively Strengthens Its Marxism Awakening of Self-Construction in the New Century." *Chinese Social Science Digest* 6 (2001): 11–16.

the study of stratification in socialist countries, believing that political processes can directly change the nature and value of social and human capital. Although this point of view has developed based on the sample cases of a number of Chinese cities from 1949 to 1993, I believe that "considering the growing role of the state in the labor market in a society with industrialized markets, the implications of China's experience go beyond China and other socialist countries."[147] The fourth perspective has examined changes in China's social stratification mechanism in the development process of a market society, believing that "changes in social stratification mechanisms have emerged from the social transformation of state socialism and are caused by the transition to a free-market system and the corresponding changes of property rights structures."[148] Many scholars in China classify the stages of market transition in China using this fourth perspective and have studied in depth the central and marginal groups involved in the transition, the income gap between mental and physical laborers, and the generational replacing of the middle class. The fifth perspective proposes a theoretical framework of social stratification based on the classification of occupations and the possession of organizational, economic, and cultural resources, holding that China's social structure is composed of ten social strata and five social economic levels. These ten social strata are: the strata of state and social administrators, managers, private entrepreneurs, professionals, clerks, individual businessmen, commercial service staff, industrial workers, agricultural laborers, and the urban and rural unemployed and underemployed. The five social economic levels are: the upper social level (top leaders, manager of large enterprises, senior professionals, and owners of large private enterprises), the upper middle level (leaders at the middle and lower levels, managers at the middle level of large enterprises, managers of middle-sized and small enterprises, intermediate professionals, and owners of medium-sized businesses), the intermediate middle level (junior professionals, owners of small businesses, clerks, individual businessmen, intermediate and senior technicians, and large households involved in agricultural operations), the lower middle level (individual laborers, general commercial service staff, workers, and farmers), and the bottom level (workers in conditions of

147 Xueguang Zhou. "Dynamic Mechanism of Stratification of National Socialism: The Case of Chinese Cities." *Foreign Sociology* 3 (2001): 26–31.
148 *Foreign Sociology* 3 (2001):15.

poverty and job insecurity, farmers, and the unemployed and underemployed).[149] A further specific explanation of these viewpoints can be found in the next chapter.

During this period, studies of stratification have become more active. More scholars with the attitude of "learning for the purpose of practice" are looking at social stratification from the standpoint of easing social conflicts and have formed, coincidentally, the following points of discussion. The first is how to assist the urban and rural poor stratum, a stratum produced by the process of social stratification. A number of researchers have pointed out that, since the 1990s, there is a growing population in both urban and rural China of poor people who come from different regions and areas, a group with a complicated composition and far-reaching influence. The problem of how to develop targeted urban anti-poverty measures is a brand new social challenge.[150] The second is how to relieve the polarization of rich and poor. Since the mid- and late 1990s, the World Bank, the National Bureau of Statistics, and some experts in this field have evaluated the Gini coefficient of China several times. It has ranged from approximately 0.35 to 0.48, indicating that China had become a country with a significant income gap,[151] making the polarization between rich and poor a public concern. A third discussion involves what social stratum structure should be like in modern China. This kind of research has pointed out that "since China's reform and opening up, an active and modern social stratum structure with a relatively reasonable structural pattern has been emerging in China. It can be said that the prototype of modern stratum structure has formed. One of its significant features is that the middle class of Chinese society is growing and social mobility mechanisms are becoming more reasonable...." However, they have also pointed out that "the present social stratum structure in China has some structural factors that might trigger social crises. This is because, in the current Chinese social stratum structure, the stratum that should be shrinking actually has not shrunk, while the stratum which should be larger is not growing. The middle class is too small while the lower middle class, composed of agricultural laborers, is too

149 Lu Xueyi, ed. *A Report on Contemporary China's Social Class and Stratum.* Beijing: Social Science Academic Press, 2002. 9.
150 Xiao Wentao. "A Research on Urban Poverty during China's Social Transformation." *Sociological Studies* 5 (1997): 42–49.
151 Lu Xueyi, ed. *A Report on Contemporary China's Social Class and Stratum.* Beijing: Social Science Academic Press, 2002. 41.

large in size."[152] Therefore, from the perspective of social structure, "only if most of the people in Chinese society enter the middle income stratum and live prosperous lives will structural stability be ensured and the whole country's long-term stability be guaranteed."[153] A fourth discussion focuses on changes in the political positions of different strata during periods of social division and reintegration, pointing out that "the social foundation of the governing party is tilting towards the strata which possesses economic and cultural resources (the stratum of managers, the stratum of professionals, and the stratum of private entrepreneurs)," while "the social foundation of the governing party among industrial workers and agricultural laborers is weakening, which cannot be underestimated because workers and farmers still account for the vast majority of the Chinese population."[154] Moreover, during the process of social division and reintegration, "the original working class, especially old workers in the original state-owned large and medium-sized enterprises who have been regarded as the 'leading class' in this group has been dramatically underrated with regards to their social positions and their interests have been seriously damaged."[155]

It is worth noting that, since the 1990s, the international sociological circle has also turned its attention to China's social stratification, focusing on the changes in social strata caused by market transition and related social equity. For instance, Professor Ni Zhiwei has generalized from the social stratification mechanisms in China after market transition to propose three main themes: market power, market stimulation, and market opportunities. These ideas have proved very influential in Western academic circles. The theory of power continuation proposed by Yanjie Bian and John Logan posits that political power has endured both in the redistribution system and in the market system. William Parish and Ethan Michelson have applied the idea of a "political market" to analyze changes in the relationships between groups. Andrew Walder has advanced that "government is manufacturer" to dissect the problem of direct government involvement in market activities.

152 Lu Xueyi eds. *A Report on Contemporary China's Social Class and Stratum.* Beijing: Social Science Academic Press, 2002. 69.
153 Li Qiang. *Chinese Social Stratification Structure in Social Transformation.* Ha'erbin: Heilongjiang Publishing House, 2002.
154 Lu Xueyi, ed. *A Report on Contemporary China's Social Class and Stratum.* Beijing: Social Science Academic Press, 2002. 38.
155 Zhang Wanli. "Twenty Years' Research on Temporary Chinese Social Class and Stratum." *Sociological Studies* 1 (2000): 37.

Another example is the local market socialism proposed by Linnan.[156] Furthermore, Deborah Davis, et al. have described nine aspects of social inequality in post-1990s China, namely, inequality between the urban and rural areas, gender inequality, regional inequality, and generational inequality, etc. They focused on comparing the differences before and after the market transition from a historical perspective and tried to demonstrate "the increasing impact of marketization on social inequality and stratification." They proposed nine principles for analyzing China's social stratification since the 1990s.[157] The discussion of China's regional social stratification has attracted the attention of international sociologists. In the past, it was rare for China's regional social stratification to be discussed in authoritative international journals, indicating the significance of research in this field.

156 Yanjie Bian ed. *Market Transformation and Social Stratification*. Beijing: SDX Joint Publishing Company, 2002. 17–28.
157 Deborah Davis. "Inequality and Stratification in China in the 1990s." *Foreign Sociology* 5 (2000): 18–30.

Chapter 16

Research on Social Stratification by Chinese Sociologists

In the previous chapter, the discussion of social stratification and stratum by Chinese theorists was introduced; in this chapter, the study of social stratification by Chinese sociologists will be explored. Since these chapters are both theoretical reviews, what differentiates this chapter from the last? I would like to point out that the analysis in this chapter is based on the work of academics specializing in sociology, and so that work is more professional and more specific than that described in the last chapter. Some viewpoints that were only briefly referred to in the last chapter will be further expanded upon here.

Looking back at the social changes that have taken place in China since the launch of the reform and opening-up policy, the most significant and fundamental change is none other than the transition of Chinese social structure. The core of this transition is the alteration of China's social stratification structure. As discussed above, the essence of social stratification is the relationships between people in possession of benefits and resources. The nature of the reforms lay in the alteration of the interest relationships between people. Therefore, reviewing the changes in social stratification facilitates an understanding of how people's interest relations have been altered by the reforms.

Since it is closely related to the reforms, social stratification has consistently been an important topic for Chinese sociologists and in sociological circles over the last thirty-plus years. This chapter does not aim to provide a comprehensive introduction to the study of social stratification since the reforms. Instead, it will mainly introduce the research achievements in Chinese sociological circles since the late 1990s. It will focus on the following four aspects: some important theories relating to the general study of social stratification, the major concepts dealing with wealth disparity, the existing research on the intermediate stratum, and the prospects for the study of social stratification.

1. A General Study of Contemporary Chinese Social Stratification Structure

Social stratification has been the dominant field in Chinese sociological studies in recent years, largely because of three key social phenomena. The first trend is the worsening polarization of wealth, which is concentrating social wealth and development opportunities into the hands of the minority rather than diffusing them throughout society, as was done in the late 1980s. Secondly, "power" is becoming increasingly decisive in the allocation of social resources and development opportunities. Finally, in top-down and bottom-up social interactions, some new structural elements are "reproduced," influencing and altering the direction of social interest relationships and social transformation. These concepts will be expounded through the work of Sun Liping, Lu Xueyi, and Li Peilin, as well as my own.

As for the analysis and explanation of the mechanisms of social structure transition and the transformation of class and stratum structure in contemporary China, sociologists have, generally speaking, formed four different conclusions. One of them is known as "Stratification Theory," which was put forward by the research group headed by Professor Lu Xueyi. Another is referred to as "Clique Theory," which I proposed. The next, developed also by me, is called "Fragmentation Theory." The fourth one is "Cleavage Theory," which was proposed by Professor Sun Liping. Apart from the above four theories, many other sociologists have suggested other explanations, such as "the double reproduction pattern of class and stratum relationships" and "relative deprivation of stratum cognition." These theories deal with other important aspects of China's social stratification mechanisms, such as the transmission of class and stratum relationships, stratum cognition, and gender stratification. A detailed explanation of these schools of thought will be given below.

1.1 "Stratification Theory"

Lu Xueyi, a researcher at the Institute of Sociology at CASS (Chinese Academy of Social Sciences), has directed a research group called "Research on the Transition of Social Structure in Contemporary China." The group explicitly states that a modernized social stratification structure is already in an embryonic form in China. "Contemporary

Chinese social stratum structure is transforming into one that is modernized and in accordance with the modern economic structure, and this new social stratum structure is tending towards stability;" and "the structure of social hierarchy in Chinese cities is transitioning into an olive-shaped structure although this transition is still a long way off for administrative regions in rural and rural-urban areas."[158]

According to empirical documentation gathered through surveys, questionnaire surveys, and case interviews carried out in different regions of China (with 11,000 samples from typical surveys and 6,000 probability sampling questionnaires) and corresponding statistical analysis, Lu and his colleagues set up a theoretical framework for contemporary Chinese social stratum structure based on the classification of occupations and the possession of organizational, economic, and cultural resources. Accordingly, they further expanded their social stratum structure to include ten strata and five ranks of social positions that applied to contemporary China. This argument is referred to as "stratification theory" in Chinese academia.

The social stratum structure they outlined consists of the following elements: ten social strata (state and social administrators, managerial personnel, private entrepreneurs, professionals and technicians, office clerks, private businessmen, commercial service staff, industrial workers, agricultural laborers, and the urban and rural unemployed and underemployed); and five social ranks (upper class, upper-middle class, middle class, lower-middle class, and lower class).

Among these elements:

a. "State and social administrators" refers to leaders or cadres who exercise administrative power in Party and government departments, public institutions, and social organizations. They accounted for about 2.1% of China's social stratum structure. (The data for this and the following elements were collected in 1999.)

b. "Managerial personnel" are senior managers (non-owners) in large and medium-sized enterprises. Approximately 1.5% of all the working population were members of this group in 1999.

c. "Private entrepreneurs" are identified as people who possess private capital and fixed assets and employ workers in operations for profit. According to current policy, only when a business has hired

158 *A Research Report on Social Strata in Contemporary China.* Ed. Lu Xueyi.

more than eight workers can its owner be called a private entrepreneur. The private entrepreneur stratum emerged after the reform and opening up; the first private entrepreneurs appeared in 1981. By 2001, the number of private enterprises had increased to 2,028,500, with 4,608,000 investors, 27,140,000 employees, a registered capital of 1821.2 billion yuan, and an operating income of 1148.4 billion yuan. According to data collected in 1999, private entrepreneurs took up 0.6% of China's social stratum structure.

d. "Professionals and technicians" made up 5.1% of the social stratum structure.

e. "Office clerks" refers to full-time office staff that assist leaders in the daily administration of Party and governmental departments, enterprises, and public institutions. They are called white-collar workers by society and are mainly made up of lower and mid-level civil servants in Party and government departments, junior management, and nonprofessional offices in enterprises and public institutions of various ownership. They constituted 4.8% of the social stratum structure.

f. "Individual businessmen" refers to those who, with a small amount of private capital, are engaged in operations like small-scale production, logistics, service provision, etc. They live on their business and take part in their business activities personally. Some of them are specialized in certain technologies or crafts, and some will take on several apprentices and hire a few assistants (no more than seven). They are called small businessmen, small employers, individual businessmen, and are referred to as "self-employed" in some countries. By 2000, this stratum had 25.71 million individual business owners with 50.71 million employees. They accounted for 4.2% in the social stratum structure.

g. "Commercial service staff" includes workers who take up nonprofessional manual or nonmanual jobs in commercial and service trades. In 1999, this stratum made up 12% of the social stratum structure, of which one-third were migrant workers.

h. "Industrial workers" refer sto those who are completely or partly engaged in manual labor, either by participating directly or by taking up support work in secondary industries (manufacture and construction). In 2000, secondary industries in China employed 162.19 million people, of which 76.4 million worked in state-owned work units, accounting for 47.1% of the total population. The majority were hired by non-state-owned enterprises, such as foreign-funded enterprises, joint-stock companies, private enterprises, individual businesses, and so on. Industrial workers made up 22.6% of the population; of these, about 30% were migrant workers.

i. "Agricultural laborers" refers to those who work on a contract basis on certain publicly owned arable lands, undertake production and operations in agriculture (forestry, animal husbandry, and fishery), and rely on agriculture as their sole or major income source. According to investigation, only 44% of employed people were agricultural laborers who were primarily engaged in agricultural labor and relied mainly on agriculture to make a living. With the progress of industrialization and urbanization, along with the implementation of relevant polices, this number will continue to shrink.

j. "Unemployed and partly unemployed people in cities and the countryside" refers to those who, though of working age, do not have regular jobs. They have been referred to as disadvantaged groups by the media in recent years and, occupying the bottom of the social ladder, constitute 3.1% of the population.[159]

1.2. "Clique Theory"

The original concept of "clique theory" was based around "interest groups," which was proposed by Professor Sun Liping, Professor Shen Yuan, and me.[160] Later, I further explained the concept of "interest groups" in a paper, holding that the term *interest clique* or *interest group* refers to "a group made up of people of similar status in relation to material interests," or "a group made up of people of similar status in relation to economic interests." Such groups are traditionally referred to as "classes" or "strata" in mainland China. So, why have I adopted the term "interest group" instead of "class" or "stratum?" There are two reasons for this. Due to the impact of terms like "historical fallacies" ("历史误区"), "mass mentality" ("大众心理"), and especially "the extension of class struggle" ("阶级斗争扩大化"), a special connotation has been imparted to the word "class," towards which people have developed a mentality and attitude of loathing

159 *Research Report of Social Strata in Contemporary China.* Ed. Lu Xueyi.
160 The prototype of four interest groups was first put forward by Sun Liping and Shen Yuan in their discussion of a research project about social structure. Reference to this can be found in "Recent Trends and Latent Dangers of Social Structural Transformation in China," published in *Strategies and Management* 5 (1998) by the above mentioned research group.

and dislike. Additionally, class refers to a group in which interest divides have been resolved and people's statuses with relation to material interests have reached a state of relative stability. Nevertheless, interest structure in China today is undergoing swift changes with constant division, disorganization, and "reintegration" of social interest groups. Therefore, the concept of class, connoting a stability of status, does not accord with the reality of present day China. In an empirical study, I analyzed a variety of data collected in the 1980s and 1990s, finding that, because social groups at that time witnessed large-scale disintegration and integration, a stable class could not be said to have existed. In line with the gains and losses of their interests since the reform, I therefore divide contemporary China's citizens into four interest groups or cliques.

First come the special beneficiaries, who have benefited the most from the reforms of the last thirty years. They are: private entrepreneurs, bosses, company presidents, senior managers, building contractors, various actors in the market, famous musicians, film stars, sports stars, etc. This group also includes those who work as managers or technical staff in foreign companies or foreign-funded enterprises. Ordinary beneficiaries occupy a second place. They have enjoyed visible benefits in both economic terms and in terms of other social resources. They are present in every social stratum as intellectuals and cadres, as well as ordinary managers, clerks, shop assistants, and farmers, etc. The interests of the third group, however, have been damaged by the reforms. This group contains those whose interests are being damaged in the present stage of the reform, even though some of them benefitted from the early stages of the reforms, such as unemployed persons and laid-off workers in cities and towns. The last group is the lowest group, which was initially called "the group suffering absolute damage to their interests." "Bottom class group" has a rather negative connotation, but this is a reality that must be faced. The gains and losses of interest are a process, while the development of class is the result of interest division. In this sense, the first group mentioned above can be called an upper class, the second a middle class, the third a lower-middle class, and the last a lower class.[161]

161 Li Qiang. *Chinese Social Stratification Structure in the Transitional Period.* 103–119.

1.3. *"Fragmentation Theory"*

"Fragmentation theory" was first put forward by me with relation to the analysis of the division and disorganization of social groups in "Clique Theory." I hold that social division and the social stratification structure of contemporary China have the features of fragmentation. People used to believe that division could only lead society into instability; however, this is absolutely not the case. A simple polarization of society will definitely bring about conflicts, but if the social division is a "fragmentation" of interests, people will enjoy diverse benefits. In that case, it seems unlikely that great conflicts of interest will emerge.

I argue that the "fragmentation" and the diversification of interests in Chinese social stratification bear the following three features. Firstly, that stratification and status groups are intertwined in a way that has given birth to diverse interest groups. The solidification of stratum structure, intertwined with the traditional status system, has resulted in the coexistence of stratum and status, with every stratum containing many status groups. Secondly, the discrepancy between residence registration and regions, intertwined with the element of the social stratum, has caused the diversification and fragmentation of people's interests. The discrepancies in economic status in different regions are much greater than those present within a social stratum. Thirdly, the differentiation of systems has mixed with social strata to generate diversified-interest groups. The diversification of social interest structure has led to the interweaving of multiple interests in society, rather than inciting social division. From the perspective of sociology, the "fragmentation" of stratum interests and social interests helps to moderate social unrest and therefore favors social stability. This has, in a way, explained why there is neither great social turbulence nor any class action, even though we are confronted with severe polarization between the rich and poor in our country.

1.4. *"Cleavage Theory"*

"Cleavage theory," which was put forward by Professor Sun Liping of the Department of Sociology at Tsinghua University, provides an understanding of the social structural features present in China today. Its fundamental argument is as follows: Chinese society witnessed a change in social resource distribution, from the "social resource diffusion" of the mid- to late 1980s to the "aggregation of social resources" in the hands of a minority since the 1990s. Disadvantaged groups have emerged from the processes of industrialization, urbanization, and marketization. Laid-off workers from state-owned enterprises and immigrant workers, among others, have been "excluded from the social structure." Their traditional social status cannot easily become a modern one. The economic reforms have not brought about corresponding social development; the fruits of reforms have not been shared by all of society; and, thus, a fractured society has come into being. "What is a society divided by cleavage? It is a society at a developmental stage at which components from different historical periods coexist while lacking organic connection to each other."[162]

Sun Liping has found that, since the mid-1990s, some crucial and fundamental changes have taken place in Chinese society. Completely different basic social structural features and forms of operational logic have emerged. Meanwhile, significant social groups that can influence and determine the basic tendencies of social development have come into being and are becoming increasingly important.

As a result, a new society with a number of distinctive characteristics is being born. One central feature is the cleavage of economic growth and social development. Economic growth has, to a large extent, been unable to bring about the natural improvement of social conditions. Another characteristic is the formation of a new logic of economic growth. This development constitutes a thought-provoking paradox: even with rapid economic growth, the majority does not benefit; without rapid economic growth, the majority will fall victim to economic stagnation. New social forces have also begun to take shape and have established new patterns of interaction, which have been exerting a powerful influence on the direction and actual process of the reforms. The logic governing the reforms has also undergone significant changes. Firstly, reform now tends to connote a competition for benefits. Secondly, the top-down reform of the early stages, which was both quite profitable and inexpensive, changed to allow diversified social forces to influence the current reform process. Thirdly, a mechanism for adjusting the measures and effects of reform appeared. For instance, some reform measures aimed at promoting social fairness

162 Sun Liping. *Cleavage: China Society since 1990s*. Beijing: Social Sciences Academic Press, 2003. 14.

have tended to produce social unfairness in practice. Some can even be twisted to become a means of corruption. Finally, an unbalanced and unequal mechanism of social forces has come into being. This is particularly exemplified by the rivalry between disadvantaged and advantaged groups (including both private economic forces and some important monopolistic sectors); there is a significant disparity in the influence these groups have on public policy, as well as their ability to make use of social opportunities. This has become a major mechanism for exacerbating the inequality of Chinese society.[163]

Sun Liping believes that three basic factors have contributed to the formation of China's "new society" since the 1990s. Firstly, since the mid-1990s, Chinese society has been gradually transforming from one that prioritizes the necessities of life to one that focuses on the acquisition of durable consumer goods. Secondly, the trend of social resource allocation has transitioned from a process of diffusion to one of re-aggregation, leading to a polarized society. Thirdly, there is the trend of globalization and China's gradually increasing level of participation in the globalization process.[164]

1.5. "The Double Reproduction Pattern of Strata Relationships"

This pattern was first expounded on by Professor Li Lulu of Renmin University of China. He holds that the composition and change of social stratification patterns are not simply determined by the rational functions of the economic mechanism. Instead, they are shaped by the social and political process and are therefore the result of a particular institutional context and transformation process. "Reproduction" is the primary feature of the relationship patterns between social strata during institutional transformation, which has two implications. First, it implies that the inheritance relationships of strata play a dominant role. Second, the reproduction pattern of social strata will continue to be reproduced in the process of institutional transformation in Chinese cities and the existing pattern of opportunities for mobility being derived from relationships will also continue. The development of the market mechanism does not lead to the "reorganization" of the patterns of relationships or the replacement

163 Sun Liping. *Transformation and Cleavage—the Transition of Chinese Social Structure since Reform.* Beijing: Tsinghua University Press, 2004.
164 Ibid.

of the stratum structure. Those groups that are initially in a favorable position are able to take advantage of their capabilities in exchanging different forms of capital, social networks, and human resources generally succeed in maintaining and continuing their superior positions in the reform of the economic system. This feature can be called "double reproduction pattern."

Based on this hypothesis, and in reference to some primary stratum frameworks, Sun Liping has proposed an analytical framework made up of five strata: advantaged authorities, ordinary administrators and clerks, professional personnel, workers and farmers, and the self-employed. Among these, the stratum of advantaged authorities includes the heads of work units and senior and middle managers. One of the features of this stratum is their power over other people, as well as resources. On the other hand, ordinary administrators and professional personnel enjoy a relatively high degree of autonomy in their work. It can be inferred that the possession of power is the core of this framework and there are multiple criteria for defining one's position in the stratum structure.

The pattern of social stratum relationships bears a number of consistent characteristics. The first is the reproducibility of advantaged authorities and professional personnel. The second is the gap between manual and nonmanual workers. That is to say, there is a conspicuous gap between advantaged authorities, professional personnel, ordinary administrators and clerks, and workers and farmers: each stratum appears to be closed to social mobility. This is related to the reproducibility of stratum structure on a larger scale. The third is the reproduction and relative mobility of strata.[165]

1.6. "Relative Deprivation in Stratum Cognition"

This theory was put forward by Dr. Liu Xin, a professor in the Department of Sociology at Fudan University, formerly of the Department of Sociology at Huazhong University of Science and Technology. Liu Xin is focused on the cognition relationship between "stratum awareness" and the way that social stratum structure in a given society can be revealed. The stratum status is established on multiple bases and any unequal allocation of social resources can find an expression in

165 Li Lulu. "Institutional Transformation and Changes in Stratification Structure-Doubled Reproduction of the Pattern of Relative Stratum Relationship." *China Social Sciences* 6 (2002): 105–118.

stratum differences. The value of social resources, to some extent, is related to social institutional arrangements, as well as people's values. The criteria for defining a stratum—in people's own minds, at least—reflect their evaluation of the importance of different social resources. If such evaluations can be combined with objective indices of social stratification in the analysis of social strata, a more accurate stratum structure can be revealed.

As part of the above idea, Liu Xin proposed the idea of "relative deprivation in stratum cognition," which means that changes in the stratification mechanism are invariably accompanied by some people becoming deprived in terms of socioeconomic and life chances. This occurs because these people have either lost their vested interests in the traditional social system or have failed to benefit significantly from the reforms. When people are relatively deprived—regardless of whether they have occupied high or low positions in objective social stratum system—they tend to regard society as being unequal. In light of questionnaire data he collected in a sample survey of the citizens of Wuhan in 1996, Liu Xin examined his proposition of "relative deprivation in stratum cognition." The results suggested that the proposition can, to a large extent, account for the discrepancies in people's understandings of social stratum in the process of social transformation in China. Those who have been reduced to a relatively deprived position—either in terms of the human resources they possess or their positions in the social stratum structure—are more likely to be aware of the stratification of society than those who hold a relatively advantaged position. In addition, this study has also come to the conclusion that, during the market transition, the economic returns that leaders obtained through the power they held was "value that appreciated on a personal level" even while "value depreciated at an organizational level."[166]

2. Studies of Wealth Disparity

The problem of wealth disparity is one of notable concern in the study of social stratification. Scholars have contributed in four main respects. First is the empirical study of the disparities between urban and rural residents in their wealth and income. Second is the

166 Liu, Xin. "The Stratum Consciousness of Chinese Urban Residents in the Transitional Period." *Sociological Studies* 3 (2001).

theoretical analysis of the worsening of wealth disparities, its social consequences, and potential countermeasures to the trend. Next follows the study of people who have been able to enter an affluent stratum. The last is the study of poor and disadvantaged groups in society.

2.1. Empirical Study of Wealth and Income Disparity between Urban and Rural Residents

The study of wealth disparity should begin with an exploration of the actual situation. By utilizing the valuable tools of social surveys and quantitative analysis, many sociologists have been able to produce a great deal of research. At present, the Gini coefficient is widely used to measure wealth disparity. The subject of measurement can be either a subject's wealth or income; the measurement of the former is more complicated, while that of the latter is much easier. Therefore, the Gini coefficient index of annual per capita household income will be employed here. Since relevant data from before the reforms began in 1979 is very difficult to acquire, the superior data of a report written by the World Bank Economic Study Group after their 1980 study of the Chinese economy will be used as a basis for comparison. According to the report, the Gini coefficient of the annual per capita household income of Chinese urban residents was 0.16 in 1979, that of rural residents was 0.33, and that of the urban and rural residents together was 0.33. Based on this, it is clear that the Gini coefficient of the annual per capita income of Chinese urban and rural residents has undergone a rapid increase since the early 1980s. The figure provided by the Chinese Academy of Social Sciences in 1988 was 0.382.[167] According to my calculations, the figure was 0.434 in 1994 and 0.4577 between 1996 and 1997. The studies of Zhao Renwei, Li Shi, etc. in 1995 put the figure at 0.46.[168] In 2004, through my analysis of the data collected in many sample surveys and comparisons with the data provided by other scholars, it was found that the Gini coefficient was not less than 0.5.[169] As far as the Gini coefficient is concerned, Chinese society

167 Zhao, Renwei and Griffin ed. *Studies on Distribution of Income of Chinese Residents*. Beijing: China Social Sciences Press, 1993.

168 Zhao, Renwei, Li Shi and Li Siqin, eds. *Restudy on Distribution of Income of Chinese Residents*, Beijing: China Financial and Economic Publishing House, 1999.

Li, Shi and Wei Zhong and Gustafson. "Distribution of Property of Chinese Urban Residents." *Economic Research Journal* 4 (2000).

169 Li Qiang. "A New Trend of Changes in Social Stratification Structure in Present China." *Jiangsu Social Sciences* 6 (2004).

has transformed from one in which equalitarianism prevailed into one with a relatively large wealth gap in the past twenty years. Such a drastic change is not common in the world.

Some scholars have pointed out that people's attitudes towards the wealth gap are also a social reality that can find three forms of expression. People's attitudes can amplify the wealth gap in their minds out of proportion to reality. Alternatively, when farmers and laid-off workers feel that they are relatively deprived, they are likely to take collective action. Finally, the decrease in absolute earnings and the severe polarization between rich and poor will intensify social contradictions and give rise to social unrest.[170]

2.2. The Theoretical Interpretation of and Countermeasures to Wealth Disparity

In order to explain the rapid widening of the wealth gap in China, many sociologists have cited the introduction of a market mechanism and market transformation. I believe that the so-called market consists of three parts: the commodities market, the labor market, and the finance and credit market (see chapter 7 of this book). The development of the commodities market in China began in the early stages of the reforms, especially with the trial price reform in cities which took place after 1987. In this experiment, the supply of commodities failed to meet the demand, causing serious inflation. The development of a labor market in China began in the mid-1990s, when the system of assigning jobs gradually came to an end in cities and the iron-rice-bowl style lifelong employment system was gradually abolished and replaced by a free employment system. On the whole, the size of China's labor force exceeded demand, which, combined with the deficiencies of the existing employment structure and an unsound employment service system, made the problem of unemployment and laid-off workers very severe. So, the immediate consequence of promoting the concept of a labor force was that of greatly increasing the size of the unemployed population. It was not until the twenty-first century that the finance and credit market began to be introduced. In recent years, the worrying instability of the stock exchange market has indicated the huge risks of the marketization of finance and credit in China. Essentially, the

170 Li Peilin. "The Psychological Impact of and Countermeasures on Wealth Gap in China." *Web of Chinese Sociology* Sep. 16, 2003.

introduction of all three types of market mechanism has led to difficulties during the market transformation; the expansion of the wealth gap is part of this problem.[171] A fair number of people have attempted to adopt the inverted U-curve theory of S. Kuznets (see chapter 7 of this book) to explain the gap between the rich and poor. Kuznets has found that in the process of market transformation, a number of rules tend to govern changes in wealth disparity. In the early stages of market transformation, the gap between the rich and poor tends to grow very quickly, but a significant narrowing generally takes place after the completion of the market transformation and establishment of a functioning market system. As for the specific case of China, some scholars hold that as China joins the ranks of the world's moderately developed countries in the first ten to twenty years of the twenty-first century, the wealth gap will shrink in size. However, other scholars hold that China's experiences have proved that the situation in China is quite different from those of other countries. The seemingly infinite size of China's labor force is a special case, so it is difficult to imagine a relative increase in laborers' incomes and a relative decrease in capital gains. Therefore, without government intervention, the wealth gap will continue to widen instead of narrow.[172]

As for the interpretation of wealth disparity, Chinese academics have contributed three distinctive theories. The first one, which I proposed, is referred to as the "two-stage market transformation theory." This theory suggests that the first stage of Chinese market transformation occurred in the 1980s, when a multitude of manual workers—who used to be a marginalized group—promptly became involved in the market economy and immediately benefitted from it. This resulted in the reversal of the wage imbalance between mental work and physical labor. The second stage of market transformation came after the 1990s, when the central social groups entered the market and instantly profited. Thus, the gap between rich and poor was rapidly widened.[173] The second theory was also put forward by me and is known as "economic and political stratification theory." This theory asserts that, before the reform and opening up, political stratification was quite conspicuous and the ubiquity of political inequality and discrimination was a serious issue. Since the introduction of a market mechanism, political

171 Li Qiang. *Social Stratification and Wealth Gap*. Xiaomen: Lu Jiang Publishing House, 2000. 79–80.
172 Li Peilin. "Globalization and 'China Experiences'." China.net.cn July 13, 2004.
173 Li Qiang. *Social Stratification and Wealth Disparity*. Xiamen: Lu Jiang Publishing House, 2000. 60–76.

differences have been reduced while economic differences have been greatly expanded. Political inequality has been substituted for economic inequality. The differences between Chinese strata have always been big, either in terms of political inequality or economic inequality. Thus, I conclude that the root causes of social inequality are deeply hidden within Chinese social structure and reflect the relationships between social groups. Political stratification and economic stratification are just two different manifestations of this inequality.[174] The third theory—"interlocking elites theory"—was presented in a paper collaboratively written by Sun Liping, Li Qiang, and Shen Yuan. This theory points out that, after the mid-1990s, elites have become increasingly interlocked in Chinese society. The interlocking of political, economic, and technical elites has brought about a visible increase in their wealth and has thereby enlarged the economic gap between elites and the low-income stratum.[175]

As for countermeasures to the problem of wealth disparity, scholars have provided suggestions concerning five aspects of the issue. One idea is to reinforce the regulation of income through taxation. The collection of income tax from high earners could be more stringently implemented while the tax burden of low-income earners and the ordinary working class could be alleviated. Another suggestion is to improve China's social security and social relief systems. A social security system capable of wider coverage could be established to include policies concerning employment, medical care, housing, pension funds, and so on. A third proposal is to create more employment opportunities. In a modern society, people's incomes are closely related to their employment; therefore, the realization of universal employment can help to reduce income differences. Alternatively, others suggest improving the education system. According to studies carried out across the world, the average level of education in a country is negatively correlated with its wealth disparity. This means that higher average levels of education tend to produce smaller wealth gaps. So, it is necessary to extend compulsory education in China to the greatest degree possible. At present, compulsory education lasts for nine years and, in the future, the goal is to extend compulsory education to cover a period of more than ten

174 Li Qiang. *Social Stratification and Wealth Disparity*. Xiaomen: Lu Jiang Publishing House, 2000. 39–59.

175 Sun Liping, Li Qiang and Shen Yuan. "Recent Trend and Latent Dangers of Social Structural Transformation in China." *Strategies and Management*. 5 (1998): 1–17.

years. The last suggestion is to strengthen the mechanism for investigating and punishing corruption. Without a sound legal system and market mechanism, corruption tends to spread. Recently, the amount of money involved in a number of sensational corruption cases has become increasingly startling. Since the wealth gap caused by corruption is the most unfair form of wealth gap and will most easily intensify social conflicts, corruption must be severely punished.

2.3. Studies of the Wealthy Stratum

In his study of private entrepreneurs in China, Zhang Houyi points out that by the end of June 2004, the number of registered private entrepreneurs in China had reached 3.34 million households with 47.14 million employees and 4214.6 billion yuan in registered capital. Compared with 1994, these three indices have respectively increased 5.96, 5.31, and 23.8 times over. He believes that the private entrepreneur group is the fastest growing social stratum in modern China and that its comprehensive quality and social status have gradually improved. The scale of its operations has expanded and its economic power has strengthened such that it has achieved extraordinary economic and social efficiency and made outstanding contributions to society. However, according to available research, private entrepreneurs themselves tend to worry a lot. This tendency has been aggravated in recent years by the appearance of the concepts of "the doctrine of original sin" ("原罪论") and "the theory of stealing state property" ("盗窃国家财产论")—both of which assume that the initial profits of private entrepreneurs have often come from unlawful sources. Zhang Houyi divides the establishment of private entrepreneurs' businesses into two types. One is establishment through internal development, referring to businesses that gradually expand in size on the basis of individual or household management. The other is establishment through external development, meaning businesses that expand by renting, contracting, and acquiring publicly owned enterprises. So, it is necessary to analyze the specific circumstances of individual cases and avoid applying vague theories of "original sin" to everything.[176]

176 Zhang Houyi. "Private Entrepreneurs are Important Components of Social Stratum Structure in China." *A Research Report on Social Strata in Contemporary China.* Beijing: China Social Sciences Academic Press, 2002.
Zhang Houyi. "The Rapidly Growing Stratum of Chinese Private Entrepreneurs." *Web of Chinese Sociology* Jan. 17, 2005.

According to some scholars, the social origins of private entrepreneurs have undergone some change since the 1990s. In contrast to the time when private entrepreneurs largely came from nonelite groups, today they are mostly from other elite groups. A mode of elite circulation has become a mode of elite reproduction. The change itself, these scholars suggest, has played a positive role in the development of that stratum. Several factors are thought to have led to this change. The increasing demand for human capital by entrepreneurs when setting up enterprises has been an important factor, as has the fact that traditional in-system advantages can be easily converted into out-of-system ones. Additionally, social relationships play an important role in the acquisition of social status and the weakening of the state work unit system has caused a lot of elites to leave.[177]

2.4. Research on Impoverished and Disadvantaged Groups

Carrying forward the tradition of researching poverty established in the early 1990s, the concept of "disadvantaged groups" has been more frequently used in research since the mid-1990s. What is a disadvantaged group? In his book, *Chinese Social Development Report by the Renmin University of China: The Disadvantaged Group and Social Support*, Professor Zheng Hangsheng defined "disadvantaged groups" as "groups that cannot maintain their basic standard of living and that of and their families through their own power or ability and require support and help from the government and society." He believed that disadvantaged groups exhibit three features. A "disadvantaged group" is made up of individuals and families who cannot meet their society's most basic living standard; they are also incapable of changing their disadvantaged status through their own efforts. Finally, such people cannot change their living conditions without the help of the government and social forces.[178]

Some scholars have discussed "disadvantaged groups" in light of globalization, suggesting that an important change resulting from resource reorganization since the 1970s has been the further division

177 Chen Guangjin. "A Study on the Private Entrepreneur Group." *Sociological Studies* 2 (2002): 7–8. "2001: A Report on the Front-line of Chinese Sociology." *Sociological Studies* 2 (2002): 1–29. Chen Guangjin. "The Evolution of Formation Mechanism of Private Entrepreneurs in China." *Web of Chinese Sociology* Jan. 26, 2005.

178 Zhen Hangsheng. *Chinese Social Development Report by the Renmin University of China: The Disadvantaged Group and Social Support.* Beijing: China Renmin UP, 2003.

of society. This has intensified against the background of globalization and a new and fairly large disadvantaged group had emerged. The characteristics of this new disadvantaged group have been thoroughly analyzed in scholarly circles. The "disadvantages" of farmers will gradually come to the surface as globalization continues. The new urban disadvantaged group is also likely to expand, including the urban newly unemployed, laid-off workers, employers in inefficient enterprises, and retired workers and their dependents. All this is separate from the existing group of "three-no individuals" (people with no source of income, no labor capacity, and no legal supporters). The lives of marginalized migrant workers in cities will also deteriorate. In order to meet the demands of the WTO and globalization, the need for a higher-quality labor force is expected to increase dramatically, leaving fewer job opportunities for lower-level laborers. This will make it harder for disadvantaged groups, who suffer from a shortage of resource and limited potential, to find employment.[179] Some research has focused on some specific disadvantaged groups, such as the marginalized "farmer-to-worker group." With systemic change, governmental policies for this group have begun to change and without governmental protection, its social status has been further marginalized.[180]

3. Studies on the New Middle Stratum since the Reform and Opening Up

The groups that have been studied most in social interest relations since the reform and opening up have mainly been the newly emerged groups that are the winners and losers in China's modernizing social structural transformation and a consequence of system reform. How should the needs and social functions of these emerging social interest groups (such as private entrepreneurs, white-collar workers, etc.) be understood? How can the financial losses and decline in social status of laid-off workers and the traditional "working class" in the reform of state-owned enterprises be evaluated? How can the social position of "migrant workers" be defined and how can

179 Wen Jun. "The Disadvantaged Group: The Impact of Globalization and Policy Options from Micro-perspectives." *Web of Chinese Sociology* Jan. 1, 2004.
180 Zhang Ruli. "From Active Marginalization to Passive Marginalization: A Study on the Behavior of Migrant Workers in Cities." *Web of Chinese Sociology* Sept. 23, 2004.

their interests be protected? A consensus has been reached on three points. Firstly, as the reform of the economic system and the process of marketization continues, the polarization of interest entities in the social structure increases. Secondly, with such polarization, more and more beneficiaries have become losers since the 1990s, which stands in contrast with the large number who benefitted in the early stages of the reforms. Thirdly, as was pointed out in the report of the Sixteenth National Congress of the Communist Party of China, a number of new social strata with distinct requirements that did not exist in the former planned economic system have appeared. The Sixteenth National Congress has identified six groups in China's newly emerging strata: "technical personnel in private science and technology enterprises, managers and technical personnel employed by foreign-funded enterprises, individual businessmen, private entrepreneurs, employees of intermediary organizations, and freelancers, etc. have appeared in the social transformation." All these new groups have been the subject of careful analysis and study by sociologists.

Since the implementation of China's reform and opening-up policies, the focus of research on the new middle stratum has been on whether a new middle stratum has emerged, where its members come from, what is the nature of its structure, its size, its behavioral characteristics, and what its future might be.

3.1 Has a New Middle Stratum Emerged?

A universally acknowledged view is that a new middle stratum has come into being in contemporary China.

A research group entitled "Research on Social Structural Change in Contemporary China" directed by Lu Xueyi holds that a middle stratum similar to those of modern Western countries has taken shape in contemporary China's social structure, but that its boundaries are not well defined. The National Social Science Fund project, "Contemporary Chinese Social Structure: Urban Social Stratification," directed by Professor Zheng Hangsheng, has found that a middle income stratum has emerged and has influenced both the new and old middle stratums in current urban social stratification. In relevant studies, I have also mentioned that, in present-day China, both the new and old middle strata are undergoing changes. The disintegration of the traditional middle stratum is expressed in intergenerational replacement—more specifically, the younger generation replacing the older generation and thus taking the position of the middle stratum. In his empirical research on the social stratification structure of contemporary Shanghai, Professor Qiu Liping has found that it can be clearly classified into five strata based on the basis of vocational evaluation. Three strata have already formed, including an upper-middle stratum (a group mainly consisting of administrative clerks and office workers), a middle stratum (a group mainly consisting of professionals and technicians), and a lower-middle stratum (a group mainly involving business personnel). Based on interviews he has conducted in Shanghai, Lu Dale also holds that a new middle class has emerged and gained much social recognition.

3.2. How Can the Middle Stratum in China Be Defined at This Stage?

In recent years, terms like "middle class" and "middle stratum" have frequently appeared in the Chinese media. Some simply equate these terms with "high income earners," "middle income earners," and "overnight millionaires." This has made the term "middle stratum" fashionable but confusing. Most researchers hold that "middle income earners" in contemporary Chinese society cannot simply be regarded as equivalent to the "new middle stratum" or the "new middle class" in an industrialized society.[181]

There is no unified definition for the new middle stratum in an industrialized society in the studies of social stratification conducted in the West, principally because the "new middle stratum is not yet a compact and simple stratum."[182] However, all researchers agree that these groups possess a number of basic characteristics. In terms of vocation, most members of this stratum group are engineers, technicians, scientific researchers, and other mental workers, including political and economic administrators and technical workers at all levels. Apart from a few senior governmental officials and senior enterprise administrators, they have become an employed labor clique and

181 Li Peilin. "Middle Earners Stratum in Sociology." *Journal of Social Science of Hunan Normal University* 4 (2003). Li Qiang. "Constitution of Chinese Middle Earners Stratum." *Journal of Social Science of Hunan Normal University* 4 (2003). Zhang Wanli. "Research Report on Chinese Middle Stratum." *Research Report on Social Structure in Modern China*. Ed. Lu Xueyi.
182 C. Wright Mills. *White Collar: The American Middle Classes*. Hangzhou: Zhejiang People's Publishing House, 1987. 84.

live on salary incomes. They tend to enjoy better economic and political treatment than manual workers; quite a large number of them have direct control of the labor process in some respect. Their political attitudes, lifestyles, personal qualities, codes of conduct, and values tend to differ greatly from those of ordinary manual workers and, to some extent, they are able to influence social ideology. They also tend to play an active role in public affairs and have influence on the management of public affairs.[183]

The researcher Li Peilin holds that the concept of a middle income stratum reflects not only income level, but also an attendant life quality, income distribution, urban and rural structure, vocational structure, and the existence of different national conditions.[184] I hold that, on the surface, the term middle class (or middle stratum) refers to a group with middle level living standards and possesses moderate assets. There is no sharp boundary between the middle, upper, and lower classes. In modern world history, the structure of the middle class has changed significantly with the transition from the "old middle class" to the "new middle class." The term "old middle class" refers to the middle stratum formed by such members of the petty bourgeoisie as small business owners and shopkeepers. Since the early twentieth century, a new middle class—a white-collar stratum engaged in management, professional and technical work, business, and office work—has become dominant in the middle class. This group is the most dominant group in the social structure of developed countries and plays the role of easing the danger of confrontation and conflict between the upper and lower classes.[185] Qiu Liping holds that the members of the modern professional group—including administrative clerks, office workers, and professional technicians—are very clearly of the middle class. In a market economy, as the control of organizations over individuals and the reliance of individuals on organizations are weakened, individuals' pursuit and worship of organizational power is reinforced. Thus, the "economic elites" and "intellectual elites," who are supposed to be the backbone of the middle class, are continuing to flock to official government positions. Such people continue to suffer derogatory

183 Zhang Wanli. "Preliminary Study on the Middle Stratum in Current China." *Jiangsu Social Sciences* 4 (2002): 85–94.
184 Li Peilin. "Middle Earners Stratum in Sociology." *Journal of Social Science of Hunan Normal University* 4 (2003): 5–7.
185 Li Qiang. "On Middle Class and Middle Stratum." *Journal of Renmin University of China* 2 (2001): 17–20.

labeling and it is therefore very difficult for them to become an independent middle class.[186] A large number of middle-class occupations will be developed further, but an independent middle class is not an inevitable consequence of this trend.

Based on her empirical study and meditation on social stratification structure in the transformation of Chinese social structure, Zhang Wanli gave, for the first time, a preliminary definition of the middle stratum in contemporary China. This definition was presented in a special report on *Research on Social Structural Change in Modern China* produced by a project research group led by Researcher Lu Xueyi in October 2001. According to this definition, the so-called middle stratum is the status group in the social stratification structure that possesses knowledge capital and vocational prestige capital. This group is mainly engaged in mental work, lives on a salary income, and has prospects of finding better-paid jobs with better working conditions. The group also holds a corresponding degree of household consumptive power, some leisure time, power over their work and those with whom they work, public awareness, public morality, and a level of concern for society commensurate with their social status. In other words, the level of their economic, political, social, and cultural status is between the upper and lower strata. Based on this definition, Zhang has also proposed seven criteria for determining whether people belong to the middle stratum.[187]

3.3. What Are the Scale, Source, and Composition of China's Middle Stratum?

At present, no official statistics can be obtained to make clear the scale of China's middle stratum. However, three estimates might be helpful in establishing a general idea. The middle stratum is thought to constitute between 13% and 15% of China's working population.[188] This would mean that it takes up 20% to 25%[189] of the total population. Finally, it is thought that the number of China's middle consumers may have reached 0.1 billion and will reach 0.2 billion within five

186 Qiu Liping. "Research on Professional Status: Indicator of Social Stratification—Shanghai Social Structure and Social Stratification." *Sociological Studies* 3 (2001): 18–33.
187 Zhang Wanli. "Research Report on Chinese Middle Stratum." *Research Report on Social Structure in Modern China*. Ed. Lu Xueyi.
188 Ibid. 254–56.
189 Xiao Wentao. "The Status Quo and Future Development of China's Middle Stratum." *Sociological Studies* 3 (2001).

years.[190] According to a prediction made by Long Yongtu, a former deputy minister of China's Foreign Trade Ministry, China's middle-income population will amount to 0.4 billion within ten years.[191]

To understand the source and composition of China's middle stratum, Zhang Wanli classifies China's middle stratum into four groups based on different economic relationships and the difference between the traditional and new middle strata. These groups include the traditional "middle stratum," including self-employed individuals such as small business proprietors and small traders. Also included are officials and intellectuals who were separated from the "middle stratum" in the planned economy. She also classifies new private business owners and township entrepreneurs who have found success as a result of the reform and opening up as being of this stratum. Finally, Zhang includes the new middle stratum that has developed as a result of the introduction of foreign investment and new high-tech talents. This group is made up of modern entrepreneurs and "white-collar workers" employed and better paid by foreign-owned companies or joint ventures whose occupational prestige, working style, and conditions are all better than those of people working in state-owned companies and traditional businesses. These two occupational groups can be regarded as the quintessence of the new middle stratum in a modern, industrial society.[192]

In his study on Shanghai's social stratification, Qiu Liping uses "occupational status" as his main criteria for ranking people, believing that there are five distinct social strata in today's Shanghai. The top stratum mainly consists of governmental officials, private enterprise owners, and foreign agents. The upper middle stratum is made up of administrative clerks or office staff. The middle stratum consists of all types of professional technicians. The lower middle stratum contains those involved in business. The lower stratum is that of workers, farmers, and community-service personnel. Essentially, Shanghai's social strata has a pyramidal shape. Respondents in Qiu's study who acknowledged that they and their family belong to the top stratum amounted to 0.5%, while 5.6% placed themselves and their families in the upper

190 See *China Information Gazette*. Mar. 28, 2000.
191 Long Yongtu. "Long Yongtu Predicts Boldly: There Will Be 400 Million Middle Class People." http://www.chubun.com/2201/12b/gb5/page42-06.htm.
192 Zhang Wanli. "Research Report on Chinese Middle Stratum." *Research Report on Social Structure in Modern China*. Ed. Lu Xueyi.

middle class, 52.7% placed themselves in the middle stratum, 35.8% in the lower middle stratum, and 5.4% in the lower stratum. The majority of the respondents regarded their families as belonging to the middle or lower middle class.[193]

I believe that the evolution of China's middle stratum can be defined according to three characteristics. First, the old middle stratum that existed prior to the reforms was composed of governmental officials who were similar to white-collar workers, intellectuals, and employees in state-owned enterprises; the new middle stratum largely comprises independent business proprietors. This replacement of the old middle stratum by the new stratum happened suddenly. The old middle stratum comprised a peer group aged between forty and fifty, whereas the new middle stratum is largely made up of young people aged around thirty. Finally, the traditional middle stratum underwent a whole-scale decline manifested in the decline of an immense, homogeneous social group made up of employees in state-owned enterprises. Therefore, the social problems that may result from the decline of China's traditional middle stratum's social status may become more acute with time.[194]

3.4. What Are the Features of the Behavior of China's Middle Stratum?

In an examination of the structural features of China's middle stratum, Zhang Wanli puts forward the theory of "multi-social division," which mainly involves the dual division of urban and rural areas, the division of regions and urban communities at different levels of urban development, the division of economic relations of different kinds of ownership, and the division of different social values and behavior patterns.[195] To explain the group characteristic of China's middle stratum, I have put forward "inter-generational replacement" theory, which suggests that the traditional middle stratum has now disintegrated; the old generation has been replaced by the young.[196]

193 Qiu, Liping. "Research on Professional Status: Indicator of Social Stratification—Shanghai Social Structure and Social Stratification." *Sociological Studies* 3 (2001).

194 Li Qiang. *Social Stratification and Wealth Disparity*. Xiamen: Lu Jiang Pulishing House, 2000. 90–91.

195 Zhang Wanli. "A Research Report on Chinese Middle Stratum." A *Research Report on Social Structure in Modern China*. Ed. Lu Xueyi.

196 Li Qiang. *Social Stratification and Wealth Disparity*. Xiaomen: Lu Jiang Publishing House, 2000. 77–100.

In an effort to understand the behavioral patterns of this group, Zhou Xiaohong puts forward the theory of "politics first and consumption next," suggesting that a nonturbulent and gradual promotion of democratic politics and a modern consumption pattern of consumer behavior will gradually develop.[197]

As for the mechanism by which the middle stratum obtains an advantaged social status, Zhang Wanli believes that the advantages held by governmental officials and intellectuals in terms of social status are related to their use of their original social relation resources, power resources, and intellectual resources in a market-oriented society.[198] Based on research he conducted in Shanxi province, David Goodman argues that members of contemporary China's middle stratum tend to maintain a good and close relationship with state and party organizations, private enterprise owners, and managers of state-owned companies alike. This is one of the preconditions for acquisition of economic resources.

Researchers have found that the returns on education for the occupational groups making up China's middle stratum are higher than those of other occupational groups, which is similar to the situation in developed countries. Liu Jingming, formerly a professor at Renmin University of China and now a professor at Tsinghua University, has explored the relationships between different types of qualification and how they have affected a person's chances of entering middle and high ranking white-collar groups in different periods of China's structural reform. Professor Liu finds that qualifications have a great influence on an individual's chances of entering a white-collar stratum. During the course of the reforms, the qualifications for entering middle and high ranking white-collar group have changed from Secondary Vocational and Technical Education to Senior Vocational and Technical Education and, finally, to standard Higher Education.[199]

3.5. What Are the Social Functions and Future of China's New Middle Stratum?

197 Zhou Xiaohong. "Middle Class: Possibility and Capability." *Collected Theses of the National Academic Symposium on Social Division and Policy Options in Present-Day China.* Vol. 2. This Symposium was held by the Research Center of Sociological Theories and Methodology of Renmin University of China and the Department of Sociology, Central China Normal University in 2002.

198 Zhang Wanli. "A Research Report on Chinese Middle Stratum." *A Research Report on Social Structure in Modern China.* Ed. Lu Xueyi.

199 Liu Jingming. "Education and the Transition of Social Stratification Structure: An Analysis of the Middle and Senior White Collar Stratum." *The Journal of Renmin University of China* 3 (2001): 2–25.

As far as social functions are concerned, it is generally acknowledged that the functions of the middle stratum include easing social conflict of interests, stabilizing society, boosting consumption, and setting good examples of conduct for the public, which has been stated in many scholars' works. According to the findings of some preliminary and small-scale empirical studies, during the present stage, China's middle stratum performs both the function of boosting consumption and normalizing the social values of the public, though only in some large and medium-sized cities. However, under the circumstances of "multi-social division," it is likely that contemporary China's middle stratum will repeat the mistakes of the Chinese Nationalist Bourgeoisie. China's middle stratum is likely to face a number of challenges and crises in its future development. These are likely to include the extrusion of the middle stratum by the increasingly polarized structure of rich and poor, the pressure of transferring surplus rural labor, and the re-employment of laborers in the modernization of urban industrial structure. The influence of the dual-division between rural and urban areas and regional differences in urbanization will also be a problem, as will the latent social taboo of wealth, pressure from Maoist ideology, and the established concept of "class analysis and struggle."[200] According to the measurements and calculations of a research project, *Research on the Transformation of Social Structure in Contemporary China,* conducted by Lu Xueyi in 1999, the middle stratum makes up 15% of China's social structure. In recent years, the number of private enterprise owners, professional scientific and technical personnel, and managerial personnel has been increasing in leaps and bounds, thanks to the rapid development of the Chinese economy. By 2002, China's middle class had come to account for 18% of the population. If an annual development rate of 1% can be maintained, the middle stratum will account for between 38% and 40% of China's population by 2020.[201]

3.6. The Heterogeneity and Identity of China's Middle Class

Sociologists, represented by C. Wright Mills, argue that the middle class, or middle stratum, is a social class with internal diversity, or heterogeneity.

200 Zhang Wanli. "A Research Report on Chinese Middle Stratum." *A Research Report on Social Structure in Modern China.* Ed. Lu Xueyi.
Zhou Xiaohong. "Middle Class: Why and How Can They Grow?". *Thesis Album of the National Academic Symposium on Social Division and Policy Options in Present-Day China.* Vol. 2.
201 Lu Xueyi. *"A Research Report on Social Structure in Contemporary China."* Ed.

According to the findings of a study on the middle class in three Chinese societies in Taiwan, Hong Kong, and Singapore, the middle classes in these burgeoning industrialized societies are internally segmented and differentiated, but have distinctive group identities.[202] A recent survey of "Shanghai's white-collar workers" shows that most interviewees feel uncertain and conservative about the title and identity of "white collar."[203] Therefore, it can be argued that the "white-collar" stratum is a newly emerging social class, which has not been politically established. This is in line with Zhang Wanli's findings in her research project, *Research on the Transformation of Social Structure in Contemporary China*. A slight difference in Zhang's findings is that the interviewees were less familiar with the term "middle class" than with the term "white collar," suggesting that "middle class" has a broader meaning than "white collar."

Generally speaking, assessments of middle class or stratum identity are closely related to three given propositions. First, the middle class's anxiety about, and nonconformity to, social status, in terms of the relationship between the lower middle class's political conservatism and its social functions. Second, the middle class and new social movements care less about the interests of a single class than bigger issues, such as major social problems that have become the focus of new social movements, including environmental protection, human rights, peace, and gender equality, etc. The "new politics" of the "new middle class" are thereby constructed based upon these issues. Thirdly, the middle stratum plays a leading role in advancing democratic ideals and launching democratic movements.[204] On the basis of these propositions, the following fundamental questions are worthy of consideration: will the practice and development of the middle class in China establish a stratum identity and set of social functions which integrates the above three propositions? Alternatively, will the middle stratum in China establish other types of stratum identity and unique social appeals?[205]

202 Xiao Xinhuang and Yin Baoshan. "The Collective Social and Political Consciousness of the Middle Class in Taiwan, Hong Kong and Singapore." *Market, Class and Politics: The Changing Chinese Society.* Eds. Liu Zhaojia, Yin Baoshan, Li Mingkun, and Huang Shaolun. Hong Kong: Inst of Asia-Pacific Studies Press of Chinese University of Hong Kong, 1999.
203 Lu Dale. "The White-Collars in Shanghai." *Social Transformation and Cultural Changes.* Hong Kong: Inst of Asia-Pacific Studies Press of The Chinese University of Hong Kong, 2001.
204 Xiao Xinhuang and Yin Baoshan. "The Collective Social and Political Consciousness of Middle Class in Taiwan, Hong Kong and Singapore." *Market, Class and Politics: The Changing Chinese Society.* Eds. Liu Zhaojia, Yin Baoshan, Li Mingkun, and Huang Shaolun. Hong Kong: Inst of Asia-Pacific Studies Press of The Chinese University of Hong Kong, 1999.
205 Zhang Wanli. "Preliminary Study on the Middle Stratum in China at Present Stage." *Jiangsu Social Sciences.* 4 (2002): 85–94.

4. New Ideas of Social Stratification in Chinese Sociology in 2008

Thus far, this chapter has been an introduction to the central ideas of the study of social stratification in the field of sociology in recent years. In 2008, what kinds of new ideas were proposed by sociologists? These ideas will be briefly introduced in this section.

We know that social stratification cannot be studied without the interpretation of some key concepts; the chosen key concepts often reflect the different basic designs of particular research projects. Wu Qingjun, Pan Yi, Chen Jingci, and Feng Shizheng have all explained the concept of "class" from different points of view. Wu Qingjun has clarified the main lines of development for the definition and analyses of "class" following Karl Marx in Western academic circles. Wu espouses the ideas of Marx and Weber as well as academic schools of thought like Marxism, Weberism, and Neo-Marxism.[206] Feng Shizheng has proposed (under the title of "Returning to Class Analysis?") that the fundamental difference between the two paradigms of class analysis and social stratification study is that the former relies upon conflict theory in its basic assumptions about social inequality, while the latter draws upon functionalist theory. So, the former pays closer attention to social deprivation and collective protest while the latter focuses upon status attainment and the market. Only by reintroducing the perspective of class can analyses of the structure and formation of social inequality in China be made more insightful.[207] Pan Yi and Chen Jingci have analyzed the unavoidable process of depoliticalization—symbolized by the disappearance of class discourse in China—and point out that, at present, the reappearance of class analysis largely extends the discourse of depoliticalization and dispels criticism about the possibility of China becoming a capitalist state as suggested in global capitalism theory.[208]

In 2008, sociologists devoted attention to many different groups, such as the elite stratum, the middle stratum, and low-income groups in a social stratification mechanism. Using two independent, large-scale national surveys, Zhu Xufeng has proposed an explanation for how elite superiority is reproduced in the social stratification structures of

206 Wu Qingjun. "An Academic Re-examination of the Concept and Theory of Class." *Chinese Journal of Sociology* 4 (2008): 59–86.
207 Feng Shizheng. "Returning to Class Analysis?—On the Paradigm Shift of Chinese Social Inequality Studies." *Sociological Studies* 5 (2008): 203–228.
208 Pan Yi and Chen Jinci. "The Disappearance of Class Discourse." *Open Times* 5 (2008): 53–60.

socialist countries undergoing reform. He believes that the transformation of policy creation in these countries can be summarized as a gradual change from an earlier system characterized by the political and administrative elite monopolizing the policy and decision-making process to a system by which the social elite participate in policy formation and the decision-making process.[209] However, Wu Zhongmin summarizes the characteristics of the elite group as being at a higher level in the social stratification system and separated from other groups by well-defined boundaries, but still open to other social groups. Without the necessary restrictions and limitations, the elite group would likely produce the following unpleasant distortions: alignment for mutual interests, self-protection, insularity, and the expansion of inefficiency.[210]

Based on data from CGSS2006, Li Peilin and Zhang Yi have measured the size of the middle class and medium income group in China at the present stage of development on the basis of income, vocation, and education. They have then compared the objective middle class and the identified middle class in terms of consistency of social attitudes, believing that the present middle class accounts for 12.1% of the Chinese population and 25.4% of urban society.[211] Tong Xin has claimed that in transnational corporations based in China, the controlling relationships higher working personnel have with their companies are influenced by the mode of post-modern capitalist production. Transnational funds have both constructed the labor market in China and affected the reconstruction of China's social structure, giving birth to a new class—the middle class.[212] Accordingly, Wu Qingjun has analyzed another working group—workers in state-owned companies—and has found that neither their group self-identity nor their class consciousness come from the production process, nor do their conflicts appear to be ones of hired labor against capital. It has been the transformation of the social system and the corruption of department managers, rather than factory discipline, that has led to this group's identification and class ideology.[213]

209 Zhu Xufeng. "Transition of Policy-Making and Elite Advantage." *Sociological Studies* 2 (2008): 69–93.
210 Wu Zhongmin. "The Essential Features of Elite Groups and Others." *Journal of the Party of the Central Committee of the C.P.C.* 2 (2008): 94–98.
211 Li Peilin and Zhang Yi. "The Scale, Self-Identity and Social Attitudes of Chinese Middle Class." *Chinese Journal of Sociology* 2 (2008): 1–19.
212 Tong Xin. "Transnational Capital and the Production of the Chinese Middle Class." *Chinese Journal of Sociology* 4 (2008): 1–19.
213 Wu Qingjun. "Workers' Group Identity and Class Consciousness in State-Owned Enterprises in the Period of Market Transition." *Sociological Studies* 6 (2008): 58–79.

I have analyzed the role of the middle class from the perspective of social structure and concluded that it has three key structural functions. It forms a buffer layer between the higher stratum and the lower stratum; it promotes conservatism; and it provides an enormous and steady domestic consumer market.[214]

Dai Weidong and Lu Jun have described different types of social mobility from the low-income group to the middle class. Based on this, they have discussed various barrier factors and their affects on the upward mobility of different groups.[215] Using analysis of survey data collected in Jiangsu province, Wang Peigang has pointed out that the objective lower-income stratum is forced to bear not only material poverty, but also inequality in social status and discrimination in social services. This has the potential to cause a unified sense of stratum identification. Their cognition of their chances and attendant hopes dramatically influence people's assessments of the gap between rich and poor.[216]

In 2008, sociologists conducted a great deal of research into social fairness; among this research, the studies of Wang Tianfu and Li Bobai are innovative. These studies have re-analyzed the concept of "redistribution" and established a new model for social stratification. They argue that the stratification structure in a socialist society is decided by the contradiction between an ideal country based on egalitarianism and a bureaucratic system based on particularism. The redistribution itself does not produce inequality, but the bureaucracy and bureaucratic political actions of the process of redistribution do. An individual's distance from the center of redistribution not only determines his or her position in the redistribution system, but also the degree to which he or she will be affected by bureaucratic politics. During a period of social transformation, the privileges of bureaucrats at the grassroots level will be maintained and the disparity in the social stratification structure will also gradually increase.[217] Wu Xiaogang has pointed out that, in recent years, employees entering the market can be divided into two groups: voluntary and involuntary employees. The

214 Li, Qiang. "The Social Responsibility of the Chinese Middle Class." *Chinese Newsweek* 47 (2008): 40–43.
215 Dai Weidong and Lu Jun. "An Analysis on the Mobility of Low Income Group to Middle Stratum." *Henan Social Sciences* 1 (2008): 78–82.
216 Wang Peigang. "On Dynamic Cognition by Each Stratum in Wealth Disparity in Present Chinese Society." *Social Science Research* 6 (2008): 115–122.
217 Wang Tianfu and Li Bobai. "An Ideal Country Based on Egalitarianism and a Bureaucratic System Based on Particularism." *Chinese Journal of Sociology* 5 (2008): 58–73.

phenomenon of employees in the private sector earning more is confined to the social group of voluntary employees entering the market later in the reform process. The effect on personal income of entering the market at a later period and the orientation of market entry form a negative correlation.[218] Li Youmei has asserted that the risk to society, as another force with decisive influence on social structure, provides a new logic beyond economic relationships in the analysis of social stratification. The diffusion and universality of risk enable the possibility of mobilization throughout the whole society and promote the production of a civil society from a new perspective and in a new arena. However, the distribution of specific risks also intensifies the differentiation of classes and strata and so may become a driving force for the production of new social conflicts.[219]

In 2008, scholars in China have paid a remarkably high degree of attention to gender equality in social stratification; their empirical research is supported by research data collected over a relatively long historical period. Wang Tianfu, Lai Yangen, and Li Bobai have adopted three sets of national data from a fifteen-year time span and developed a new theoretical model for gender stratification in a socialist society. They have determined that the farther one is from the redistribution center, the more one is likely to suffer serious gender discrimination. As China's social transformation progresses, discrimination against women—who were originally protected by the redistribution system and held higher social status than at present—will gradually become an increasingly serious issue.[220] Wu Yuxiao and Wu Xiaogang have discovered that, in the 1980s, the total level of gender segregation in nonagricultural professions had an upward trajectory but in the 1990s, began to decline. This was because, in different historical periods, the main factors influencing professional gender segregation were different.[221] Based on three sets of data covering China in a number of different historical periods, Li Chunling and Li Shi have pointed out that the income gap between the genders has been increasing steadily and strikingly over the past

218 Wu Xiaogang. "Voluntary and Involuntary Mobility for Jobs and Income Inequality in Urban China during 1993–2000." *Sociological Studies* 6 (2008): 33–57.

219 Li Youmei. "From Wealth Distribution to Risk Distribution." *Chinese Journal of Sociology* 6 (2008): 1–14.

220 Wang Tianfu, et al. "Income Gap for Different Genders in Cities and Its Evolution: 1995-2003." *Sociological Studies* 2 (2008): 23–53.

221 Wu Yuxiao and Wu Xiaogang. "1982–2000: A Study on Gender Segregation in Nonagricultural Occupations in China." *Chinese Journal of Sociology* 6 (2008): 128–152.

twenty years. In the first ten years of the reforms, the market mechanism was the main cause of the enlarging gender income gap. In the last ten years, however, gender discrimination has become the most important factor affecting the income gap between the genders.[222]

5. The Prospects for Social Stratification Research

At the turn of the century, the study of social stratification in China is very active because dramatic changes in China's social structure have taken place. So, in order to look far ahead into the future development of the study of stratification, we should analyze the trends of stratification structure in Chinese society. In my opinion, there are likely to be several important tendencies in the transformation of social stratification. These trends are very deserving of further in-depth study.

Research on changes in the whole social structure is necessary. In this new century, the overall context of social structural changes will gradually become clear. Although it might take at least forty years for Chinese society to change from a pyramid-shaped social structure into a spindle-shaped one, there is still an obvious tendency for the group at social bottom to shrink with each passing year. So, sociology is expected to bring forth sociological theories with Chinese characteristics.

The dramatic changes in stratification structure have tended to solidify. Since the twenty-first century began, the stratification structure of Chinese society has already tended towards comparative stability; the boundaries between different strata have gradually come into being and mobility between the strata has decreased. Internal stratum identities have been reinforced and the specific lifestyles of each stratum have also taken shape. So, scholars will increasingly concern themselves with the study of models of stratification.

Because of the widening gap between rich and poor, the study of this trend will remain very important. People will come to care more about the social consequences of the polarization of wealth and poverty. Unequal wealth distribution has always had the potential to cause social instability. This is especially so in Chinese society, which has a tradition of "equally sharing the wealth of the rich and the poor," so unequal wealth distribution and social instability are twin sisters. From

222 Li Chunling and Li Shi. "Market Competition or Gender Discrimination: The Trend of Widening Income Gap for Different Genders and Its Causes." *Sociological Studies* 2 (2008): 94–117.

the perspective of decision-making consultation, we should study the mechanism by which inequality might trigger social instability and implement measures to avoid the occurrence of such instability.

Recently, quite a few Chinese scholars have emphasized that when the per capita GDP of a nation ranges from 1,000 to 3,000 USD, its social structure is subjected to a high degree of strain and social contradictions are highly intensified. China is currently at this exact point, so the question of how the Chinese sociological community should study and deal with such a problem, both in theory and in practice, is a new problem produced by the process of reforms and construction.

The next few decades will be a period during which the middle stratum will develop faster than ever. In China, there is not a unified middle stratum at present; the different parts of the middle stratum vary greatly and their dissimilarities far outweigh their uniformity in terms of economic interest, lifestyle, and education. This means that, in China, it is extremely difficult to form a unified middle stratum. The middle stratum in China does not have unified interest requirements. Because the internal components of the middle stratum in China are extremely complicated, the study of its different internal components will also become a very important topic.

For many years, scholars have focused on the study of the wealthy stratum and private entrepreneurs in China. In a socialist country, the study of the wealthy stratum involves a fundamental problem: how property and wealth should be distributed between the different strata of a society. This question is also very important to the Chinese public. Thirty years have passed since China started its reforms; the wealthy older generation will pass its property on to a younger generation. The study of inheritance laws and inheritance taxes will therefore be appropriate.

Because of the polarization of Chinese society, the problems of the lower stratum have become more prominently discussed. Recently, researchers in this field have begun to advocate building a more comprehensive social security system and, above all, enhancing this system in rural areas. So far, China has not been able to build a unified social security system covering the whole rural population, which means that a strategy of diversified security corresponding to different circumstances is necessary. This must include a social relief system, a social insurance system, a form of social appeasement, and an updated household residency system. In some comparatively wealthy areas, a five-securities support system could be implemented to include a welfare system for the disabled, collective welfare, and programs for the public good. At the same time, the mutual assistance system in rural areas must also be strongly advocated. At present, the main task in cities is to continue to propel the comprehensive implementation of the three guarantees, including the guarantee of basic support for the unemployed, unemployment insurance, and a minimum subsistence guarantee system for urban citizens. In summary, the social security of the lower stratum will be of common concern to sociologists.

Chapter 17

Epilogue: A Forecast of Changes in China's Social Stratification Structure for the Next Thirty Years

By providing a foundation for the introduction of the basic theories of social stratification and social structure, and the methodologies for their study, this book has described the ways in which social stratification has changed in the People's Republic of China over the past sixty years. Particular attention has been paid to the changes in social structure in the past three decades of reform and opening up. People may ask: what will social stratification and social structure be like in the future? In particular, how will social structure change in the decades to come—for example, in the next thirty years in China?

It is not easy to predict the future of China. Over thirty years ago when China's reforms started, people could never have envisaged that China would develop in the way that it has. However, the conditions today are much more conducive than those of thirty years ago to forecasting how the Chinese society will progress. The phrase "more conducive conditions" refers to China's current social conditions. Currently, Chinese society operates regularly and normally. If a society does not operate normally, according to the law of social development for a period of time—such as with the Cultural Revolution—the development of society cannot be predicted on the basis of this period. Even Chairman Mao Zedong did not expect the "Cultural Revolution" to drag on for such a long time, so no one could have predicted the future of China during the Cultural Revolution. Similarly, even China's reforms and opening up have been far outside the norm. From this perspective, regularization, legislation, and the establishment of rules represent the most significant progress that Chinese society has made over the past thirty years. Therefore, it is possible to offer a prediction of how the Chinese society will develop in the next thirty years.

In this chapter, I will attempt to predict how China's social structure will change in the next thirty years. Although China's social structure is very complex—because it includes economic structures, political structures, occupational structures, and social organizational structures, etc.—the most important of these is the structure of the social position of various social groups in Chinese society, or the social structure of class and stratum. The social positions of groups in Chinese society traditionally form a pyramid-shaped social structure—a structure with a large bottom and small top. In chapter 11, I used the ISEI (International Socio-Economic Index of Occupational Status) method to analyze Chinese people's socio-economic statuses based on data from the 2000 census. This study revealed that Chinese people's socioeconomic statuses form an inverted T-shaped structure. People with ISEI scores of 23 and below comprised 64.7% of the population, forming the huge bottom in the inverted T-shaped structure. The remaining 35.3% represented a number of socio-economic status groups. Each group was very small, but layer upon layer they produced the perpendicular column in the inverted T-shaped structure. There is a sharp divide between those with ISEI scores of 24 and those who scored 90, a sharp division between the lower and higher scoring groups. It is clear that Chinese society is divided by enormous social differences.

This inverted T-shaped social structure is very uncommon. Compared with the traditional pyramid structure, the sharp division of the inverted T-shaped structure is more dramatic, like a right angle. There is no buffer or transition zone between the lower stratum and other strata. It is an either-or dichotomous structure. How has this come about? Though there are many reasons, in China, the most direct cause has been the household residence registration system, which separates urban and rural citizens. As has been mentioned above, the inverted T-shaped social structure, as well as the social tensions it causes, can be used to explain the various social contradictions and problems in contemporary Chinese society. For instance, the expanding gap between rich and poor is the visible result of the inverted T-shaped social structure. Furthermore, current social concerns such as "issues concerning agriculture, the countryside and farmers," "house demolition," "grievance petitioners," "mass disturbances," etc., are all related to this social structure.

This kind of social structure indicates that there is still a long way to go before China can achieve a modern social structure dominated by a middle stratum. It is, therefore, vital to analyze potential alterations to the inverted T-shaped social structure in order to explore the developmental trends of China's social structure over the next thirty years. An attempt will be made to explore the possible alteration of social structure in three aspects.

To begin with, changes are likely to take place in the industrial structure. The concern here relates to how the ratio of labor and the functions of status in all industries will change. We know that the basic principle for predicting structural change is the principle of inertia, meaning that the expansion or extension of the curve of change based on historical data may predict the future according to the facts of the past. Thus, an analysis will be made of the changes in labor in three industries in China that have taken place between 1978 and 2008. In 1978, the proportion of employees in primary industries, secondary industries, and tertiary industries were 70.7%, 17.6%, and 11.7%, respectively. In 2008, 29 years later, these proportions were 39.6%, 27.2%, and 33.2%. This represents a reduction of 31% in the proportion of agricultural laborers in the population; if the proportion of agricultural laborers continues to shrink at this speed over the next thirty years, the proportion of China's agricultural laborers will decrease to 8.6%. Of course, this prediction is subject to variation as predicted data cannot accurately represent future realities. What can be confidently asserted is that the proportion of China's agricultural laborers will continue to decrease dramatically and that the inverted T-shaped social structure will change radically. The proportion of laborers in secondary manufacturing industries has increased by 9.6% in the past thirty years; the rate of this increase is not great and the development of manufacturing has its own cycle. Therefore, it can be estimated that the increase in labor in manufacturing will be minimal in the next thirty years. If calculated according to the 9% increase of the last thirty years, the proportion of employees in secondary industries might reach one-third of all the employees in China in 2038. The proportion of employees in tertiary industries has risen the fastest, by about 21%. It is estimated that this proportion will increase greatly in the next thirty years and it has been predicted that employees in tertiary industries will account for more than 50% of the working population. Occupational structure is restricted by industrial structure. Therefore, the basic characteristics of the occupational structure in China indicate that the number of white-collar workers will outstrip the number of blue-collar workers by 2038, producing a modern social structure. However, this is not enough to change China's industrial structure; it is also necessary to innovate within other systems in order to create a harmonious society. For instance, as mentioned above, the proportion of agricultural laborers will drop to 8.6% by 2038, which will necessitate a corresponding adjustment in the household residence registration system. Otherwise, conflicts will reoccur when the household residence registration system cannot keep pace with these changes. An example of such a failure was the way in which the proportion of agricultural laborers accounts for 39.6% of the population, while the proportion of those registered officially as farmers has remained at 54.32% at the end of 2008.

Secondly, the changes in the proportions of social groups in various types of economic work units will be analyzed on the basis of statistics collected in urban areas. In 1978, almost all urban employees in China worked in state-owned or collective work units, among which the proportion of employees in state-owned work units was 78.3% and the proportion of employees in collective work units was 21.53%; only 0.15% were self-employed. By 2008, however, the proportion of employees in state-owned work units had decreased to 21.34% and the proportion of employees in collective work units had fallen to 2.19%. In contrast, 27.55% workers were employed in private enterprises, limited liability enterprises, joint stock enterprise, or joint-stock cooperative units. The other 5.37% worked in foreign-funded enterprises and enterprises funded by Hong Kong, Macao, and Taiwan. Urban self-employment accounted for 11.95%. The statistics above show great changes in the structure of employment, indicating a significant decline in the proportion of employees working in state-owned and collective work units. Over the next thirty years the proportion of employees in state-owned and collective work units will continue to decrease very slowly, but the declining trend will continue and the proportion is estimated to fall to 10%. At the same time, the vast majority of urban employees will join the private sector in limited liability companies, and joint stock companies. Such industries will employ 50% of the working population. While the proportion of self-employed individuals is likely to rise to over 20%, the proportion of employees in foreign-funded enterprises and enterprises funded by Hong Kong, Macao, and Taiwan might rise to over 10%. It seems that the planned economy has weakened while the private sector's strength has greatly increased. While some believe that this has been detrimental to the interests of the general public, in fact, that has not been the case. In a society dominated by the private economy, the government's economic monopoly will be weakened and the availability of small and intermediate quantities of capital will greatly increase. As mentioned above, the proportion of the self-employed alone might exceed 20% in 2038. Meanwhile, the rise of the Internet seems to be conducive to the development of self-employment. In a country like China with a population of over 1.3 billion, it is more beneficial for the public if capital is distributed rather than monopolized. If it is monopolized, the profits of those possessing small quantities of capital will inevitably be damaged. From a long-term perspective, it is more beneficial for citizens if a society is built to allow those possessing small and intermediate sums of capital to play a dominant role; this kind of society is referred to as an olive- or spindle-shaped society. In such a society, the middle class is the dominant social group.

Finally, changes in the educational structure of Chinese population will be analyzed. The educational structure is affected by many factors; the most prominent change has been one in which groups have been receiving higher education. One of the main indicators of the distribution of higher education is the gross enrollment rate of the current population (statistics relating to enrollment generally refer to the section of the population aged between eighteen and twenty). In 1978, only 85.6 million students were admitted to Chinese universities and the gross enrollment rate was 1%. The gross enrollment rate was 3.4% in 1990. By the late 1990s, the gross enrollment rate of higher education had risen rapidly; the gross enrollment rate rose to 10.5% in 1999, 15.0% in 2002, 21.0% in 2005, and 23.0% in 2007. This rate is projected to rise to 25% in 2010, according to experts. Based on the report of Zhou Ji, the former Minister for Education, the gross enrollment rate will reach 40% in 2020. Taking this rate into consideration, the gross enrollment rate is likely to exceed 55%, or even 60%, by 2038. If the proportion rises to over 60%, what international standards will it meet? Currently, countries with a gross enrollment rate in higher education of greater than 60% include Australia with 74%, Belgium with 61%, Denmark with 67%, Finland with 87%, the Russian Federation with 65%, Sweden with 82%, the UK with 63%, and the U.S. with 83%. In summary, very few countries reach this number and even some developed countries do not reach 60%: France has a rate of 55% and Germany, 50%. The status quo urges us to give the following issues serious consideration. The development of China's colleges and universities should be tightly controlled over the next thirty years because a corresponding higher level of occupational structure is necessary to match the number of well-educated graduates. Will China need so many university graduates within the next thirty years? Germany, a well-developed country, has a gross enrollment rate of 50% and its approach has been to focus on the cultivation of skilled workers to meet the needs of actual production operations. Additionally, people who have received higher education will account for the majority of employees—and when the majority in a society receives higher education, changes take place in its politics, economy, and culture. Changes will also take place in its organizations and operational methods, as well as even its social attitudes and values. So, in this sense, systemic reform in China is indeed a long-term historical task.

Index of Terms

Index of Names and Titles